S0-BZU-062

ACKNOWLEDGMENTS

We owe much to our editors at the *Wall Street Journal* for the genesis of this book. In the fall of 1985, as reporters assigned to cover science and medicine, we separately were pursuing feature stories on what appeared to be major advances in genetics. One of us (J.E.B.) was exploring new discoveries of genes that predispose individuals to coronary heart disease. The other (M.W.) was pursuing a story on the innovative study by population geneticists in Utah of large Mormon families to reveal the hereditary aspects of common disorders such as cancer. During casual conversations over coffee we began to realize that the developments we were reporting separately were springing from the same basic advance in molecular biology, a newfound ability to identify previously hidden genes.

It appeared that the best way to report the nature of this advance and its unprecedented implications for society was in a series of stories for the *Wall Street Journal*. The idea appealed to our immediate news editor, Neil Ulman, who helped us organize a mass of material into five front-page stories, prodded us through those moments when we felt we might have taken on more than we could handle, and edited our copy into the kind of clear, readable prose for which the *Journal* is known. The series, which also included an article by our colleague, Alan Otten, appeared between February 3 and March 19, 1986. (Otten's excellent article on genetic screening inspired our further research in that area.) The series, and the reader response to it, led to this book. In addition to Neil, we would like to thank the *Journal*'s managing editor, Norman Pearlstine, and deputy managing editor, Paul Steiger, who not only encouraged us in this endeavor but also allowed us to combine our research for the book with our reporting for the *Journal* of the rapidly breaking developments in molecular genetics.

We also want to acknowledge, with deep gratitude, those scientists who so graciously and patiently spent hours explaining their research in terms that our unscientific minds could understand. Among the scientists we especially would like to thank are Nancy Wexler of the Hereditary Disease Foundation, Mark Skolnick of the University of Utah, Louis Kunkel of Harvard Medical School, Jan Breslow of Rockefeller University, Ray White of the University of Utah, Bert Vogelstein of Johns Hopkins School of Medicine, and Alfred Knudson of Fox Chase Cancer Center, all of whom were kind enough to review portions of our manuscript (although we alone are responsible for its accuracy).

Over the two years we worked on the book, there were many, many others who took our calls, suggested ideas, and shared what we know was very precious time. They included:

David Botstein, James Gusella, David Housman, Robert Weinberg, Victor McKusick, Philip Reilly, Marc Lappé, Michael Conneally, Herbert Pardes, Miron Baron, Janice Egeland, John Minna, Ted Dryja, Milton Wexler, Webster Cavenee, Eric Lander, Randall Burt, C. Robert Cloninger, Scott Grundy, Antonio Gotto, and Neil Holtzman.

We are especially grateful to our friends and family; to Ben Patrusky, a fellow science reporter whose insight into the scientific process kept us on track; to Walt Bogdanich, Hank Gilman and Barry Meier, colleagues who repeatedly assured us that the book would be read; to Debby and Allan Ostrega; Elysa and Arthur Goldblatt; Florence and Charles Blatt; and Steve Savage, whose support is impossible to measure but is highly valued; and to Rachel Waldholz, who had to give up many weekends with her Daddy and was greatly missed in return.

We want to thank our editor at Simon & Schuster, Bob Bender, whose enthusiasm for the project propelled it to its conclusion. Bob's astute judgment and the help provided by copy editor Patty O'Connell are evident in the following pages. And, of course, there would have been no book at all had it not been for our agent, Barbara Lowenstein, who helped us enlarge our initial ideas.

Simon and Schuster

New York

London

Toronto

Sydney

Tokyo

Singapore

Genome

The Story of the
Most Astonishing Scientific
Adventure of Our Time—
The Attempt to Map
All the Genes in the
Human Body

Jerry E. Bishop & Michael Waldholz

toExcel

San Jose   New York   Lincoln   Shanghai

Genome

All Rights Reserved. Copyright © 1990, 1999
by Jerry E. Bishop and Michael Waldholz

No part of this book may be reproduced or transmitted in
any form or by any means, graphic, electronic, or mechanical,
including photocopying, recording, taping, or by any
information storage or retrieval system, without the
permission in writing from the publisher.

Pulished by toExcel,
an imprint of iUniverse.com, Inc.

For information address:

iUniverse.com, Inc.
620 North 48th Street
Suite 201
Lincoln, NE 68504-3467
www.iUniverse.com

ISBN: 1-58348-740-9

Printed in the United States of America

DEDICATION

To Ruth Altman and Betty Holcomb,
whose genetic predispositions for patience,
love, and tolerance were
severely tried during the writing of
this book and were found to be
strong and enduring.

TABLE OF CONTENTS

INTRODUCTION • Breakthrough   15

CHAPTER 1 • From Curse to Crusade   29

CHAPTER 2 • Alta   49

CHAPTER 3 • Worcester   69

CHAPTER 4 • El Mal   81

CHAPTER 5 • Bruce Bryer and Reverse Genetics   103

CHAPTER 6 • Dystrophin   121

CHAPTER 7 • Two Hits   132

CHAPTER 8 • Cancer Unleashed   154

CHAPTER 9 • Genes and the Heart   179

CHAPTER 10 • The Map   200

CHAPTER 11 • "Siss im Blut"   225

CHAPTER 12 • Genes in a Bottle   249

CHAPTER 13 • Predictive Medicine   267

CHAPTER 14 • A Niche in Society   285

CHAPTER 15 • Choices   307

Notes   323

INDEX   337

EPILOGUE   353

*To wrest from nature
the secrets which have perplexed
philosophers in all ages,
to track to their sources
the causes of disease . . .
these are our ambitions.*

Sir William Osler

Breakthrough

i t is difficult, even hazardous, to attempt to describe a scientific breakthrough while it is still in progress. Most discoveries are made in bits and pieces, from scientists working in different places at different times. Invariably, contemporary commentators, even the scientists themselves, will misjudge the significance of some of the events they are observing, overlooking events that later prove of pivotal importance while overemphasizing developments that ultimately prove to be of merely peripheral interest. Only years later does the historian have the luxury of looking back to pinpoint the particular experiments and identify the scientists that played key roles in the breakthrough.

Nevertheless, there occasionally is a scientific breakthrough of such overwhelming implications for society that its importance is immediately recognized. Such is the case with the rapidly developing ability of a new science—molecular genetics—to locate and identify individual human genes.

No one, not even the scientists themselves, can say exactly which of hundreds of experiments over the last decade and a half opened the door to this powerful new technology. Many of the developments described in this book begin somewhat arbitrarily in the mid-1970s. Since then, the number of genes known to man has multiplied more than tenfold. Today, new gene discoveries are being made at an almost geometrically progressing rate. Yet these newly identified genes constitute only a tiny fraction of the 50,000 to 100,000 genes that make up the human genome, the genetic endowment that lies in the nucleus of every human cell. And so far, every forecast of how soon every human gene will be identified has been exposed as shortsighted by an unanticipated improvement in the technology of molecular genetics. A task that scientists fifteen years ago thought would take centuries to complete—mapping the location of every gene—now appears likely to be accomplished in only a few decades.

While the gene discoveries haven't been shrouded in secrecy, their full import hasn't been widely discussed—or even appreciated. Like the atomic bomb physicists in World War II, the molecular geneticists have been so engrossed in their adventure that few have had the time or inclination to step out of the laboratory and contemplate the magnitude of the research. And few of the discoveries have been sufficiently dramatic in and of themselves to warrant the national headlines and public attention that stimulate debate.

The first discovery born from the new science—the identification of the gene that causes the very rare Huntington's disease—made the front page of many U.S. newspapers in 1983, as did the discovery in 1987 of the gene that causes muscular dystrophy. But the 1989 identification of the gene that causes cystic fibrosis was shoved off many front pages by the news that baseball great Pete Rose was banned from baseball for gambling. Indeed by 1989, new genes were being found at such a frequent rate that the mass media paid less heed when scientists identified the genes that render people susceptible to coronary heart disease or multiple sclerosis, and they gave little or no note to the identification of genes responsible for producing exotic enzymes with unpronounceable names.

Few in the public recognized, therefore, the swelling cascade of gene discoveries. As a result, sociologists, ethicists, politicians, and editorial writers, all of whom were quick to expound on the issues surrounding, say, the transfer of a gene from one species to another, were largely silent about the revolution taking place in the field of gene discovery.

At most, pundits have limited themselves to hailing only the most immediate and obvious impact of gene identification, that of the conquest of three thousand known genetic diseases, ranging from such well-known disorders as cystic fibrosis and muscular dystrophy to such obscure-sounding diseases as neurofibromatosis and xeroderma pigmentosum. By no means is this an unimportant consequence of the new science. The ability to reach into the human genome and pluck out the single gene that renders its possessor retarded, crippled, or destined for an early death is of immeasurable importance to tens of thousands of humans who suffer genetic disorders. Identification of a defective gene reveals the biochemical pathways through which the defect wreaks havoc on the human mind and body. From there it is only a matter of time, perhaps a short time, until scientists find a way to block the pathway by developing new medicines to cure or prevent a disorder. Few could quarrel with such advances in medical science.

But unmasking the identity of genes that render people *susceptible* to any of a host of chronic and crippling diseases is an entirely different matter. These aberrant genes do not, in and of themselves, cause disease. By and large, their impact on an individual's health is minimal until the person is plunged into a harmful environment. The person who inherits genes that interfere with the normal processing of cholesterol, for example, may be little affected unless he or she habitually consumes a high-fat diet. Because such a person is genetically unable to handle an overload of fat efficiently, the arteries begin clogging up with cholesterol-laden deposits. The ultimate result is a heart attack, perhaps a fatal one, early in life. Similarly, a person who inherits a genetic susceptibility to cancer may never develop a malignancy. But if that person smokes cigarettes, eats (or fails to eat) certain foods, or is exposed to carcinogenic chemicals, the chances of developing cancer are several times higher than the cancer risk faced by the genetically nonsusceptible person.

The list of common diseases that have roots in this kind of genetic soil is growing almost daily. As of this writing, it includes colon and breast cancer, Alzheimer's disease, multiple sclerosis, diabetes, schizophrenia, depression, at least one form of alcoholism, and even some types of criminal behavior. How many more human ills will be added to the list is unknown, although some contend that almost every disorder compromising a full and healthy four score and ten years of life can be traced in one way or another to a genetic vulnerability.

Identifying these susceptibility genes allows physiologists to understand the underlying biochemical causes of major human ills. This knowledge opens the door for pharmaceutical chemists to devise new therapies to treat these ailments, as in the development of a potent new type of drug that reduces cholesterol levels in the blood much more quickly than previous anti-cholesterol medicines. There already are hints that recent discoveries of susceptibility genes will lead to the creation of drugs to halt the relentless progression of the nerve damage characterizing multiple sclerosis, to correct the biochemical derangements of the brain that cause schizophrenia, or even to repair genetic defects that predispose some people to cancer.[1]

At the very least, finding these aberrant genes permits those who have inherited them to avoid dangerous environments; the person who knows he is genetically susceptible to coronary heart disease can be taught from a very early age to avoid high-fat foods, while the alcoholism-prone individual can be warned to avoid alcohol.

It is possible, and would seem desirable, to know early in life, perhaps at birth, if an individual is genetically susceptible to one of these devastating disorders. Avoidance of risky environments could be taught early in childhood and thus would become a natural part of the individual's life-style without the discomfort and difficulties of giving up cherished habits. Early detection of genetic susceptibility also could alert parents and physicians to watch for symptoms of incipient disease in a child, so that treatment could begin before too much damage is done (as, for example, might be possible in diabetes).

But the new technology has ramifications far beyond the prevention and alleviation of disease. Once a defective or aberrant gene has been identified, routine testing of individuals for the gene is relatively simple, involving only a few drops of blood. As recently as 1987, a genetic test took as long as two weeks. Today, if a laboratory technician knows which gene to look for, he or she can identify it in as little as two hours. This "while-you-wait" test is now done manually. As use of the test expands, it almost certainly will be automated, and results will be available even sooner.

It is highly likely that within a decade tests for a variety of aberrant genes will be cheap and easy enough to permit testing of large numbers of people. Initially, only those persons who are at risk of inheriting a defective gene might be tested. For example, anyone who had a parent die prematurely of a heart attack might be tested—indeed, might want to be tested—to see if he or she had inherited one of the several defective genes that can render one susceptible to coronary heart disease.

As the list of known defective genes grows, there will be mounting pressure for mass screening of the population, at least of the newborn population, to pinpoint anyone predisposed to future illnesses. There is ample precedent for such mass genetic screening of newborn infants. The advent in the 1960s of a simple blood test for a genetic disorder known as phenylketonuria (PKU) led almost every state in the United States to mandate testing of all newborns for the disorder. There was

a good reason for such laws. If a PKU-affected infant could be identified at birth, he or she could be placed on a special diet to prevent
the mental retardation that characterized the disorder. Automated tests
to screen newborns for at least three other similar "inborn errors of
metabolism" are being developed. It is routine these days to test
newborn black infants for sickle-cell trait and newborn Jewish infants
for Tay-Sachs disease, both of which are inherited.

If mass screening can be justified for these relatively rare genetic
disorders, then screening newborns for susceptibility to such common
diseases as diabetes, schizophrenia, coronary heart disease, or cancer
would seem even more worthwhile.

Of course, society might decide to use such tests in other ways.
There are circumstances where the interests of society in knowing an
individual's genetic susceptibility would be paramount. It would seem
too risky for an airline to permit a person with a genetic tendency for
alcoholism, or for a premature heart attack, for that matter, to take
command of a wide-body jet with its 350 passengers—or for a trucking
company to permit such a person to roam the highways in a fifty-ton
truck. A corporate board of directors might be considered irresponsible
to stockholders should it elect a president and chief executive who
might be genetically predisposed to manic-depression or Alzheimer's
disease. A police force could hardly risk hiring and arming a young
man or woman who was genetically predisposed to schizophrenia.
Almost certainly voters, or at least the press, will demand to know the
genetic profile of presidential candidates, while opposition senators
may well inquire into the genetic predispositions of presidential nom-
inees to the cabinet and the Supreme Court.

But, the first uses of the new gene identification technology haven't
been the diagnosis of either the adult or the newborn. Instead, the
technology has been used for genetic diagnosis of the unborn. Almost
immediately upon the discovery of the genes that cause muscular
dystrophy and cystic fibrosis, a precise prenatal test was offered to
pregnant women who, by virtue of having already borne one child
with either of these disorders, were known to have a high risk of
bearing a second afflicted child. For these women and their husbands,
the precision of the new prenatal genetic tests has been of immeasurable relief. Often it has allowed them to proceed with a pregnancy
in full confidence that they will not bear a second totally disabled child
who may be doomed to die before or during the second decade of life.
The advent of the tests has allowed many couples the opportunity to
have children they wouldn't otherwise have borne.

Prenatal testing for severely crippling, life-shortening disorders that inflict an almost unbearable burden on a family is one thing. Prenatal testing for susceptibility genes whose effects may not be manifest for years is another matter. Within the foreseeable future, young couples will have the power to determine as early as the eighth week of pregnancy whether their future child will be subjected to the nightmare of schizophrenia or manic-depression before age twenty, or will be threatened with cancer or multiple sclerosis before age forty.

We may think it too farfetched to believe that young couples in the future might actually choose whether to continue a pregnancy on the basis of genetic tests that predict a child's incapacities, deficiencies, or susceptibilities in the distant future. But social mores change rapidly in the face of new technology. The advent of techniques like amniocentesis and more recently chorionic villus sampling for just a limited number of chromosomal defects has led to an explosion in prenatal testing in the last decade. As a result, pregnancies are routinely terminated today for reasons that would have left previous generations aghast. Genetic counselors say it's not unusual to encounter young couples who, for whatever reason they may formally state, are actually seeking a prenatal test solely to determine the sex of the fetus. And sociologists say younger couples today are showing an increasing tendency to expect—and demand—only "perfect" children.

(There is, in fact, evidence that doctors soon will be able to test a human embryo for certain genes within hours after an in vitro, or test-tube, fertilization. Selecting an embryo for implantation into the mother on the basis of its genetic makeup may become more acceptable to future parents than terminating a pregnancy, considering the controversy over abortion.)

By 1989, the rate of new gene discoveries was such that at least one top molecular biologist predicted his colleagues soon would be pinpointing the kinds of genes that control intelligence, behavior, and even musical, artistic, or athletic skills. "There is absolutely no doubt in my mind that within the next ten to fifteen years the new technology will lead to the discovery of genes that are strong determinants of human behavior," said Robert Weinberg, one of the nation's foremost molecular biologists. He expects that behavioral characteristics, such as a talent for music-writing, will prove to be affected by a half dozen to a dozen genes. "When that happens, the debate over the relative importance of nature versus nurture in the developing human will be cast in an entirely new light."

Weinberg is a researcher into the mechanics of cancer-causing genes at the Whitehead Institute, the prestigious molecular biology center at Massachusetts Institute of Technology. Widely respected as a perceptive scientist, he is one of the first practitioners of molecular genetics to worry aloud about the "profound effects" of the gene-identifying science, calling its impact on society potentially "very corrosive."

"Right now our society runs on the premise that everyone has a biologically equal chance to be anything he or she wants," he said in an interview in his cramped office on the MIT campus in Cambridge, Massachusetts. "But what will happen when, in fact, the scientists find strong evidence that everyone's fate is greatly affected by the inheritance of a group of very specific and identifiable genes. There are many who scoff at such a possibility. But the trajectory of scientific advances makes me believe it's inevitable. How we handle this new information is something we as a society ought to begin thinking about and discussing as soon as possible."

Indeed, by late 1989 a handful of social ethicists were beginning to discuss among themselves their fear that the gene discoveries would lead to the creation of a new social stratum called the biological underclass. People identified as having certain genetic weaknesses, they argued, might be discriminated against by employers, they might have difficulty getting health and life insurance. Businesses, for instance, might be less willing to hire people predisposed to illnesses that could drive up the employer's health insurance costs. Employers might want to begin screening prospective workers to detect their genetic susceptibilities.

The ethicists sprinkled their talk with such new dark-sounding terms as "genetic discrimination," "genetically unemployable," and "genetic labeling." And they began calling on the scientists to come out of their labs and begin discussing how best to keep a person's individual genome—his or her personal genetic profile—private, while still providing the benefits of preventive care and treatment.

It would be easy to dismiss such predictions as being in the realm of science fiction or, at the least, as being issues that can be dealt with in some future century. The fact is that the identification of human genes is accelerating at a rate that not even the molecular geneticists themselves would have predicted five years ago. These scientists are now asking the congresses and parliaments in industrialized nations to fund an organized international effort to locate and identify human genes. Akin to the Manhattan Project that led to the atomic bomb and the Apollo Project that landed men on the moon, the Human

Genome Project has as its goal nothing less than the "mapping" as rapidly as possible of every one of the estimated 50,000 to 100,000 human genes.

Whether the mapping of the genome proceeds as a formally organized effort funded by various governments or continues as a loose collection of independent research projects, it is inevitable that within a single generation the result will encompass the location and identification of the large majority of human genes.

The technology of gene identification or "mapping" cannot be stopped any more than the technology of the automobile, the machine gun, or the atomic bomb was stopped. Its uses, however, can be controlled. But to guide and control the uses of a new technology requires society to understand the technology and be aware of its potential.

The mapping of the genome is the climax of an adventure that actually began in 1900 with one of the most unusual coincidences in the history of science. Publication of Darwin's theory of evolution less than half a century earlier had ignited a revolution in botany. Plants that could be bred rapidly in controlled environments offered the quickest means to confirm that nature did, in fact, conserve mutations that enhanced survival of the species, discarding those that detracted from survival. There was no question, at the time, that characteristics of parents were passed on to offspring—that had been assumed by even the most primitive societies—but little thought had been given to exactly how this inheritance took place.

In 1900, at the peak of this botanical golden age, three botanists—Karl Correns in Germany, Erich von Tschermak in Austria and Hugo De Vries in Holland—working independently and completely unaware of each other's research, discovered that inheritance of traits followed a remarkable and predictable pattern. The pattern was unmistakable evidence that each inherited trait in an offspring was determined by two heredity units, one from each parent.

In an act that has drawn admiration from scientists ever since, all three searched the botanical literature and discovered that an Austrian monk named Gregor Mendel had published a largely unnoticed paper in 1866 reaching the same conclusion.[2] All three dutifully credited Mendel with primacy in the discovery of the laws of heredity. Today, every high school biology student in the world is taught how Mendelian laws determine whether a child will inherit his mother's brown eyes or his father's blue eyes.

In 1904, an American scientist, Walter Sutton, concluded that these mysterious heredity units were contained in the sausage-shaped structures in the central nucleus of every cell. These structures are called chromosomes, from the Greek words for "color body," because early geneticists had to stain them with special dyes to make them visible under the microscope. Like the heredity units, the chromosomes come in pairs, with one half of each pair being inherited from one parent and the other half from the other parent. (It is now known, after years of confusion, that humans have twenty-three pairs of chromosomes.) In 1909 a Danish biologist, Wilhelm Johannsen, coined the word *genes*, from the Greek for "giving birth to," as a name for these invisible heredity units carried by the chromosomes.

The effort to find a gene and determine what it does and how it works was launched in 1906 when the American Thomas Hunt Morgan began his famous studies of the chromosomes of fruit flies. It soon became clear that almost every biochemical characteristic in every living creature was determined by the genes. The genes, in other words, were the basis of all life. By 1941, two other Americans, George W. Beadle and Edward L. Tatum, had shown that the genes' function was to produce proteins, the structural components of all living matter, as well as the enzymes that carry out the infinite variety of second-by-second chemical reactions that make life possible.

Despite these remarkable discoveries, the nature of the genes remained a mystery. No one knew what a gene looked like, how it worked, or how the cell managed to replicate its genes in order to pass a complement on to its offspring. By the 1940s, however, a series of discoveries began suggesting that the genes were composed of an acid found in the nuclei of cells. This nucleic acid was rich in a sugar called deoxyribose and hence was known as deoxyribonucleic acid, or DNA.

The secret of the genes was suddenly revealed in 1953 when British physicist Francis Crick and his young American colleague, biologist James Watson, determined the physical structure of DNA. The structure resembles a spiraling molecular ladder: two sugar-phosphate strands linked together at regular intervals by "rungs" of simple molecules. Each rung consists of two opposing molecules called base nucleotides, or simply bases. The "ladder" is twisted into the now-famous double helix.

There are only four kinds of bases, A (for adenine), T (thymine), G (guanine), and C (cytosine). But they are dotted along each strand by the hundreds of millions. More astonishing is the fact that everywhere there is an A base on one strand, it is coupled to a T base on the opposing strand. Each C base on one strand is, without fail, paired with a G base on the opposing strand.

Each chromosome is composed of an incredibly long but tightly coiled piece of double-stranded DNA. Genes, it seemed clear, are segments of DNA. Thousands of genes are aligned along each chromosome's DNA.

Each gene, as Beadle and Tatum had discovered in 1941, is the blueprint the cell used to assemble a protein. Proteins are long necklaces of simple amino acid molecules, each protein consisting of a distinct sequence of amino acids.

When the structure of DNA was unveiled, it became evident that the sequence of the bases in a gene specified the sequence of amino acids the cell is to follow in assembling that gene's particular protein. The bases, in other words, constitute an alphabet of only four letters, G A T and C. Each base "word" specifies a particular amino acid and a complete sentence of hundreds or even thousands of base "words" specifies a protein.

In the early 1960s the genetic code was broken. It was discovered that only three base "letters" were needed to specify an amino acid. The "word" CTG, for instance, orders the amino acid leucine to be laid in at that spot in the newly forming protein, while the word GAT specifies aspartic acid. Since there are four base "letters" in the alphabet, there are sixty-four possible three-letter words in the genetic

The Double Helix

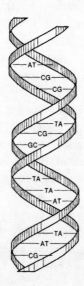

language. There are, however, only twenty amino acids. Hence, there is a redundancy in the code; five other "words" besides CTG can specify leucine, for instance.

But the structure of DNA also revealed another genetic secret, how during cell division the DNA copies itself—replicates—to produce a new set of genes for the daughter cells. The secret lay in the pairing of the bases, A to T and G to C. At replication, the two strands of DNA zip apart down the middle. A new strand begins to form on each of the originals, laying in an A wherever there was an opposing T, a T where there was an A, a G to a C and a C to a G. The end result is two new double-stranded lengths of DNA, identical in base sequences to the originals.

With the genetic code broken and the secret of gene replication revealed, biochemists began developing the tools of gene manipulation. In the mid-1970s enzymes were discovered that could snip pieces

DNA replicating itself

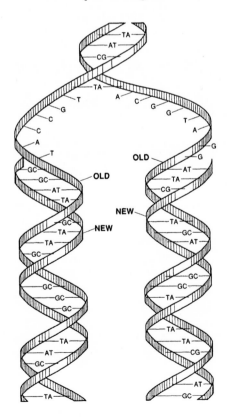

of DNA out of one organism's genome and splice it into the genome of another organism. The transfer of human genes into the genetic apparatus of bacteria and yeasts opened the era of genetic engineering, whereby fermenting of genetically altered microbes led to the production of literally gallons of once-precious human proteins for medical use.

As marvelous as these genetic engineering feats seemed at the time, they were limited. Genetic engineers could manipulate only the genes whose protein products already were known. For example, protein chemists in the 1960s determined the amino acid sequence of insulin, a fairly simple human protein. When the genetic code was broken, it was possible to deduce the makeup of the snippet of DNA— the gene—that caused cells to produce the particular sequence of amino acids that comprised insulin. With the tools of genetic engineering it was possible then to fish blindly in a pool of DNA snippets and find the one piece that conformed to the amino acid sequence of insulin. Hence, the human insulin gene was quickly isolated and inserted into bacteria that proceeded to produce vats of human insulin. The human insulin avoided the allergic reactions that limit the use of beef or pork insulin for many diabetics.

But where the insulin gene came from was anyone's guess, for what neither the genetic engineers nor the molecular biologists could do was to peek into the chromosomes of a human cell and pinpoint the location of a gene. They hadn't the vaguest idea of where the insulin gene lay among the tens of thousands of genes contained in the twenty-three pairs of chromosomes.

Thus, in the late 1970s, geneticists and molecular biologists were frustrated in the decades-long hunt for the human gene. It was as though they stood on the steps of some huge Alexandrian library where the accumulated genetic knowledge of three billion years of evolution was stored. They knew the architecture of the library with its twenty-three pairs of wings. They knew the alphabet used to write its 50,000 to 100,000 tomes. They even knew the titles of a few of the genetic tomes. But they had no way of finding which books lay in each wing and on each shelf. What they desperately needed was a map.

Then, almost unexpectedly, several lines of biological and medical research began to converge, and the means of drawing the genetic map became evident. The "genomic library" was suddenly accessible. Humans were about to gain unprecedented mastery of their genetic destinies.

In retrospect, from our 1990 viewing stand, we can see that the threads of research leading to this breakthrough stretch back in many directions in time and space, to a crippled child in Spokane, to an impromptu lecture in a Utah ski resort, to a young woman with heart disease in Detroit, and to a simple question about cancer asked by a pediatrician in Texas. One of the threads reaches back to an apartment in Los Angeles in 1968.

CHAPTER 1

From Curse to Crusade

milton Wexler was a Hollywood psychoanalyst who counted a number of movie stars among his friends and patients. His two daughters, Alice and Nancy, were all that any father could ask for: bright, attractive, and full of promise. In 1968, Alice, the older of the two women, was twenty-five and deep into graduate studies in Latin American history at Indiana University. Twenty-two-year-old Nancy, who would play a key role in the coming genetics revolution, was following her father's profession. She held a bachelor's degree from Harvard's Radcliffe College, had studied briefly under Sigmund Freud's daughter, Anna, in England, and had just returned from Jamaica, where, under a Fulbright scholarship, she had spent the academic year studying the mental health problems of the poor. That fall she would begin her doctoral studies in psychology at the University of Michigan.

The two women had come to Los Angeles to celebrate their father's sixtieth birthday. But Wexler had asked them to his apartment to talk of their fifty-three-year-old mother and his ex-wife, Leonore. As the girls sat on his bed, the shades drawn against the fierce California summer sun, Wexler described recent events in Leonore's life. A calm and rational man who had spent a career choosing his words carefully, he struggled to control his emotions.

A few months before, Leonore had to report for jury duty, he told them. She had gotten up early, dressed carefully and impeccably as was her habit, and driven downtown, putting her car in the parking lot near the courthouse. As she was crossing the street, she was approached by a policeman. "Aren't you ashamed of drinking so early in the morning?" he admonished her. Only after her anger at such rudeness subsided a few moments later did she realize that she had been walking erratically, weaving in a manner that suggested drunkenness. Alarmed, she called her ex-husband, who, trying to keep her calm, suggested she continue on to court and then come over to his office.

This much, while distressing, wasn't totally surprising to Nancy and Alice. When they were young, Leonore had been an attentive, loving mother, but she also suffered bouts of moodiness, depression, and erratic behavior. "I remember my mother was a lousy driver," Nancy recalls. "Once when I was a teenager, sitting in the car with her going down Sunset Boulevard, she kept driving in this sort of herky-jerky way, pulling and pushing her foot on the gas pedal, as if she was some sort of Sunday driver who didn't know what she was doing. I lost my temper and yelled at her. It was cruel thing for a kid to do, but, of course, back then, I didn't know. She was so upset she stopped the car and told me to get out, which was a very unusual thing for her to do."

Such behavior seemed to afflict Leonore more frequently as the years passed. While Nancy was in Jamaica, Leonore had come to visit her. "I was shocked by the change in her," Nancy says. "We slept together in the same bed. Before we fell asleep, as we talked, her arms and legs seemed unusually restless; I'd never seen that before." Moreover, "she had become so infantile she couldn't go off on her own. She couldn't shop. She seemed so absolutely lost inside herself."

Shortly after Leonore called her ex-husband in a near panic, Wexler called a colleague, a neurologist, who hurried over to the psychoanalyst's office as soon as Leonore arrived. After listening to Leonore's story and the history of her symptoms, his diagnosis was quick and certain.

Leonore had Huntington's disease.

"It shocked me," Milton Wexler says. "All I felt was horror. All I could think about was my two beautiful daughters, both working on their doctorates, and I got sick."

That afternoon in the bedroom Wexler told his daughters of the neurologist's verdict. Huntington's disease was incurable and untreatable. It would slowly but relentlessly destroy their mother's mind, steal control of her body, and, eventually, take away her life.

Then Wexler had to broach the matter that had sickened him so. Huntington's disease is a genetic disease. Leonore had apparently inherited the Huntington's disease gene from her father and there was a fifty-fifty chance she had unwittingly passed it along to each of her daughters. And, what's more, there was no way to know if they had inherited it.

"Imagine what it must be like telling your grown children something like that," Nancy says two decades later. "Alice and I immediately decided we would never have children, that we'd never knowingly take the risk our parents unknowingly had taken. We said it boldly to Dad, as if that at least was something we could do. Dad said that if he had known, he wouldn't have taken the gamble of having us, but that he wasn't sorry that we were here. We were stunned, too stunned to cry, and, I think, mostly worried about our mother and each other." Then, with a brashness of youth and, most certainly, in an effort to reassure their father, the young women declared, one after the other, "Well, fifty-fifty isn't so bad."

Among the three thousand or so known human genetic diseases, Huntington's has its own peculiar tragedy. It is a genetic time bomb. The inheritance of the Huntington's disease gene is implacably fixed at the moment of conception, but the symptoms of the disease don't become manifest until the carrier approaches or reaches middle age, usually years after he or she has borne children. Thus, the defect is preserved and unwittingly passed on through the generations. It is a rare disease, with about 25,000 American victims afflicted at any one time, and another 125,000 who, like Nancy and Alice Wexler, are at risk of having the gene and who may yet develop the disease.

"It is the most terrifying disease on the face of the Earth," says Milton Wexler, "because its victim is doomed to absolute dementia as terrible as Alzheimer's disease, a loss of physical control akin to muscular dystrophy, and a wasting of the body as bad as the very worst of cancers."

The disease is named after George Huntington, a mid-nineteenth-century physician who practiced medicine on the eastern tip of Long Island. In 1872, in the only scientific paper he ever published, titled "On Chorea," Huntington described a scene he encountered while driving along the road from Amagansett to East Hampton, the area that today is the playland of the well-to-do and the literati but was farmland at the time. "We suddenly came upon two women, mother and daughter, both tall, thin, almost cadaverous, both bowing, twisting, grimacing," he wrote.[1] At the time, such uncontrolled movement was diagnosed as an adult form of Saint Vitus' dance. But Huntington was aware that the frenetic movement and staggering gait of the two women was far different from the Saint Vitus' dance seen in children who temporarily developed uncontrolled movements during a fever (now known to be a temporary result of bacterial infections).

As a child, Huntington often had accompanied his grandfather and later his father, both physicians who had practiced in the same area, as they made their rounds. He had seen similar patients then and knew that the disease progressed "gradually but surely, increasing by degrees and often occupying years in its development until the hapless sufferer is but a quivering wreck of his former self."[2]

Then, in an observation that was to enshrine his name permanently in the literature of medicine, he noted that the strange chorea (the medical term for purposeless movements) appeared to run in families, "an heirloom from generations away back in the dim past." He observed that, when either parent had the disease, "one or more offspring almost invariably suffer from it, if they live to adult age. But if these children go through life without it, the thread is broken and the grandchildren and great-grandchildren are free from the disease."[3]

In 1968, almost a century later, little else had been learned about Huntington's disease. It was known that the inheritance of the mysterious gene from parent to child followed the pattern of what the Austrian monk Gregor Mendel called a dominant gene (or "hereditary factor," in Mendel's words) in pea plant experiments. It was later shown that, as with Mendel's peas, human traits are controlled by two factors, or genes, one coming from the father and one coming from the mother. In his plants, Mendel showed that in some cases an inherited trait will be expressed because of the presence of a single dominant gene from just one of the parents, while in other cases two recessive genes, one from each parent, are required for a trait to be expressed. (If brown eyes is a trait in humans that is dominant, only one brown-eyed gene is needed from one parent for the trait to be expressed, whereas blue eyes require the inheritance of two recessive genes, one from each parent.)[4]

Beyond this slender fact, Huntington's disease remained an enigma. The purpose and functioning of the defective dominant gene and its normal counterpart was a complete mystery. Try as they might, scientists could find no biochemical abnormality in the blood, urine, or tissues of Huntington's disease victims that even hinted at how the

How Recessive Inheritance Works

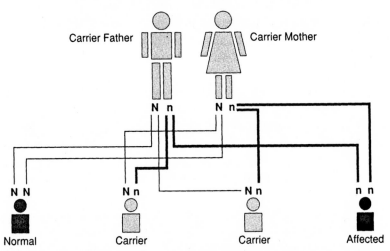

Both parents usually unaffected, carry a normal gene (N) which is generally sufficient for normal function despite the presence of its faulty counterparts (n)

How Dominant Inheritance Works

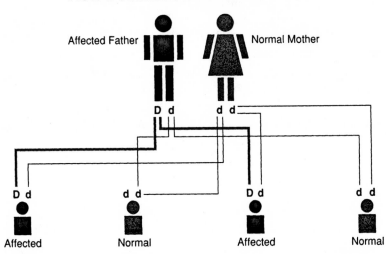

One affected parent has a single faulty gene (D) which dominates its normal counterpart (d)

gene carried out its lethal work. The anatomical and biochemical abnormalities seen in the brain at autopsy gave no clues. And there wasn't any way to find where the gene resided on the twenty-three pairs of human chromosomes. Without even one of these discriminating factors, there was no test that could tell Nancy or Alice whether they had inherited their mother's deadly, defective copy of the unknown gene, in which case they, too, would develop the disease in mid-life—or whether they had inherited her normal copy, in which case, in Huntington's words, the thread was broken.

While Leonore was still lucid, Nancy and Alice questioned her about her family background, hoping, in part, that piecing together the genesis of the disease might somehow lift the veil of uncertainty over their own future. Gradually, it emerged that Leonore's father, a Russian Jew who changed the family name to Sabin from Maziowieski when he passed through Ellis Island, had come to America at the turn of the century, while still an adolescent. The family had always assumed he left to escape the persecution and poverty that oppressed the Jews in Eastern Europe. But Wexler now strongly suspects her grandfather's immigration was a conscious or unconscious attempt to escape the tragic malady that afflicted the Maziowieski clan.

Leonore was the youngest of Sabin's four children and his only daughter. By the time Leonore was born, Sabin must have been showing signs of mental and physical instability, since, in 1921 (when his daughter was only six years old), he was institutionalized in a state hospital on Long Island. Seven years later he died.

By the time Milton Wexler and Leonore Sabin met, however, her father's fate was a submerged family secret. Leonore was a bright, vivacious young woman who loved to dance. She had studied for a master's degree in biology at Columbia University and, when they met, she was teaching high school in the Bronx. Milton Wexler was tall and dapper, stylish in dress and manner. He already had a thriving law practice on Wall Street and, for the times, was doing quite well. (Soon afterward, Wexler abandoned his practice and followed some friends into what he felt would be the more exciting field of Freudian psychology, eventually receiving a doctoral degree from Columbia University.)

Wexler later told his daughters that, on several occasions, while he was courting Leonore in Brooklyn, Leonore's mother and brothers took him aside as though they wanted to divulge some private family matter, but they seemed unable to specify exactly what it was.

In 1950, when the Wexlers were living in Topeka, where Wexler researched schizophrenia at the famous Menninger Clinic, word came that Leonore's eldest brother, Jessie, forty-eight years old, was "having trouble" and had to quit work. The three Sabin "boys" had been delightful uncles, Nancy recalls. In their youth they had been fun-loving, and, fascinated with "swing" and jazz, all played musical instruments. When they were older, Paul and Seymour formed the Paul Sabin Orchestra and played at the Tavern on the Green and the Hotel Delmonico in New York City, and at posh hotels in Miami. Jessie fancied himself an amateur magician and delighted his nieces with fabulous tricks, spinning coins around his fingers at top speed and making them magically appear out of his ears, his nose, and his pockets.

Nancy and Alice remember, however, that on their occasional trips to Brooklyn from Topeka, their uncle Jessie began to seem "odd." He spoke with a slur and shuffled when he walked. "I remember on one visit, he seemed to be having trouble with his magic, the tricks didn't work, the coins fell to the floor while his fingers danced and twisted," Nancy notes.[5]

Jessie soon sought a neurologist who, after examining him, asked to see the other two Sabin brothers. All three, he concluded, suffered a "progressive degeneration of the brain... of unknown etiology... with poor prognosis." Huntington's chorea wasn't mentioned at first, and the family outwardly behaved as if the strange disorder was a coincidental occurrence instead of what it was, a frightful trick played on all three brothers at birth.

In fact, while the news surprised Milton, it almost demolished Leonore. Growing up, Leonore had somehow come to believe that her father's disease, which she knew to be Huntington's, only affected males. But when Jessie's illness was finally described as Huntington's, a neighbor of the Wexlers' in Topeka, a neurologist, told them that men and women were equally at risk. Leonore suddenly realized that not only were her brothers likely to experience the worst that was possible from the family curse, but that she and her young daughters also were at risk. Neither Milton nor Leonore discussed this possibility, burying it as if that could make it go away. Nonetheless, Nancy believed that from then on her mother's "equanimity about herself and her children's genetic safety was smashed."[6]

Realizing he would soon have the burden of supporting his brothers-in-law, Wexler abandoned his beloved research at the Menninger and moved to Los Angeles to open a private practice in psychoanalysis. One by one, the Sabin brothers died of the disease. Milton and Leonore continued to keep the nature of the curse in the shadows, never telling Nancy and Alice. "I had known there was something the matter in my mother's family, something that had hit all her brothers, but I didn't know what it was," Nancy says. "My mother was always so hidden about it, it was very hush-hush."

Milton Wexler does recall briefly reading about Huntington's disease during this period, but he convinced himself that his wife and children had escaped the disease, since Leonore showed no signs of being affected. "As a rational man steeped in science, it's almost unbelievable to me now that I could have ignored the possibility," he says. "But denial is common with the disease. It's a way to survive." Wexler later viewed Leonore's erratic moods and deepening depression over the years as the consequence of mental conflicts traceable to the death of her father in her formative childhood years, compounded by the early deaths of her brothers and the sudden fatal heart attack her mother suffered soon after the Sabin brothers died.

For Leonore, "things began unraveling," Nancy recalls, "but they were so subtle, so insidious, no one knew what was going on. As I grew older it became clear that something was wrong. She snuggled my sister and me like we were toys or pets, at some elemental level unaware of who we were or what we were about. She listened but often did not respond and did not know how to make it better. She cuddled and kissed and fussed but she also grew softer, sadder, silent, more listless, vague and undefined. It was as if some dark subterranean river was taking her away from us.

"In retrospect, I do not know if the ominous gene was beginning to take hold or if she was mainly numbed and in shock," Nancy says.[7]

As Leonore drifted, Milton's career blossomed. His life was filled with famous people and stimulating research. He began to write screenplays. Coming home at night was "deadly," he recalls, and in 1964, the Wexlers separated and then divorced. "I was convinced beyond doubt that, if there was any danger, she had escaped it by then," says Wexler.

But, on that hot afternoon in 1968 when Milton revealed Leonore had Huntington's disease, the family was forced finally to face facts. Nancy and Alice grieved for their mother, for themselves, and for the children they vowed they would never have. And they began to experience the spectre that hangs perpetually over the children of Huntington's disease victims. A few days after that devastating session with her father, Nancy was at a friend's house helping in the kitchen. She dropped a carton of eggs, splattering them over the floor. There was a sudden silence in the room. "No one said a word but I know everyone was thinking what I was thinking: 'Is this the beginning? Do I have it, too?' I started watching and worrying, and to this day, any moment of forgetfulness or physical slip can get me going," she says. Alice agrees: "There isn't a day I don't think about it."

Wexler, however, was not one to passively accept his daughters' uncertain fate. That August afternoon, "before we could even focus on what it meant, he was telling us, 'Let's fight this thing,'" Nancy says. "He told us how he'd already begun to contact scientists and was making plans to raise money." That September as he drove his daughters back east to pursue their graduate studies, Wexler talked of how he could marshal his talents to solve the mysteries of Huntington's disease.

At the time, such a hope did not seem unreasonable. By 1968 four Nobel prizes had been awarded within the decade for discoveries relating to the genes. Newspapers and magazines were full of articles hailing unprecedented breakthroughs in the understanding of the genes and the possibility that the genetic diseases might well be eliminated by replacing bad genes with good ones, a feat later called gene therapy.

What these breathless accounts of gene manipulations glossed over, however, was the primitive nature and severe limitations of the advances. Reports of gene cloning and gene engineering usually failed to mention that the trick was being accomplished only with a few genes whose molecular structure was already known, such as the gene for insulin, the first gene to be engineered. Scientists were still unable to pull off the trick that would be needed to solve the mystery of Huntington's disease and hundreds of other genetic diseases, that of identifying the exact chemical makeup of unknown genes—and the vast majority of genes remained hidden. If DNA was the chemical encyclopedia of life, the molecular biologists had learned to read only, "See Dick and Jane run."

As Wexler quickly learned, Huntington's disease was of little interest to molecular biologists or anyone else, for that matter. The disorder had received a modicum of public notice when it first incapacitated and then killed folksinger Woody Guthrie, who had gained fame for his ballads about the Depression and the oppressed working class, and who is known to a new generation as the father of Arlo Guthrie. When Woody died, a foundation called the Committee to Combat Huntington's Chorea was formed by his widow, Marjorie. By 1968 she was campaigning, not very successfully, to raise money for research into the disease. Wexler decided to set up a California chapter of Guthrie's organization and, aided by the friends he had made among the wealthy and the famous, he began trying to raise money.

He also began making the rounds of scientists who might advise a fledgling foundation on how to tackle the problem of Huntington's disease. He learned that outside of neurological circles, few scientists, including geneticists and biochemists, were more than dimly aware of the disease. Indeed, research on it was practically nonexistent. The federal government's medical research arm, the National Institutes of Health (NIH), was being heavily funded with money from a Congress eager to see cancer and heart disease conquered, but not a single federal cent was earmarked for Huntington's disease and little more for any other inherited disorder.

There were reasons for this lack of interest. Few Americans were threatened by Huntington's. The disorder never struck outside the families that had been haunted by it for generations. Only a few dozen such families were known in the United States and, like the Sabins, they tended to keep their affliction a secret. By contrast, other genetic diseases, such as muscular dystrophy or cystic fibrosis, were not only more visible and more common but they also could strike unexpectedly. About one-third of all cases of muscular dystrophy are new mutations arising spontaneously at conception when a quirk of nature somehow damages the genes, while cystic fibrosis seems to strike without warning because the parents are unaware that they both carry the recessive gene.

Perhaps the biggest obstacle of all was the lack of interest among medical scientists. Huntington's disease was not a disease where a scientist could make his or her reputation. Other genetic diseases were attracting scientists because they had a foundation of research that offered a good chance of making significant discoveries. Biochemical abnormalities had been found in both cystic fibrosis and muscular dystrophy, showing research potential that an astute young scientist might be able to exploit and thus gain a name for himself.

But, given the dearth of knowledge, any scientist tackling Huntington's disease would have to start from scratch—and there were no clues as to *where* to start. Moreover, the study of inherited diseases like Huntington's focuses heavily on comparisons of family members who inherit the disease with those who don't. This is time-consuming enough in diseases that are apparent at birth; for a genetic disease that strikes in adulthood, it seemed more than anyone could ask. A geneticist tackling Huntington's disease would have to wait two or perhaps three decades to see which of a victim's children inherited the Huntington's gene and which escaped it. In short, a scientist could expect to spend a lifetime studying the disease and its victims without the slightest prospect of making even a small niche for himself in medical history.

Milton Wexler set out to remedy the state of indifference. He was uniquely qualified to take on what appeared to be a hopeless cause, being highly motivated by the fear that the disease would someday claim his daughters. But he also considered himself primarily a scientist and, while he never mentioned it, he deeply regretted abandoning his career in research. He wanted to get back to science, and Huntington's gave him the chance.

Wexler was the kind of idealistic twentieth-century American who believed that problems, given the proper effort, were meant to be solved, the kind who believed that Americans, given their ingenuity and resources, could conquer space and overcome disease.

"I may be an optimist or a fool," says Wexler. "But I know that solutions are found somewhere between stubbornness and grandiosity."

During his days at the Menninger, for example, he had boasted to his colleagues that he could cure schizophrenia and, to prove it, had challenged them to give him their toughest case of the mental illness. They presented him with Nedda, a backwoods woman in her late thirties who had been hospitalized for years, and who, as Wexler himself described her, "was weird beyond belief." For months, Wexler tried every bit of psychoanalytic technique available, without success. She would paint her face with lipstick and tie her hair with dozens of bows made of toilet paper. She also spoke in what psychologists call "word salad," a sort of gibberish. Later he was to write, in one of several scientific papers he published about the case, that she put him through "hell-fire," that she would assault him with "the violent language of a jungle tigress," with "crude sexual overtures," and even with physical violence, once ripping the shirt off his back.[8]

One day, in the midst of one of these rampages, Nedda kicked Wexler in the groin. His instant response was to slap her hard across the face. Before he could even regret his unprofessional act, Nedda, stunned, asked him, "Why did you do that?" Those were the first coherent words she had spoken to him. The slap led Wexler to believe that Nedda required rigid restrictions, that for years she'd lacked the kind of limits placed on her behavior that adults normally give their children. By following a course of treatment that included strict discipline and rewards for good behavior, Wexler was able, within months, to get Nedda out of the hospital and into a semblance of normal life. Though her treatment initially was the result of serendipity, Wexler had a lesson for life: Stubborn confidence can pay off.

To launch his battle under the aegis of Marjorie Guthrie's committee, Wexler's first step was to form a board of scientific advisors. One of the scientists Wexler corralled was Seymour Benzer, a geneticist at the California Institute of Technology in nearby Pasadena. Benzer and several other prominent scientists in Southern California were invited to present their work at a gathering that Wexler hoped would lead him to scientists whose area of research might parallel the problems found in Huntington's. One of Benzer's students had been studying the genetics of fruit flies. Benzer explained the work, giving what Milton thought was a "terrific" lecture on a strange genetic mutant called "drop-dead fruit flies," whose late-onset disease seemed to mimic Huntington's. The mature adults often deteriorated suddenly, flopping about out of control before dying. X rays of the fruit flies' brains showed that they had "cheeseholes," in which sections of the brains mysteriously withered away.

"I was very naive, but I thought well, here we are, we've got an animal model for the disease," says Wexler. "If we could just figure out why the cheeseholes formed in fruit flies, we'd be able to know what caused the similar problem in people and we'd be on the road to finding a treatment."

But in a long talk after the meeting Benzer and several others urged the pyschoanalyst to pursue a different tack, to seek out young, ambitious scientists who hadn't yet made their mark. "You've got to find researchers who are desperate for funding, who are schooled in the latest scientific techniques but who are young enough to reject the accepted notions," Benzer told him.

Taking this advice to heart, Wexler hired a Caltech graduate student, Ronald Konopka, to make the rounds of laboratories in order to spot aspiring young scientists who were casting about for an unoccupied research niche that would bring them both funds and a chance at fame. In a ten-day marathon around the country, young Konopka, a tall, gangling, laconic fellow, traveled to some of the country's top biology labs, including those at Massachusetts Institute of Technology, Harvard, Yale, and Columbia. He was a great listener and was awed by the talent he saw. Sitting in the labs of the newly minted Ph.D.s, Konopka would ask them to discuss their work and, then, after listening for quite a while, would get up to leave, often without even telling them why he'd come. Only when he was pressed would Konopka explain the foundation's efforts to recruit young scientists. Frequently he offered the young scientists free transportation and thousand-dollar honoraria to come to a meeting in California to discuss their work, even though it was unrelated to Huntington's disease.

Right from the start, Wexler began to win a reputation that quickly spread from lab to lab.

Several years after joining with Marjorie Guthrie, he created his own foundation, calling it the Hereditary Disease Foundation, an attempt to net as broad a range of scientists as possible. This schism with Guthrie came after Wexler helped organize a fund-raising concert at the Hollywood Bowl, which included folksingers Richie Havens, Joan Baez, Pete Seeger, and Guthrie's son, Arlo. Marjorie Guthrie wanted the bulk of the money raised to help expand state chapters. Her strategy, like that of other disease organizations, was to create a large enough following to lobby Congress for money to back research. But Wexler wanted what little money they raised to serve as seed money to encourage scientists to consider studying Huntington's disease. After it became clear that Guthrie and Wexler had different philosophies, he spun his chapter off into a separate organization, although he and Guthrie would often continue to lobby for funds together.

Wexler's foundation used its first funds to back a series of workshops for the young scientists. Those attending the Huntington's disease workshops practiced the traditional method of sharing research— presenting a planned lecture with dozens of slides documenting their findings, followed by a brief, formal question-and-answer period. But Wexler soon became frustrated with this formalized approach, which seemed to stifle spontaneous give-and-take among the scientists present.

Finally, at one early workshop, an exasperated Wexler chastised his guest speakers and ordered them to ignore their written talks and their slides. Instead, he described to them the process he had used his entire professional career to goad hidden truths from his patients. This psychoanalytic process, he explained, was based on a willingness to "spin out ideas without concern for their quality."

"It was unheard of at the time," says Michael Conneally, a geneticist at Indiana University Medical Center, who was one of a handful of researchers already working in Huntington's. "We began to look forward to the workshops because it was so different. Milton would spend hours figuring out things like what kind of table we should sit at, or which restaurant would be most conducive to this sort of free-floating exchange."

A second strategy wasn't planned, but was pursued vigorously only when its impact on the scientists became apparent. Wexler's practice and life-style in Los Angeles brought him in contact with some of

Hollywood's most glamorous and famous stars. Many became his friends, and offered to help raise money for the fledgling foundation. (In the 1980s, Wexler wrote a successful screenplay, *The Man Who Loved Women*, with one of his friends, the director Blake Edwards.) At one of the earliest workshops held in Los Angeles, the scientists were invited to a dinner with several movie stars. Among those who attended parties for the scientists in later years were Carol Burnett, Ann Landers, Candice Bergen, Gregory Peck, Jennifer Jones and her husband Norton Simon (the wealthy industrialist), and Blake Edwards's wife, Julie Andrews. At one fund-raising party, Cary Grant served as the doorman.

"A lot of us were in our late twenties or early thirties and we'd spent our entire adult lives at the university, mostly in our labs," recalls Allan Tobin, a biologist at the University of California in Los Angeles. "To have this elegant and successful psychoanalyst interested in our ideas, and then to rub shoulders, so to speak, with these famous people was inspiring."

The Wexlers also brought something else to the foundation that scientists soon found captivating and irresistible—Nancy. From the beginning the Wexlers would introduce the scientists, most of whom confined their studies to microscope slides and tissue specimens, to people affected by the disease. The result was often riveting. But when the scientists learned that Nancy, who was about their age and also pursuing a career in science, was at risk, the impact often was even greater. "Over the years," says Conneally, "Nancy became this emotional glue that held us all together."

Nancy initially wasn't involved in her father's effort to organize an attack on Huntington's disease. After learning of her risk, she threw herself into her graduate work at the University of Michigan and, for a time, tried not to think about the danger. "I didn't tell anyone, because I was worried what they would think, that they might not want to associate with me," she says.

Then, in 1970, Leonore tried to commit suicide. "It was horrible," says Nancy. "It hit me for the first time how desperate she was, because suicide was so absolutely foreign to anything anyone in my family would ever consider." Returning to Los Angeles, Nancy Wexler found that her mother had taken a massive overdose of sleeping pills. Then, she lay in her bed facing photographs of Alice and Nancy, as if to say good-bye. She had been saved by the housekeeper, who had noticed Leonore's bedroom lights still on, even though it was late. When Nancy Wexler found out that her mother was furious at being saved, she realized, "My God, this is going to get worse."

It was then that Nancy threw herself into the disease work with a passion that matched her father's. "At first, I think I did it to do something, to feel as if I was taking some mastery over this calamity that had befallen us," she says. Her father was still formally associated with Marjorie Guthrie, so Nancy started a Michigan chapter of Guthrie's organization. She began to organize Huntington's disease families in the Detroit area, traveling from Ann Arbor to help counsel the families and find them financial support, medical help, and custodial care for their affected relatives.

One of the families included a middle-aged auto worker who lived in a trailer with his wife, who had Huntington's disease, and three children. "One day, he said he was going into the hospital for a hernia operation and he told me, 'I'm not coming out,' and he didn't," she recalls. "I think life for him was just too overwhelming. Working with them and the other families was terribly sad, but I felt absolutely compelled to do it," she says. At first the experience was draining and she felt increasingly depressed. More than once, the depression made her so tired she found herself falling asleep at the wheel driving back to school.

But the families were also therapeutic. As affected people and their families talked of their fear, anger, and desperation, she began to feel a kind of relief. "It was almost magical, as if having others say aloud what I'd been thinking, but not accepting, suddenly made my feelings manageable," she says. "Instead of just seeing Huntington's as my disease, I began to see it from my perspective as a psychologist, too." For the first months after she began organizing the families, she "had these two separate lives, but slowly they came together," she recalls. Eventually, her life became "totally immersed" in Huntington's disease and she decided to do her doctoral thesis on the psychological impact of the disease on Huntington's families. She also took a greater interest in her father's workshops, although, she says, she rarely understood the technicalities of the science being discussed.

For the first few years after that hot August afternoon in Los Angeles when she had learned of her family's medical history, Nancy had felt out of balance, as if she needed to grab hold of something solid after being tripped. "I remember walking along the beach after Dad told us about my mother and looking up at the palm trees and thinking, 'What does this strange disease called Huntington's have to do with us?'" she says. "It was as if some alien thing had invaded our lives and had suddenly changed it."

She felt as if the life her family had built upon the rock-hard foundation of middle-class values, education, and common sense was crumbling. Perhaps, she thought, it really never had existed at all. In its place was the sense that nothing could be trusted. The only thing, she found, that began to give her relief and a sense of stability was "the fight." As the years passed she found that she could stay upright by grabbing onto the fight, by waging war against the terrible disease that had invaded her family's lives.

Slowly, requests for grants to pursue particular aspects of Huntington's disease began to trickle into both the Hereditary Disease Foundation and the NIH. In 1972, the foundation held one of its workshops at a meeting organized by the World Federation in the Neurology of Huntington's Chorea in Ohio. As the invited speakers reeled off the results—or lack of results—of their research, it became clear that almost no progress had been made in unraveling the mysteries of Huntington's disease.

But then Ramon Avila Giron stepped to the podium.

Avila Giron was a physician in Venezuela who had devoted his doctoral thesis to a Venezuelan family that had an unusual number of Huntington's victims. Avila Giron's presentation included a grainy black-and-white film he had made in a remote village on the shores of Lake Maracaibo, a huge baylike body of water on the northern coast of South America.

Suddenly the audience was hypnotized. The film showed people, obviously afflicted with the staggering gait and uncontrolled movements of Huntington's disease, going about the village, neither ostracized nor treated as deranged or drunk by the other, unaffected villagers. Says Nancy Wexler: "Most of us with Huntington's in our family were used to seeing one member, maybe two, with the disease. Here we were seeing what looked like dozens of people in one village with the problem. It was eerie. Our first thoughts were that it was just strange. But several of us began to realize the villagers might be worth studying, although we couldn't quite say why."

Several of those present were skeptical of Avila Giron's claim that all the Huntington's victims seen in the movie were members of the same family. Nonetheless, the foundation board decided that someone should go to Venezuela and check out his claim that numerous families in the dozen or so villages sprinkled along Lake Maracaibo were afflicted with the disease.

However, little came of the discovery until 1976, when two separate events occurred that made a full-scale investigation of the Venezuelan families not only a possibility but potentially the breakthrough Huntington's researchers had been hoping for.

First, Congress finally recognized Huntington's disease as worthy of federal funding. For several years, Huntington's disease families in New Jersey had urged their congressman to get an appropriation for Huntington's disease research. The congressman dutifully introduced a bill creating a federal Huntington's Disease Commission. But the proposal, lumped in with similar bills to study other relatively rare diseases, was vetoed each time by President Gerald Ford. Finally, it was tacked onto a nurse training bill that had considerable support, and was passed.

The commission was created with Marjorie Guthrie as chairman, Milton Wexler as vice chairman for research, and Nancy as deputy director. Within six months the director quit and Nancy, who by then was teaching at the New School for Social Research in New York, was made executive director on her thirtieth birthday. "It was all very incestuous," she says.

The commission's job was to recommend avenues of research that could be funded by the NIH's National Institute of Neurological Disorders and Stroke, or NINDS. As the commission cast about for new ideas to recommend, the Wexlers and their advisors heard word of a pivotal discovery in Texas about a totally unrelated disorder. Two medical geneticists at the University of Texas Southwestern Medical School in Dallas, Michael Brown and Joseph Goldstein, had found the genetic defect that caused familial hypercholesterolemia, or FH, an inherited disease that leads to early, severe heart disease (discussed later in Chapter 9). The keys to the FH discovery were children who had inherited two defective copies of a gene, one from each parent, and were thus called homozygous for the defective gene. Their symptoms were twice as severe as their parents', each of whom had only one defective copy of the gene (and thus were heterozygous for the gene).

The discovery fired the Huntington's Disease Commission's hopes. If a child who was homozygous for the Huntington's disease gene could be found, his or her disease might be doubly severe and occur earlier in life. The biochemical pathways by which the gene inflicted its damage might stand out more clearly and offer undiscovered clues to the disease.

But the number of known Huntington's disease families in the United States was small. The odds were astronomical against a man and a woman, both of whom were unwittingly carrying a single copy of the Huntington's gene, actually meeting, falling in love, and marrying. And the odds against them bearing a homozygous child, if they had children, were four times as great.

"As soon as the [Brown and Goldstein] paper was published, a number of us had the same idea," says Nancy Wexler. "If there is a homozygote anywhere in the world, the child probably is somewhere among the Venezuelan villagers along Lake Maracaibo." If it was true, as Avila Giron had said five years previously, that all the Huntington's victims in the village were members of the same extended family, then it was possible, in fact likely, there had been some inbreeding, given the remoteness of the village. If so, there was a chance that an occasional homozygote for Huntington's disease had been born.

Indeed, by 1978, the pursuit of the Huntington's gene had become an obsession for Wexler, both as a career and a crusade. On Mother's Day that year Leonore died. The last few years and months of her life were truly a nightmare for the Wexler family, and each new tragic episode fueled Nancy's impatience for, if not an answer, at least a research avenue worth pursuing. In a not-yet-published memoir Nancy Wexler wrote in 1989, called "50/50: Genetic Roulette," one can read, in her eloquent and straightforward prose describing Leonore's progressive demise, Nancy's mounting anger, and the images that stalked her as she and her father rummaged through the closet of existing science for a clue to solving the disease:

> My elegant mother who was always dressed in clothes which perfectly fit her size 5 figure, to whom personal hygiene was synonymous with character—my mother was in cotton "wash and wear," drenched after every meal with spilled food. New shoes would wear out the moment you put them on her feet from her incessant movements. In one nursing home she used to sit in a chair in the narrow space between her bed and the wall. No matter where you put the chair, the force of her movements would edge it back against the wall until her head began bashing into the hard plaster. She had a bloodied scab on the back of her head from hitting the wall but you couldn't wean her from the chair or secure it in place.
>
> When my mother was sitting relaxed, her fingers—which I always will remember as long and beautiful—kept up a constant motion as if she were playing a sad tune on a silent piano. Her face twisted, her toes jumped. When she walked, her left side sagged below her right. Her legs would suddenly buckle as if

an unexpected blow had hit her unaware in the back of the knees. . . .

One of the worst torments was not being able to understand my mother as her control over her speech began to deteriorate. At first she just sounded drunk, as if she had a large potato in her mouth. Then her speech became impossible to decipher. I would try to intuit what she might be talking about and give her options from which to choose. If I grasped a word or sentence, I would be sure to repeat it, even if I really didn't understand all she was saying, so that at least she could feel that she had connected. Her isolation behind garbled speech seemed unbearable to me. . . .

At the end she looked like a prisoner from Dachau, so thin that if she lay sideways in bed you almost couldn't see her. Her limpid brown eyes still haunt me with their sadness and their fear. She had worn a bald spot on the back of her head from rubbing against the pillow and even lambswool padding could not prevent her fragile skin from breaking down into bedsores from her incessant movements against the sheets.

Wexler writes that at Leonore's funeral the family read some letters she had written many years earlier to a childhood friend:

They were letters from the early days of her marriage, the coming of war, and when she was pregnant with Alice. They were cheerful, exuberant, witty letters, intelligent and thoughtful letters. They were the correspondence of someone I didn't remember and Alice scarcely knew. The letters reconstructed before our eyes a woman who had been vibrant and alive and a fighter. For the first time, now that she was freed from her pain and suffering, now that it was finally over, we could allow ourselves to feel the momentousness of the loss. We could afford to remember her when she was healthy and miss her.[9]

The Huntington's Disease Commission came out with its final report in early 1978, just a few months before Leonore Wexler died in a Los Angeles nursing home, her frail body tied down to keep her from hurting herself. In the report, the commission urged an expedition to study the Venezuelan villagers and search for a homozygote.

But the expedition to Venezuela was about to gain a new urgency. Six months before the trip, the Wexlers' foundation held one of its workshops at Marina del Rey, the home to many of Los Angeles's most glamorous yachts. The meeting was designed to update its grantees and potential grantees on the status of research into the inherited diseases. Early in the workshop, Allan Tobin, the Harvard-trained molecular biologist who had become the foundation's research director, went through a laundry list of promising research avenues that might be pursued in the inherited diseases, particularly in Huntington's disease. At one point Tobin, a tall, broad-shouldered man with a bush of curly hair, began tugging at his curls in obvious excitement, and, to the bafflement of his audience, began to drop such unfamiliar words and phrases as restriction enzymes, polymorphisms, gene libraries, genetic maps, and Southern blots. The audience was baffled.

"I finally told him," Nancy Wexler recalls, "to stop free-associating aloud in front of people who had no idea what he was talking about and tell us what he meant."

Tobin had maintained ties with a circle of molecular biologists at Harvard and was aware that a new theory was stirring unusual excitement among biologists at nearby Massachusetts Institute of Technology. A few days after the Marina del Rey meeting Tobin called David Housman at MIT. Tobin and Housman had known each other during Tobin's days at Harvard, even sharing the same baby-sitter. Tobin earlier had invited Housman to lecture at UCLA the following month, and now Housman agreed to fill his host in on the new MIT research at that time.

Housman, although only thirty-three years old, was one of MIT's senior researchers in the fledgling field of biotechnology. Over dinner during his visit to Los Angeles, he conceded to Tobin that he knew little about Huntington's disease. But he was familiar with unpublished work underway at MIT in the laboratory of David Botstein, down the hall from his own lab. As Housman talked, Tobin realized that the search for a Huntington's disease homozygote might become secondary, because all the Venezuelan families could hold the secret to the disease.

CHAPTER 2

Alta

What excited Housman and Tobin were reports of a new technique for locating genes including, possibly, the Huntington's disease gene. Although nothing yet had been published on the new technique, it was being openly discussed at MIT, and it was known that an experiment to test the technique was underway at the University of Massachusetts medical school in Worcester. If it proved workable, it would be the breakthrough that geneticists had sought for half a century. Research on the inherited diseases would be changed in a fundamental way.

The rumors and reports of a breakthrough circulating on the molecular biology grapevine in the spring of 1979 were traceable to what was almost a chance remark made at a ski resort in Utah a year earlier. Each April, as the spring thaws begin to discourage skiers, a handful of biologists and geneticists at the University of Utah take their most promising graduate students to a lodge at Alta, Utah, high in the Wasatch Mountains that tower over Salt Lake City. During the two-day gathering, the students describe their particular research projects in short, informal presentations, followed by discussions between the students and the senior scientists present, including distinguished researchers from other institutions who are sometimes invited to participate. The meeting is a highly prized opportunity for the students to learn from the ideas of more experienced scientists and a chance for the senior scientists to catch up on what their colleagues in other laboratories are doing. All in all, the meeting is an opportunity for some high-caliber scientific kibitzing.

The students and many of the Utah faculty were anticipating the April 1978 gathering in particular. Utah molecular biologist John Roth had invited two outsiders to the Alta meeting that year, molecular biologist David Botstein of MIT and biochemist Ronald W. Davis of Stanford University. Roth had an ulterior motive for inviting Botstein and Davis. Each June the three of them taught a three-week postgraduate course in molecular biology at the Cold Spring Harbor Biological Laboratory on the northern shore of New York's Long Island, a once-famous center for genetics research that was regaining its prominence under the directorship of Nobel laureate James Watson, the codiscoverer of the structure of DNA. Roth felt the Alta meeting would give the three a chance to get together privately and finalize the curriculum for the Cold Spring Harbor course two months hence.

Both Botstein and Davis were rising young stars in the new sciences spawned by Watson and Crick's discovery of the molecular structure of DNA in 1953. Davis was among the new generation of biochemists who were perfecting and using the tools, the enzymes, by which DNA could be manipulated—snipped, cloned, transferred, and welded. Botstein was among the new breed of molecular biologists who, aided by the biochemists' enzymatic tools, were rapidly uncovering the secrets of how DNA functioned in the living cell.

The students were looking forward especially to the encounter with Botstein. The MIT biologist was gaining a reputation somewhat akin to that of a rock star among the young scientists. A large, barrel-chested man with jet-black curly hair and a gap-toothed smile, Botstein had a combination of self-confidence and garrulousness that graduate students rarely saw in their mentors. He loved to talk and relished long, often detailed accounts of his theories and concepts, of his scientific conquests, and of other scientists' flaws. He had the knack of making his listener, whether a friend or a momentary acquaintance, feel as though he, Botstein, was confiding some special new insight or bit of gossip to him alone and for the first time. To the delight of the young researchers, and often to the chagrin of their scientific elders, Botstein eschewed the sometimes forced disinterest and false politeness with which scientists critiqued each other's work. He didn't hesitate to label those scientists whose work he disdained as "parasites, bullshitters, and jerks." Moreover, his ad-lib monologues—which could be triggered by anything from a simple, mundane question to the delivery of a full-blown scientific paper—often were brilliant, for Botstein was one of those rare "conceptualizers," a scientist who could take a single fact and in a matter of minutes show his fascinated listeners how it fit into a much larger picture of scientific enterprise. Few walked away from an encounter with Botstein without a stimulating new appreciation of their own research.

For Davis and Botstein the Alta meeting was a chance to relax. "We were supposed to comment on the grad students' research, give some advice, that sort of stuff," Botstein recalls. "Nobody expected it to be much of anything but a nice, pretty, mellow place to go for a few days."

Nevertheless, on the second afternoon there occurred one of those rare coincidences that sometimes change the course of science, when the right remark is made at the right moment midst the right combination of people. The triggering event was a long-winded, complex statistical argument by a graduate student studying under Mark Skolnick of the University of Utah.

Mark Skolnick was a population geneticist, a discipline that interested Botstein and Davis only peripherally. Population geneticists study families to see if a particular trait, including a disease trait, is being passed from one generation to another in the pattern dictated by Mendel's laws of inheritance. If a disease afflicts certain members of a family in succeeding generations in the Mendelian pattern— Mendelian segregation, in the geneticists' jargon—that is prima facie evidence that the culprit is a deleterious gene, even though one doesn't know which gene it is or how it works.

In some diseases, such as Huntington's, where a single dominant gene causes the disease and there is a 50 percent chance each child will inherit the deleterious gene, the Mendelian pattern is so obvious that it can be spotted by studying the incidence of the disease in only three generations or even two generations, provided the disease strikes a sufficient number of the second generation (as it did in the Sabin family).

But if the disease is caused by a recessive gene, in which case the child has to inherit two copies of the deleterious gene, one from each parent, it is more difficult to spot the Mendelian pattern, since there is only a 25 percent chance that any child would inherit the two copies. To spot the segregation of recessive genes requires studying perhaps three, or even four generations, and studying aunts, uncles, and cousins, as well as parents, grandparents, brothers, and sisters. For the population geneticist, the larger the family the more "informative" it is for a suspect gene.

In the 1970s population geneticists like Skolnick were beginning to tackle an even more subtle type of inherited disorder. There was increasing evidence of genes that don't, themselves, cause disease but instead render a person susceptible to a disease. There are families that seem haunted by a higher-than-normal risk of any one of several of the more common diseases, such as cancer, coronary heart disease, depression, or schizophrenia. Yet, the disease itself doesn't segregate in the classical Mendelian pattern. A well-known example is breast cancer. Women whose mothers had breast cancer have a 10 percent higher-than-normal risk of developing breast cancer. Although higher than that faced by the average woman, this risk isn't as high as the Mendelian laws would predict for a recessive gene (a 25 percent higher risk) or a dominant gene (a 50 percent higher risk).

To population geneticists the 10 percent higher breast cancer risk for these daughters suggested they have inherited a gene or combination of genes that renders them vulnerable to some agent in the environment, perhaps radiation, chemicals, viruses, diet, or some unknown factor, that could trigger the malignancy. Only a few of the susceptible daughters might ever be exposed to the environmental agent under just the "right" circumstances and only these would develop breast cancer. The other daughters would be luckier, either because they failed to inherit the breast-cancer susceptibility genes or they were never exposed to the environmental agent.

To spot the segregation of such predisposing genes, the population geneticists needed huge families where blood relationships, back to great-grandparents and out to third, fourth, and fifth cousins, were well documented. "If you want to see whether a susceptibility to breast cancer is inherited, the first thing you do is look at sisters," Skolnick explains. "If there are only two or three sisters [in a family], there's little likelihood you'll see more than one case of cancer—even if something is being inherited—since cancer is just not that common. But if there are five, six, seven, or even nine sisters, the chance of seeing a cluster of breast cancer increases. And if the same clustering occurs among sisters in other branches of the family, and if you are able to look at three, four, five, or six branches—well, then you can begin to make some statistical assumptions that the breast cancer isn't due to chance but to something that's being inherited."

Almost no population in the world has better-documented genealogies than the Mormons of Utah, descendents of the five thousand pioneers who followed Brigham Young across the plains and mountains 140 years ago to the valley of the Great Salt Lake to create what he called "a Kingdom of God."

"From the very beginning, the Mormon leaders did two things for which geneticists will be forever thankful," says Skolnick. First, they encouraged large families in order to build the church quickly. For a century, Mormon families averaged eight children each, and even today their birth rate is twice the national average. And second, the founders established a church practice of keeping track of ancestors. As a result, the Genealogical Society of Utah—an organization affiliated with the Church of Jesus Christ of Latter-Day Saints, the formal name of the Mormon church—maintains genealogies going back a century and a half on thousands of Mormon families.

In the early 1970s the Mormon church agreed to give University of Utah geneticists access to its genealogical records. The university hired Skolnick in 1974 to computerize the vast network of Mormon pedigrees. Skolnick had spent the previous two years working in Parma, Italy, under Luca Cavalli-Sforza, a renowned geneticist who was trying to organize the pedigrees of several long-established Parma parishes—a collection of local villages—in an effort to trace ordinary genetic differences in a fairly homogenous population. Skolnick had a knack for computer programming and one day, while puzzling over how a computer could be programmed to use the pedigrees, happened to notice the web of an Italian fisherman's net. It inspired him to design a computer program that could flit through the intricate matings of an extensive family, following routes akin to the pattern of a fisherman's net. The program gave Cavalli-Sforza the tool he needed to track the passage of genes through the Parma families, and the grateful Italian, on hearing of Utah's access to the Mormon pedigrees, recommended his student and his fisherman's net computer program to organize the genealogical records in Salt Lake City.

Skolnick leapt at the chance. Although only thirty-two years old, he had been mapping his career goals for years, ever since he was in high school in San Mateo, California. His father was a psychologist, and among his friends was Joshua Lederberg, already a world-renowned geneticist and Nobel prize laureate. In dozens of conversations with Lederberg, the young Skolnick's interest was piqued by the idea of following the "flow" of genes through mass populations. Lederberg took the thoughtful Skolnick under his wing, and the two talked often about how man had evolved over the years through the conservation from generation to generation of natural mutations.

Genes, explained Lederberg, appeared to undergo deviations during cell replications. While some of these deviations produced severe illness and handicaps to the individual, others could produce more efficient and healthier humans. The species naturally "conserved" these good mutations simply by the Darwinian law of natural selection; the healthy mutants survived adversity better than their contemporaries. Skolnick was convinced early on that he could combine his interests in computers and science to help clarify how this evolving system was continuing in the present.

"I couldn't believe my luck at being at Utah at that particular time," Skolnick says. "I felt there was no place in the world as suited as Utah for discovering the genetics of disease, for determining if susceptibility to common diseases could be inherited."

Skolnick's hope was to check the genealogies against hospital records of breast cancer patients and uncover extended families who, over four generations, might have a higher-than-normal incidence of breast cancer. A study of a Mormon family already known to suffer hereditary breast cancer convinced the young geneticist that the Utah genealogies could reveal the existence of a common breast-cancer susceptibility gene if there was one. Discovering such evidence would be a coup for any aspiring geneticist. But Skolnick also quickly realized that the search for a breast-cancer susceptibility gene would be a protracted, tedious job and a long shot, even with the aid of the remarkably complete Mormon genealogies. And Utah, like every other university, pressures its scientific faculty to bring in a steady stream of research grants to pay the salaries of its scientists and their graduate students.

"I was convinced that the real contributions of the Mormon families would be in attacking the common diseases [such as cancer]," he recalls. "But, in science, sometimes you have to tackle what's possible to prove that you are on firm theoretical ground. Dreams don't win you research grants, results do."

To get such results and establish his reputation, Skolnick decided to use the Mormon genealogies first to tackle a genetic disease that in many ways epitomized the problems known to frustrate geneticists in most inherited diseases. These research results were the basis of the Alta lecture that, four years later, Botstein and Davis listened to with polite interest.

The disease was hemochromatosis, from the Greek *haima* for blood and *chromatos* for color. The nineteenth-century French and German physicians who first recognized the disease found the liver and other tissues of afflicted people loaded with an iron-rich pigment they thought was derived from blood. It is now known to be a disorder of iron metabolism in which the tissues absorb and retain large quantities of iron from the diet, so much so that many victims develop a peculiar bronze-colored skin. The iron overload leads eventually to cirrhosis of the liver, diabetes, and heart damage, among other problems.

Hemochromatosis usually doesn't develop until a person reaches middle age. At first, it was thought the disorder was the result of years of alcoholism, excessive intake of iron or some other heavy metal, vitamin deficiencies, or the accumulated effects of any of a number of other environmental agents. But in the late 1930s, as doctors noticed that the disorder frequently occurred in siblings of afflicted persons and in children of parents who were cousins (consanguineous marriages), evidence accumulated that hemochromatosis could be inherited.[1]

A forty-five-year-long controversy broke out over just how the disorder was inherited. When researchers tested the younger descendents of a patient with hemochromatosis, they found that, while one or two might have overt hemochromatosis, others only had higher-than-normal levels of iron in their tissues without having the disease. Some geneticists argued that these children and in some cases even grandchildren of the afflicted person were in the latent or early stages of hemochromatosis and would develop overt disease symptoms when they reached or passed middle age. If this were true, then hemochromatosis was due to inheritance of a single dominant gene.

On the other hand, countered other geneticists, the disease could be caused by a recessive gene. Those who developed the full-blown disease, they argued, were homozygous for the recessive hemochromatosis gene, that is they had inherited two copies of the defective gene, one copy from each parent. Their relatives who merely had higher-than-normal iron levels were heterozygous for the recessive gene, having inherited only one copy of it. The one copy was enough to cause a mildly abnormal iron metabolism but not enough to cause full-blown hemochromatosis.

Resolving the issue was of medical as well as scientific import. It would determine if doctors should treat the children and relatives of a person who had hemochromatosis. If hemochromatosis was caused by a dominant gene, it meant that those young descendents with abnormal iron levels were destined to develop the cirrhosis and other fatal consequences as they reached middle age. These consequences might be postponed, perhaps for years, by monitoring these young people's iron levels and periodically bleeding them when their iron levels got too high.

But if the disorder were caused by a recessive gene, those young offspring and relatives who had merely mildly high iron levels, indicating inheritance of only a single recessive copy of the gene, would be unlikely to develop the full-blown disease and wouldn't need monitoring or bleeding.

One way to resolve the dominant vs. recessive controversy, of course, was to wait until the grandchildren and great-grandchildren of an afflicted person reached middle age and see whether those who had high iron levels as youths developed overt hemochromatosis (as predicted by the dominant gene advocates), or whether they remained free of the disease (as predicted by the recessive gene theory). But this would take decades.

Moreover, this approach was complicated by several problems. Hemochromatosis is several times more common in men than women, presumably because, throughout their fertile years, women lose a considerable amount of iron each month during menstruation, so effects of the iron overload are postponed. Thus, tests of iron levels in any sister, daughter, or granddaughter under age fifty would be uninformative and probably misleading. Body iron levels also can be affected by alcoholism and diet, particularly the consumption of modern iron-fortified breads and breakfast cereals. If and when some of the grandchildren and great-grandchildren developed hemochromatosis, it would be uncertain whether it was because they inherited a dominant gene or because their drinking and eating habits had aggravated the partial expression of a recessive gene.

A quicker way to resolve the controversy would be to find a way to distinguish those offspring and relatives who had inherited the hemochromatosis gene from those who hadn't *before* they developed the full-blown disorder. Unfortunately, there was no clue to identity of the hemochromatosis gene itself. The unknown gene was hidden somewhere among the tens of thousands of human genes spread amongst twenty-three pairs of chromosomes, and the genes were far beyond the resolution of even the most powerful electron microscope. Despite the discovery of the structure of DNA, no one yet had figured out how to sift through the long strands of DNA and determine where one gene stopped and another gene began. Even if someone had figured out where genes began and ended, there would have been no way to know which particular gene was the cause of hemochromatosis, since no one knew how the gene caused the fatal iron overload. Indeed, even in the mid-1970s, more than two decades after the Watson-Crick discovery of the double helix, the possibility of ferreting unknown genes out of the human genome was considered a feat that *might* be accomplished sometime within a century.

Occasionally, however, geneticists had been able to track the passage of an unknown gene through the generations of a family with the aid of a genetic marker. A marker is some visible or detectable trait that is inherited along with the unknown gene. An example of a genetic marker might be eye color, an easily seen trait. If one were to look at two or three generations of a family and notice that everyone who developed a particular disease had blue eyes, while those who remained free of the disease had green eyes, then one might conclude that the disease gene was being inherited along with the gene for blue eyes. In such a case the blue-eye gene and the disease gene are so closely "linked" that they travel together as they are passed from parent to offspring. The blue-eye gene then would be a genetic marker for the disease gene. If the scientists knew where in the maze of genes the marker was located, then the approximate location of the gene in question would be known, too.

Although geneticists have occasionally checked a family to see if eye color is linked to a genetic disorder, it has rarely proven to be a useful genetic marker. One reason is that there is too little variation in eye color among humans. In some major ethnic or racial groups, for instance, brown eyes are predominant. Even if a deleterious gene were linked to the brown-eye gene in such groups, there would be no way to trace its inheritance, since there would be no way to tell whether a child's brown eyes came from the father or the mother.

To be useful, a genetic marker has to take one form in the father and a different form in the mother. Only then can the geneticist look at a child and tell whether he inherited the marker—and the linked deleterious gene—from his father or from his mother. In other words, a useful genetic marker is a genetic trait that occurs in a variety of forms, or morphologies, in humans; it has to be a polymorphic trait, one that can have many forms.

Genetic markers, however, are hard to find. Genes have a habit of becoming "unlinked," that is, separating from each other. This occurs when a new germ cell—an egg in the female, a sperm in the male—is being formed. Unlike the rest of the body's cells, a germ cell has one-half of each of the twenty-three pairs of human chromosomes, since, at fertilization, the sperm and egg each contribute their one-half of a complete set of chromosomes to the newly conceived being. Thus, when the germ cell is being formed, it "picks" pieces from the bearer's paired chromosomes, somewhat like a diner in a

Chinese restaurant picking some dishes from column A and other dishes from column B. The rules the germ cells follow in this process of recombination, as it's called, are presently unknown and the recombination is unpredictable. It never occurs exactly the same way twice, which explains why siblings are different from each other even though they inherit their genes from the same two parents.

The phenomenon tells population geneticists much about the relationship of a marker to a gene. As a result of recombination, a genetic marker that appears linked to an unknown disease gene in one or two generations suddenly is no longer linked in some members of the third and fourth generations. The deleterious gene and the marker gene separate at some point in this recombination phenomenon. The closer the marker gene lies physically to the disease gene, the more likely the two genes are to survive the recombination process still linked together. Thus gene mapping is a search for markers tightly linked to a disease.

The chances of finding a genetic marker that is both polymorphic and tightly linked to the gene of concern are extremely low. Consequently, useful genetic markers are rare. One of the first linkages of a gene to a human genetic marker was found in a family in the mid-1930s when color blindness, which is genetically determined, was found to be linked to the hemophilia gene. That is, males in the family found to have hemophilia were also color blind.

In the case of hemochromatosis, the geneticists had looked vainly for a genetic marker that was linked to the hemochromatosis gene so they could track its inheritance through the generations and settle the dominant vs. recessive issue In 1975 French geneticists reported hints that a complex of a dozen or more genes seemed to accompany the hemochromatosis gene as it was passed from parent to offspring. The genes in this complex could not, themselves, be identified. But they produced a set of proteins that lay on the surface of cells. These proteins were detectable and, thus, could serve as markers. These particular cell-surface proteins were first discovered on the tiny white blood cells called lymphocytes and are known collectively as human lymphocyte (or leukocyte) antigens, or simply the HLA complex.

Because there are more than a dozen of these HLA genes proteins and because many of them have many forms, almost no two humans possess the same combination of HLA genes. They are almost as polymorphic as fingerprints. And, equally important to the hemochromatosis researchers, a blood test can determine how closely any two persons' human lymphocyte antigens match—or mismatch.

Other geneticists immediately seized on the French suggestion that the HLA complex might be a genetic marker for hemochromatosis. Evidence began to accumulate that, in a particular family, certain HLA proteins seemed to be shared by those who had the disease or an iron overload, while the same HLA proteins often were missing from their relatives who had normal iron levels. This hinted that certain HLA genes were relatively tightly linked to the mysterious hemochromatosis gene and usually (although not always) accompanied it whenever it was passed on to an offspring.

Several University of Utah physicians, particularly Corwin Q. Edwards and George E. Cartwright, long had been interested in the hemochromatosis mystery and knew of several cases of the disease among the Mormons. With Skolnick and his fisherman's net computer program available, they decided to use the Mormon genealogies to settle the hemochromatosis inheritance issue.[2]

In a study that would serve as a model for later gene searches, the Utah researchers tracked down more than 250 relatives of 10 hemochromatosis patients, including third and fourth cousins. In one case, 117 near and distant relatives of one patient were located. Blood samples were taken to test each relative for iron levels and HLA proteins. The tests uncovered 55 relatives of hemochromatosis patients who themselves had abnormally high iron loads, seven of whom had such high levels they were considered to have full-blown hemochromatosis.

Skolnick's computer program tested for linkage between the HLA genes and hemochromatosis. Skolnick and a graduate student, Kerry Kravitz, set up computer models of dominant and recessive inheritance of the HLA genes to see which model best explained the pattern of iron overload in the families. As the results rolled in, it became increasingly clear that the hemochromatosis gene was recessive. The geneticists could identify twenty Mormons with high iron levels who clearly had inherited two copies of the hemochromatosis gene, one from each parent, and thus were homozygous for the gene. Many of these already had been diagnosed with hemochromatosis. But in eight cases—six women under the age of menopause and two men in their twenties—the disease hadn't yet progressed to the point where there was detectable damage to the liver. These undiagnosed relatives, being homozygous for the gene, were clearly going to develop the full-blown disease eventually. All the male homozygotes and all but two of the women were immediately bled to deplete their iron stores and, it was hoped, stave off the inevitable damage.

For another 145 relatives, many with abnormal iron levels, the inheritance of the HLA marker showed they could have inherited only one copy of the gene. Only five needed precautionary treatment for excessive iron levels. While 140 of these heterozygotes were in little danger, the knowledge that they possessed one copy of the deleterious gene was a warning to be on the watch for the disease among their relatives and, especially, their children and grandchildren.

The analysis turned up an unsettling fact about hemochromatosis: If the Mormon population is any guide to the general population, the recessive hemochromatosis gene—but not necessarily the disease—is forty times more common than previously suspected.

There was one discovery in the hemochromatosis story that seemed of little more than academic importance at the time. It hadn't been known exactly where the hemochromatosis gene lay among the twenty-three pairs of human chromosomes when geneticists first began their effort to understand the pattern of inheritance of the disease. But the genes for the HLA complex were known to lie on chromosome number 6 and appeared tightly linked to the hemochromatosis gene. Thus, it seemed that the hemochromatosis gene also must lie on chromosome 6, close to the HLA genes.

This "mapping" of the hemochromatosis gene to chromosome 6 with the aid of the polymorphic HLA marker, was a preview of what, a decade later, would become one of the most ambitious projects ever attempted in biology, the mapping of the entire human genome.

At the time of the Alta meeting, Skolnick and his colleagues were in the final stages of writing up their hemochromatosis study for publication. It seemed appropriate to summarize it at the Alta retreat. Recollections more than a decade later are dim on exactly who said what in the few moments after Kerry Kravitz, Skolnick's graduate student, had presented the hemochromatosis story in, what seemed to Botstein, somewhat tedious detail.

Skolnick took the floor briefly, however, and made clear to Botstein the power a closely linked marker gave researchers in discovering a previously hidden gene. Then, Jon Hill, another of Skolnick's graduate students, lamented how frustrated the Utah researchers were in their search for a breast-cancer susceptibility gene because they lacked polymorphic markers like the HLA complex. Whatever was said, the remark suddenly galvanized Botstein and Davis.

"I remember there was this long silence in which Davis and Botstein just looked at one another," Skolnick says. "Then they suddenly started talking about transposons, Southern blots, deletions, and restriction enzymes. Finally, Botstein looked at me and said something like, 'Theoretically, we can solve your problem, we can give you markers, probably markers spread all over the genome. Nobody has specifically looked for them or [when they saw them] thought about them as anything but junk. But they exist. It's not proven, but I think they exist over all the chromosomes. Moreover, they can be identified and their locations on the chromosomes pinpointed.'

"Botstein is a truly inspired thinker and he just went on explaining this concept that he was creating right there in front of us," Skolnick adds. As Botstein talked, he frequently turned to Davis for detail and elaboration of some of the more technical aspects of the concept.

Skolnick and his graduate students had little understanding of the details, but more than a decade later he remembered the shiver down his spine as he thought of the powerful tool these men were describing. He could only wonder, Were they right?

As molecular biologists slowly had unraveled the secrets of DNA over the years since 1953, Botstein explained, they occasionally encountered short stretches of human DNA that seemed to vary from person to person. These short stretches, in other words, were polymorphic. The variations occurred in the sequence of the DNA bases.

By 1978 molecular biologists learned how to determine the sequence of the bases on a given segment of DNA: they learned how to "read" DNA. Although it was tedious work, they did such "sequencing" whenever they were trying to understand the function and purpose of some interesting DNA segment. The base sequence is the same in all humans for 99.9 percent of the DNA. But during these sequencing efforts the biologists occasionally encountered short stretches of DNA where the base sequence differed from person to person. Some of these variations presumably were in genes and spelled the difference between individual humans. But many variations seemed to have little to do with genes, and their purpose, if any, was unknown. The biologists, being more interested in basic, general principles that applied to all humans rather than in the minor differences between humans, had paid little attention to these polymorphic variations in base sequences and, in fact, often found them more of a nuisance to their experiments than a source of new insight.

The polymorphisms, as the variations came to be called, could be any of several kinds. One person's DNA might lack one or more bases that exist in the same stretch of his neighbor's DNA. Or John might have a sequence of a few bases that is repeated several times, whereas in the same stretch of Mary's DNA, the same sequence occurs only once.

Such polymorphic differences in base sequences, Botstein and Davis were willing to bet, were sprinkled up and down the DNA of each of the chromosomes. If the variations were as frequent as Botstein and Davis suspected, then most unknown genes probably lay near at least one of the polymorphic stretches and probably traveled in tandem with the marker as it moved from parent to child.

In short, the base-sequence polymorphisms could be the ideal polymorphic genetic markers. All that was needed was a way to spot these polymorphisms in each member of a family.

Botstein had one more surprise for the incredulous geneticists. There was a way to identify the base-sequence polymorphisms. Only a few years earlier, the biochemists had uncovered one of the dark secrets of bacteria. Bacteria have a nasty habit of transferring genes among themselves. If there is a mutant gene that protects one of the microbes against, say, a powerful antibiotic, it isn't long before the new mutant gene is being transferred to neighboring bacteria, and a colony of antibiotic-resistant bacteria emerges to confound humans.

To take on a new gene, however, a bacterium first has to snip open its own DNA and then splice in the new gene. Each type of bacteria has its own favorite spot for inserting new genes. Thus, each type produces an enzyme that will snip its DNA at exactly the same spot each time.

The biochemists were uncovering a growing number of these DNA-snipping enzymes in bacteria, each one of which cut DNA at a different but predictable spot. Known as restriction enzymes, these chemical scissors had become vital tools of the exploding field of commercial genetic engineering, allowing the gene engineers to snip out genes from human DNA and splice them into bacteria for mass production of human proteins that could be used as medicine, such as insulin for diabetes or interferon for some forms of cancer.

In the test tube as well as in the bacteria, a restriction enzyme scouts along a strand of DNA until it comes to a specific sequence of bases. It then severs the DNA, always cutting it at exactly the same spot in the sequence. For instance, in the streptomyces bacteria, an enzyme called SAC-1 always zeroes in on a sequence of bases reading G A G C T C, and whenever it finds such a sequence, it severs the DNA between the T and final C.

The restriction enzymes could be the keys to finding the DNA polymorphisms, Botstein told the geneticists. A decade later no one recalls exactly the examples Botstein used, but here is an oversimplified version of the concept he tried to get across:

Suppose the polymorphism is a sequence of four bases, T G A T, that is repeated a different number of times in each person. As described previously, the sequence is repeated three times in John's DNA but it occurs only once in Mary's DNA:

Suppose the T G A T polymorphism lies between two "cutting sites" of the SAC-1 enzyme. The stretch of DNA with the polymorphism from each person is then "digested" with the SAC-1 restriction enzyme that settles on G A G C T C and cuts the sequence between the T and final C:

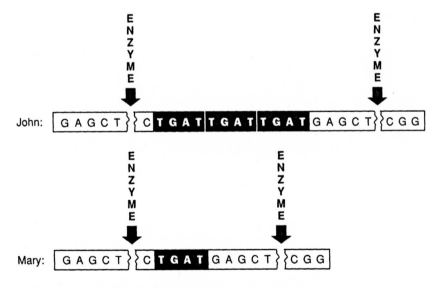

The length of the fragment snipped out by the enzyme from John's DNA would be longer than the fragment snipped out of Mary's DNA. John and Mary's DNA could be distinguished by the length of these fragments cut out by the SAC-1 enzyme, a form of identification as specific as a fingerprint.

There were other variations in base sequences that would similarly produce fragments of different lengths when the DNA of different individuals was exposed to the same restriction enzyme. The ultimate result would be the same whatever the case: If two individuals were polymorphic for the base sequences along the same stretch of DNA, the restriction enzyme would scissor out fragments of different lengths. And there were simple laboratory techniques for comparing lengths of DNA fragments, Botstein explained.

With these techniques one could use a DNA polymorphism as a marker to track the inheritance of a gene from a parent, say the father, to the child. The father, of course, has two copies of the polymorphism, one inherited from each of his parents. If his DNA is extracted and digested with a restriction enzyme, it will yield two fragments, a short one and a long one, representing his two versions of the polymorphism.

Now, suppose it's known that the father has a defective gene that travels with the longer fragment and a normal version of the gene that travels with the shorter fragment. One could then take the child's DNA and digest it with the same restriction enzyme. If one of the child's fragments is the same length as the father's longer fragment, then the child obviously inherited the longer version of the father's polymorphism—and the defective gene linked to it. If, however, one of the child's fragments is as short as the father's shorter fragment, then the child must have inherited the father's polymorphism that is linked to the normal version of the gene.

(The child's DNA, of course, would yield two restriction fragments but the other fragment would be the one inherited from the mother. A similar comparison of the mother's and the child's restriction-cut fragments could reveal which of the mother's two copies of the poly-morphism the child inherited.)

Although the possibilities of Botstein's on-the-spot theory sent a pulse of adrenaline through Skolnick, he was skeptical that Botstein and Davis, right there before him, had solved one of the greatest obstacles facing gene discovery. Moreover, Skolnick had to concede that the technology was over his head. "I spent the next few days and weeks talking to everyone," he says. "I had very little knowledge of molecular biology and I couldn't figure out how Botstein was going to find these fragments or exactly what they were."

At one point, Skolnick called his father's friend, Joshua Lederberg, who by then had become president of the prestigious Rockefeller University in New York. "I said I had been talking to Botstein and that we had a really great idea that would revolutionize human gene mapping. I was excited and I wasn't sure I was making sense. He interrupted me and told me what Botstein's idea was. Lederberg thinks very fast, and even though he hadn't heard about the idea he put two and two together. That was when I was convinced the idea was worth pursuing."

Botstein, who reviewed his impromptu theorizing on the plane from Utah to Boston, also grew increasingly intrigued by the possibility of using the polymorphic DNA sequence variations as genetic markers. But he was in the midst of numerous projects and left the notion alone. Meanwhile, Skolnick kept collaring everyone he came across, testing the possibility that Botstein and Davis were right. "I had this religious feeling about it all," Skolnick remembers. At a meeting at the National Institutes of Health, Skolnick discussed the idea with Maurice Fox, a biologist at MIT who, it turns out, had been Botstein's mentor and had offices and a laboratory near Botstein's.

Back in Cambridge a few days later, Fox had Botstein describe the newborn theory to him. But Botstein told Fox that he hadn't any intention of taking on the effort to find the DNA polymorphisms himself, even though he was sure the polymorphisms were there. "I regarded it as an idea, a theory. I thought it was a neat idea and I understood right away its importance. I wanted to see it carried out but I didn't want to actually search for the damn things myself," Botstein recalls. Fox, however, had a candidate, a former graduate student who had both the talent and the expertise to track down the DNA polymorphisms.

Ray White, who had studied for his doctorate under Fox a year after Botstein, had recently finished his postdoctoral studies at Stanford and had landed a position with the University of Massachusetts medical school at Worcester, an hour's drive west of the MIT campus in Cambridge. Fox knew White was restless and casting about for a more exciting project.

"Fox called me and told me about Botstein's idea," White recalls. "I thought about it a long time—overnight—and called Botstein the next day. He said, 'I'm glad you called; you were the second person I was going to call.' I never asked him who the first person was." The two met, Botstein outlined his concept and how he thought the DNA polymorphisms could be found. "Neither Botstein nor Davis were willing to commit their labs to the project," White says. "I had the time, the inclination, and the technology, so I took it on."

White had planned to vacation in Oregon, where his parents lived. He was an accomplished pilot and flew his own plane across the continent. On the way back, he stopped off at Salt Lake City and spoke with Skolnick, who described the potential power the Mormon families would provide in any search for new markers.

Back in Massachusetts the first order of business for White was finding money to fund the project. By early fall White, having consulted frequently with Botstein throughout the summer, completed and sent in a proposal to the National Institutes of Health. There was still a lingering public distrust of any experiments involving the laboratory manipulation of genes. The NIH grant had to be approved not only by the scientific advisory committees but by the institutes' Recombinant DNA Advisory Committee (RAC), a procedure that would delay the grant for several months.

By this time the biologists needed a name for the polymorphisms they would be searching for. DNA polymorphisms seemed too vague. They settled on "restriction-fragment-length polymorphisms." In publication, the name quickly would be reduced to RFLPs, given the scientific journals' practice of reducing multi-word nouns to initials after the first reference. Verbally, RFLPs degenerated to "riflips."

Others in a similar situation might have waited for White to find a riflip, and then published the results of his experiment with, perhaps, an ending comment that the riflips might be used as markers for unknown genes. But, true to form, Botstein saw far broader implications in the riflip concept. If riflips could be found at regular intervals along the chromosomes, they could serve as starting points for locating any gene, deleterious or normal. The riflips could serve as benchmarks for genetic surveyors in the same way those little concrete geodetic markers spread around the nation serve as starting points for land surveyors. If geneticists discovered that a particular gene they were seeking was linked to a particular riflip, and the location of the riflip was known, then they would know the approximate location of the gene. The first and main task, then, would be to develop a riflip "map" of the chromosomes.

The young researchers decided they had better publish the concept of a riflip map before someone else took credit for it. Since the scientists had been talking it up to almost everyone who would listen, the threat that someone else might claim it as his own was very real. Neither Botstein nor White nor Davis was conversant with the finer points of using genetic markers in families to locate and track genes. This was the province of the population geneticists. Skolnick, the young Utah geneticist, was the natural choice to carry this out.

White's grant proposal was circulated among the four authors—Botstein, White, Skolnick, and Davis—each adding and refining sections within his own expertise. White and Botstein outlined the techniques for uncovering a riflip.[3] They then calculated that, to survive the recombination phenomenon, a riflip and a gene would have to be within ten million bases of each other. At that distance, there would be a 90 percent chance that the riflip and gene would stay linked as they were passed from one generation to the next. Thus, it would be useful to pinpoint at least 150 riflips spaced roughly every twenty million bases along the twenty-three pairs of chromosomes.

For his part, Skolnick calculated that, if the riflip and the unknown gene were tightly linked, the gene could be spotted by studying the inheritance of the riflip in as few as three five-child families, or fifteen people. If the linkage was loose—the riflip and the gene more likely to separate—it might require studying a riflip's distribution in as many as three hundred relatives of a family before the gene could be pinpointed. Obviously, the Utah Mormons with their extensive, documented genealogies would be the logical group in which to begin using the riflips to pinpoint unknown genes, particularly deleterious genes.

Although it would be months before their final manuscript was ready for publication, Botstein, White, and Skolnick were talking openly and freely about their concept, and word was spreading rapidly. That's how Allan Tobin of the Hereditary Disease Foundation had heard about the concept in Los Angeles.

At a gathering of molecular biologists in the summer of 1979 at the Banbury Center, the conference center located near and associated with Watson's Cold Spring Harbor laboratory on Long Island, Botstein and White gave a full oral presentation of the riflip hypothesis. It was, White recalls, met with some skepticism.

"Among the more kosher molecular biologists [at the Banbury meeting], there was a lot of bitching and grumbling about it [the hypothesis]," White says. "Several people were quite annoyed that we would be so presumptuous" as to propose mapping the entire human genome, he recalls. Some of their fellow molecular biologists argued that there wouldn't be enough variation, or polymorphism, in the riflips for them to be useful genetic markers, while others said there would be too much variation. Such a reaction "was the first hint that we were really on to something," he adds.

By November, the paper was in final form and was submitted to the *American Journal of Human Genetics*. After revisions in January, the full paper appeared in the journal's July 1980 issue. It began: "We describe a new basis for construction of a genetic linkage map of the human genome."[4]

In the paper, Botstein and his colleagues only briefly broached possible uses of a riflip map. It would be useful in prenatal diagnoses, they suggested. Moreover, with riflips as genetic markers, "parents whose pedigrees might indicate the possibility of their carrying a deleterious allele [gene] could determine prior to pregnancy whether or not they actually carry the allele and, consequently whether amniocentesis might be necessary."[5] Persons whose family histories hint that they might have inherited a gene for some disease, like cancer, that shows up later in life "ought to be able to determine in advance of obvious symptoms whether they are at risk of the disease," the authors noted. And the riflips could help biologists uncover defective genes that so far had remained invisible.

If anything, Botstein, White, Skolnick, and Davis were short-sighted in their 1980 landmark paper. Within months of its publication, molecular biologists and geneticists around the world were teaming up and launching searches for riflips linked to major genetic diseases. The attempt to map each of the 50,000 to 100,000 human genes, a feat that was considered to be a century in the future, was now underway.

CHAPTER 3

Worcester

On October 16, 1979, nine months before the Botstein-White-Skolnick-Davis paper was published, Arlene Wyman, the post-doctoral student in White's laboratory in Worcester, held a piece of developed X-ray film up to the light and stared intently at a sprinkling of black smudges on the film. The previous day Wyman had attached radioactive tracers to DNA fragments from seventeen individuals and covered the tagged fragments with a piece of unexposed X-ray film. The next morning the black smudges in the developed film, caused by the faint radioactivity of the tracers, showed the positions of the DNA fragments relative to each other. If the smudges were side by side, it would mean all seventeen fragments were about the same length. But if any of the smudges were out of line, it would mean the fragments were of different lengths.

"We were using very large fragments, fifteen thousand bases long and up, so we couldn't distinguish short differences in length," Wyman recalls. Nevertheless, Wyman thought she saw some slight differences in the positions of four of the smudges. "It was hard to tell whether the differences were real when I held the film up to the light," she says. "But I did see this kind of funny result. It was interesting but worrisome. It could have meant that we were just getting sloppy in our experiments." But experiments in the following weeks showed the length differences were real. The same section of DNA taken from each of four individuals and subjected to the same restriction enzyme had produced fragments of different lengths. It meant this particular section of DNA varied from individual to individual, it was polymorphic. Wyman was looking at a restriction-fragment-length polymorphism, one of Botstein's riflips.

In retrospect, the Wyman-White experiment in 1979 has proven to be a pivotal moment in the genetics revolution now underway. The techniques White and his student worked out in those crucial months would be used to serve as a template for the flurry of gene searches undertaken in the years to follow. The techniques also would become the tools for making the riflip map conceived at Alta and, ultimately, the map of the entire human genome. Indeed, many of the prenatal diagnoses for genetic diseases that are done almost routinely today are derived from the techniques first used in the Worcester laboratory a decade ago.

Wyman was not the first person to see a riflip. As Botstein had noted at the Alta gathering, the molecular biologists had stumbled across them occasionally in their manipulations of DNA. (While Botstein, White, and Skolnick were describing the riflip concept to their colleagues, across the Atlantic a noted British scientist, Sir Walter Bodmer, broached the same idea in a brief article he wrote with Ellen Solomon concerning the use of blood-related genes as markers.)[1] Botstein and Davis themselves, in different experiments, had found riflips in the DNA of yeasts although the term riflip wasn't used. And in at least three instances, human DNA polymorphisms linked to genetic diseases had been uncovered. These latter reports concerned the globin genes, a complex of genes that produce the protein portions of hemoglobin, the oxygen-carrying component of the red blood cells. Defects in these globin genes lead to a variety of inherited blood diseases, the best-known of which is sickle-cell anemia, the severe blood disorder that most often afflicts people of African descent.

In early 1978, about the time of the Alta meeting, Tom Maniatas and his colleagues at MIT reported that they had been probing through the DNA from a globin gene of one individual. As is always the case, the individual had two copies of the gene, one inherited from his father and the other from his mother. When the MIT researchers snipped fragments out of each copy with a restriction enzyme, they found that the fragment from the maternal copy of the gene was of a different length than the fragment scissored out of the paternal copy.

The parents, in other words, had been polymorphic for this particular genetic fragment.

About the same time, Y.W. Kan and Andrees Dozy at the University of California, San Francisco, reported they had found a length difference in the DNA fragments cut out of the beta-globin gene. When a restriction enzyme was used to slice up the beta-globin gene of most Americans, the enzyme would cut out a fragment of DNA only seventy-six hundred bases long. But in many people of African descent, including many who carry the sickle-cell trait, the same restriction enzyme applied to the beta-globin gene snipped out a fragment almost twice as long, thirteen thousand bases. This fragment-length difference—a riflip—was used by other researchers in 1978 to determine whether an unborn child had inherited a defective beta-globin gene for the sickle-cell trait, the first use of a riflip for a prenatal diagnosis. And in England, Alec Jeffries, who was developing a technique for identifying individuals by their DNA "fingerprints," was finding differences among individuals in the sequences of the DNA bases in their globin genes.

In these instances, however, the biologists were dealing with fragment-length polymorphisms identified with known genes. But in their 1980 paper Botstein-White-Skolnick-Davis would propose something entirely different, a map-making expedition into the human genome searching for "anonymous" riflips. A surveyor doesn't care whether a landmark is a mountain peak or a street corner as long as he knows where it is in relation to other landmarks. In the same sense, the genetic cartographers didn't care whether a riflip was part of a gene or whether it had some other function or no function at all. The most important property of these nameless riflips would be their locations on the chromosomes and their distances from each other. Once 150 or so riflips were located at regular intervals along the chromosomes, one would have a crude map of the genetic continent.

No one before had ever attempted to find an anonymous riflip; the known riflips had been found more or less as a by-product of studying specific genes. There were no guidelines telling how one went about finding a riflip for the sake of finding a riflip. And there was no assurance that the riflips existed in the numbers and distributions that Botstein and Davis had postulated at Alta.

In the summer after the Alta meeting, White, conferring frequently with Botstein, began devising the strategy to find an anonymous riflip. It would be a crucial experiment in that, if successful, it would prove the concept Botstein and Davis had sketched out at Alta. It also would be a crucial experiment for White. White was just at the point in the early career of a young scientist where the choice of a research project could seal his professional fate for the next few years, if not for his entire career.

White hadn't intended originally to pursue molecular biology. His father was a dentist in a small rural town outside Eugene, Oregon, and the family's plan was that Ray would eventually enter medicine. But by his sophomore year at the University of Oregon, White had second thoughts. He turned to his faculty advisor, George Streisinger, a research scientist of some note, who had set up one of the country's first molecular biology institutes at a major university.

"I was talking to George in his office and telling him that this medicine stuff just wasn't right for me," White recalls. "He asked me if I had ever considered research, and before I could answer, a shadow passed by the door and George yells, 'Hey Frank, do you have any room in your lab?'"

The passing scientist, molecular biologist Frank Stahl, professed that he did, and the following fall White joined Stahl's laboratory as an assistant. He soon found himself addicted to the attempts to wrest the secrets of DNA. "I became what they call a lab rat," he says, noting that his grades "went to zilch" as he ignored everything else but his lab experiments. "It was a truly wonderful time in my life; nothing I had ever done really captured me so."

Before White graduated, Streisinger asked him where he planned to go to graduate school. "I hadn't really thought that being a researcher was something you could do for a living," he says. "Being an academic, an intellectual, was a much more sophisticated thing than what my family was about." Both Stahl and Streisinger suggested that, if White was interested in pursuing research in molecular biology, the best person to study under was Maury Fox at MIT.

White stayed with Fox six years, becoming proficient in the language and techniques of the rapidly developing field of DNA research. In 1971, he went to Stanford University to do postdoctoral research with David Hogeness. By this time, the young Oregonian had become exceptionally self-assured. "I went to Hogeness's lab with the expressed idea of disproving some of his work in cell replication of drosophila," he says, pausing slightly. "I didn't." But while there, White had a ringside seat to one of the major advances in molecular biology when biochemists Stanley Cohen of Stanford and Herbert Boyer of the University of California, Berkeley, successfully spliced a gene from one bacterium to another. Their historic experiment opened the entire field of genetic engineering and, for researchers like White, provided the splicing tools that would enable them to manipulate DNA with the dexterity of a genetic tailor.

In 1975, White was offered his first full-time professorial job at the University of Massachusetts. At the time Botstein contacted him, White was deep into—and becoming bored with—determining the sequence of bases in the DNA of blowflies.

Botstein's proposal that White take on the search for the first anonymous riflip was welcome but risky. If the strategy he and Botstein worked out was correct and the experiment paid off, White would be a jump ahead of any competition to develop a riflip map. The revolutionary nature of the riflip map and its importance to medical genetics would assure a steady and ample stream of research funds and a large laboratory, either at the University of Massachusetts or some more prestigious university. On the other hand, if the strategy failed or if the riflips were rare, White could spend several fruitless years in obscurity, scratching, like most scientists, for research funds.

For White, the adventure of the riflip search outweighed its risks. Says Wyman: "Ray would sit around and talk to his colleagues about this idea he had heard from Botstein; he was very excited about it." White's enthusiasm infected his lab assistant. "It sounded more interesting to me than the genes of blowflies so I volunteered my services," Wyman says.

At thirty-two, Wyman was making an unusual second round of postdoctoral training. A chemistry undergraduate major at Cornell University, she had moved into molecular biology during her graduate work at Harvard Medical School in Boston. But in 1974, when she was ready for her postdoctoral training, "molecular biology wasn't really rolling. The restriction enzymes had just been discovered and sequencing [of the bases in DNA] was just a vague idea," she says. So her postdoctoral research at Harvard was in developmental biology, the study of the genes in the developing embryo. "I took a couple of years off to teach, but by 1978 I realized I was out of touch" with a sudden splurge of developments in molecular biology. "I decided to go back and take a second postdoc and learn about [gene] cloning." She landed a spot in White's laboratory at Worcester. "He was doing DNA sequencing and had made a nice reputation for himself."

White, she quickly discovered, "is easy to work for. He's supportive, congenial, quick, and bright," Wyman says. "White also is a real presence, a big presence, dominating but not domineering." In a way, she adds, "Ray is on the same side of the spectrum as Botstein." The difference, says Wyman, who has worked with both men, is that Botstein is "right out there, very explosive" while White "is more mannerly. But both are very powerful personalities."

For their riflip search Wyman and White faced a formidable task. They would need a probe that would swim through an entire complement of human DNA and land on a unique sequence of bases that existed in all humans at only one spot on only one chromosome. In other words, if every human's genome was regarded as a gigantic library, the probe would seek out exactly the same book and page in every human.

No one knew what "text" (sequence of bases) was printed on this unique page but it had to have one key property: There had to be slight differences in certain "words" that would distinguish one individual's "page" from another's, that is, the page had to be polymorphic.

The probe would be a fragment of DNA that would "hybridize" or cling to the hidden riflip. Hybridization is a tool that is as central to molecular biology as a saw is to carpentry. It relies on nature's inviolate rule that a T base in one strand of DNA always binds to an A base on the opposing strand and the G base always binds to a C base.

If a molecular biologist has a fragment of single-stranded DNA in hand from, say John, and he wants to find the same fragment in Mary's DNA, he first makes a complementary copy of John's fragment. If John's fragment has a base sequence reading ATTCGAT etc., the biologist chemically synthesizes a complementary fragment that reads TAAGCTA etc., following the rule of T to A and G to C. This complementary fragment, called cDNA, becomes his probe.

The biologist then takes his cDNA probe and drops it into a flask containing a jumble of single strands of Mary's DNA. The probe—TAAGCTA etc.—will seek out and hybridize only with a fragment of Mary's DNA that reads ATTCGAT etc. The result is a double-stranded piece of "hybrid" DNA (hence the term hybridization). Like a molecular fishing hook, the cDNA probe has plucked out a fragment of Mary's DNA that has exactly the same base sequence as John's fragment.

To find a DNA probe that would seek out and hybridize with such a postulated riflip, White and Wyman turned to one of the unsung feats of molecular biology, a DNA library. Several farsighted biologists had realized in the early 1970s that it would be useful if they and their fellow researchers worked with standardized pieces of DNA, thereby

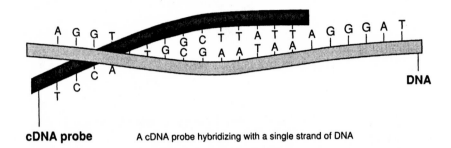

cDNA probe   A cDNA probe hybridizing with a single strand of DNA

avoiding a source of confusion when comparing results of experiments in different laboratories. To this end they created what they called DNA libraries where fragments of DNA from a known source were stored. Researchers from other laboratories could literally check out copies of human DNA segments of known identity, take the copies back to their own laboratories, and use the copies as probes to identify anonymous pieces of DNA.

One of the first of the DNA libraries was created by Tom Maniatas and his colleagues at MIT. They had taken an individual's DNA and snipped it into fragments each about fifteen to twenty thousand base pairs long. Each fragment was then inserted into a tiny virus, a so-called phage, that infects bacteria. Once inside the bacterium, the phage with its cargo of human DNA, is faithfully reproduced—cloned, as the biologists say—by its bacterial host. Maniatas's DNA library, then, consisted of a collection of bacteria peacefully multiplying away in laboratory dishes, unwittingly Xeroxing the fragments of human DNA.

White had spent the summer and early fall working out the proposal to the NIH for a grant to carry out the riflip search and doing some preliminary experiments at MIT. On a chilly fall day, before the grant had come through, he arrived at his Worcester lab and handed Wyman a glass vial containing the bacterial hosts of an entire human genome from Maniatas's library. Wyman dutifully distributed the bacteria into small nutrient-filled "wells" of plastic trays. Each well, in effect, contained a different fragment of human DNA. Wyman carefully labeled each well with her initials and a number.

If Botstein and Davis had been correct at Alta, somewhere among the hundreds of wells were the wells that contained the fragments of DNA with the unique polymorphic sequences, the riflips. If only one of the riflips could be found, it could be used to make a cDNA probe that would seek out and hybridize with the same polymorphic sequences—the same riflip—in the DNA of others.

There were hints, however, that they might encounter a problem that could require months of tedious work to overcome, if it could be overcome at all. Over the years of manipulating DNA, biologists kept running across short sequences about 250 bases long. They would find such a short sequence in one spot on a strand of DNA and then a little farther down the strand they would find the same sequence repeated. No one knew the purpose of these short sequences or why they were duplicated every so often. It was as though some mischievous type-setter kept dropping in a four-letter word just to frustrate the molecular biologists.

There were hints that these repetitious sequences, as they were called, were sprinkled up and down the entire genome. Any cDNA probe that contained copies of one or more of these short repetitious sequences would be useless as a probe for a riflip. If such a probe was tossed into a pool of DNA, it would hybridize indiscriminately with any other fragment of DNA that had the same repetitious sequence. It was entirely possible that every one of the wells held fragments containing these repetitious sequences, that none of the fragments could be used as a riflip probe.

Thus, Wyman and White first had to eliminate all the library fragments that contained the repetitious sequences. "This had never been done before and nobody knew what fraction [of the library fragments] were free of these repetitious sequences," Wyman says. "So we took a shot in the dark. We figured that the repetitious sequences were spread around kind of randomly, so there should be some fragments that didn't have them. There were." To eliminate the fragments containing the bothersome repetitious sequences, they prepared a short stretch of cDNA containing a complementary copy of the repetitious sequence, thus creating a probe for the bothersome four-letter word. They produced multiple copies of the probe and applied one to each of the hundreds of library fragments. Whenever the probe hybridized with a library fragment, the fragment was set aside. Hundreds of fragments were thus crossed off as useless for riflip probes.

"I don't remember how many usable fragments we ended up with, maybe eight or ten, certainly less than fifty," White says.

The work often was frustrating although not boring, Wyman recalls. Sometimes, the human DNA library fragments were contaminated with DNA from bacterial viruses, and produced confusing results. The restriction enzymes didn't always make clean cuts in the DNA. A special type of blotting paper that could soak up the DNA fragments off a sheet of gelatin had to be found. "It seems like every few weeks some problem would come up," she says. "It took months to learn all the mistakes. It might take a whole month just to learn that something isn't going to work."

By spring Wyman and White had singled out five of the DNA library fragments to use for probes for riflips. Each probe was a unique sequence of fifteen to twenty thousand bases that wasn't repeated anywhere else in the genome. Each, in other words, was a unique genomic page. When tossed into a pool of DNA from any individual, each probe would hybridize with a similar unique sequence in that individual's DNA.

They now had come to the crucial part of the experiment. Were any of the unique sequences picked out by the five probes polymorphic? They would test each of the five probes against the DNA of several individuals to see if any one of the probes plucked out a sequence that, when cut with a restriction enzyme, would produce different lengths from different individuals.

Thanks to the geneticists in Utah, they had on hand blood cells donated by fifty-six Mormons. Wyman began the tedious task of performing a Southern blot, named after the Scots scientist, Edward M. Southern, who invented it in 1975. She began taking the DNA from each of the Mormons' blood cells and cutting it into fragments with a restriction enzyme. The DNA fragments from each Mormon were placed on a sheet of gelatin, to which an electric current was applied. The current forced the shorter fragments to move through the gelatin faster than the longer pieces. Once the fragments were separated by length, they were soaked up on a piece of blotting paper, carefully preserving their relative positions. (See illustration on opposing page.)

A library probe, tagged with a radioactive tracer, was then placed on the blotting paper and allowed time to hybridize with any fragment of matching sequence. Probes and fragments that failed to hybridize were washed off. A piece of X-ray film was left lying on the blotting paper overnight, allowing the faint radioactivity from the probe to expose the film. The dark smudges on the developed film would show the relative positions of the fragment-probe hybrids. The farther apart the smudges, the greater the difference in length.

On that mid-October morning Wyman was examining the results of a Southern blot on the restriction fragments from several Mormons. The fragments were those that had hybridized with a probe taken from well number 18, tray number 1016 and, hence, the probe was labeled pAW1016-18 ("p" for probe and AW for Arlene Wyman).

Making a Southern blot:

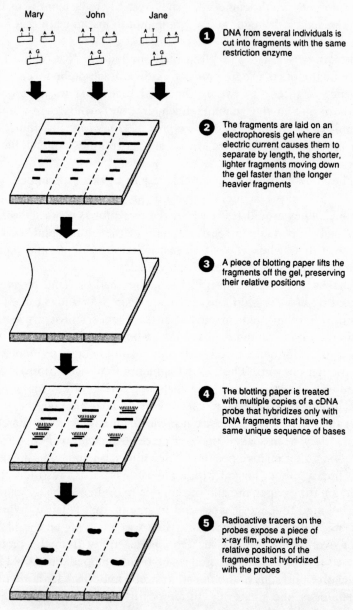

Mary   John   Jane

1. DNA from several individuals is cut into fragments with the same restriction enzyme

2. The fragments are laid on an electrophoresis gel where an electric current causes them to separate by length, the shorter, lighter fragments moving down the gel faster than the longer heavier fragments

3. A piece of blotting paper lifts the fragments off the gel, preserving their relative positions

4. The blotting paper is treated with multiple copies of a cDNA probe that hybridizes only with DNA fragments that have the same unique sequence of bases

5. Radioactive tracers on the probes expose a piece of x-ray film, showing the relative positions of the fragments that hybridized with the probes

The example above shows that Mary, John and Jane all possessed two copies of a piece of DNA with a unique sequence of bases, as identified by the cDNA probe. Small differences in the base sequences among the three, however, caused the restriction enzyme to cut each piece of DNA at a different point, producing different lengths of fragments, showing this unique segment of DNA is polymorphic; it is a restriction-fragment-length polymorphism, RFLP

Later, in a more refined version, the probe would be known simply as pAW101. Because the fragments were long, fifteen to thirty thousand bases, it was difficult to detect length differences of only a few hundred bases. Nevertheless, to Wyman's practiced eye there seemed to be a slight difference in the positions of the smudges representing fragments from four of the Mormons, a hint that the fragments might be of different lengths. However, the smudges representing fragments from thirteen other Mormons seemed to be side by side, indicating they were all of the same length.

Wyman and White were more worried than pleased by the hint of a length polymorphism. Once before "we had gotten real excited" by a similar result, only to discover the length difference was caused by the failure of the restriction enzyme to completely cut the DNA of one person, Wyman recalls.

They began repeating the experiment, leaving the fragments on the gelatin sheet for longer times and allowing them to separate by greater distances. By early December, "we were sure we had a polymorphism," Wyman says. The smudges were showing wide variations among individuals in the lengths of the fragments that hybridized to the pAW101 probe.

The fragments from some individuals were as short as fourteen thousand bases, while the fragments from others were as long as twenty-nine thousand bases. Among the fifty-six Mormons, Wyman and White could distinguish eight different restriction fragment lengths (later expanded to thirty different lengths). In a large majority of the Mormons—forty-four of the fifty-six—the maternal copy of the

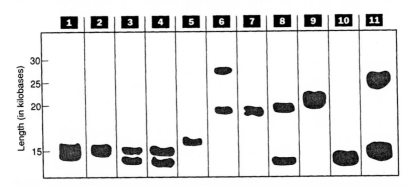

Southern blots made by Wyman and White of fragments cut by a restriction enzyme from the DNA of eleven individuals. The differences in lengths of the fragments show that this region of human DNA varies from individual to individual and, thus, is polymorphic. There are two fragments for each individual, one inherited from the father and one from the mother. (Adapted from Wyman and White, *PNAS*, 77:6757, 1980.)

riflip was of a different length than the paternal copy, a feature that would make the riflip an ideal genetic marker for determining whether a gene linked to the riflip had been inherited from the father or the mother.

Later they would discover that the length differences were due to a short sequence of bases buried in the fragments that was repeated in tandem a different number of times in different individuals. The short sequence might be repeated only a few times in one individual and a few hundred times in another. Known as a variable number tandem repeat (VNTR), it would be found to be the most common type of riflip in the genome.

In mid-December Wyman, in the ninth month of her first pregnancy, took a three-month maternity leave. White moved into the laboratory and, with a technician, finished the experiment. In mid-March, Wyman returned to the lab, occasionally bringing her newborn boy with her, and she and White began writing up the experiment for publication. The paper, titled "A Highly Polymorphic Locus in Human DNA," appeared in the November 1980 issue of the *Proceedings of the National Academy of Sciences*.[2]

"The human geneticists paid more attention to the paper than the molecular biologists," says White. "The molecular biologists were kind of laid back about it, saying it wasn't molecular biology." The Wyman-White experiment, confirming the concept laid out in the Botstein-White-Skolnick-Davis paper published a few months earlier, impressed the geneticists, who were in desperate need of DNA markers for the major genetic diseases.

The month the paper was published, White moved to the Howard Hughes Medical Institute at the University of Utah where, promised a steady flow of research funds, a large laboratory, and access to pedigrees and the DNA of fifty Mormon families, he began working on the riflip map that Botstein and Davis had envisioned at the Alta gathering two and half years earlier. Wyman, who was married to a faculty member at Brandeis University in Waltham, Massachusetts, moved to Botstein's laboratory at MIT, where a few months later she tracked down the location of the riflip to which probe pAW101 hybridized. It lies on the lower tip of chromosome 14.

CHAPTER 4

El Mal

Lake Maracaibo, despite its name, is an enormous saltwater gulf that opens like a mouth in the northern face of South America, swallowing in the Caribbean Sea. Along its shores poor villagers live together on subsistence fishing. The smallest of these communities are the isolated *pueblas de agua,* or water villages, where several dozen wood and tin–roofed houses rise on wooden stilts above the shallow waters of sheltered lagoons. Behind the water villages, the shoreline is a thick and marshy forest, uninhabitable and impassable to travelers. Day and night the fishermen work the lake in long narrow skiffs called *chalanas.* At home the women tend to large families; the number of children typically reaching a half dozen to a dozen and sometimes more. Except for some modern conveniences such as manufactured clothing and small outboard engines for the skiffs, life in the pueblas has changed little in centuries. That is, except for the invasion almost two hundred years ago by "El Mal." Translated from Spanish, *El Mal* means the sickness, taken literally, "the bad."

Local legend traces the invasion to German and Spanish sailors, whose trading ships in the early 1800s regularly anchored off the large port city of Maracaibo to the north. The merchant ships bartered finished goods for stores of berri-berri nuts, which grew in the jungles and whose extract the Europeans used for tanning leather. For reasons of commerce and sport, the sailors often would sail southward from Maracaibo, stopping off at the pueblas. Some of the sailors inevitably slept with the village women, some married and stayed, and others moved on. One of the sailors, so the story goes, left behind El Mal.

It was to these same pueblas that Nancy Wexler was first drawn in July 1979. For El Mal is Huntington's disease, the same horrible illness that stalked Wexler's family. If the legend is correct, the isolated fishermen and their families unwittingly received the mutant Huntington's gene from one of the European sailors. But it is also possible that a normal gene in one of the Venezuelan villagers was mutated by a spontaneous genetic accident of the type that might occur only once in a hundred million births—and that the villager passed it along to his or her descendents. In either event, the Huntington's disease gene was to become an unending source of poison to a single Venezuelan family, tainting their stream of DNA as it coursed through generation after generation after generation.

By the time Wexler arrived in Venezuela, the descendents had grown into a family tree of over 7,000 people, 2,600 of whom were still alive in 1979. Nearly 100 people living in a half dozen or so larger towns and small water villages already were afflicted with El Mal. Another 1,100 children and young adults had a 50 percent risk of developing it by the time they reached middle age, since at least one of their parents carried the mutant gene.[1] For the Venezuelan families living along Lake Maracaibo, the disease was as terrifying and devastating when it struck as it was to Milton Wexler when he first learned the fate his daughters might have inherited from his wife.

But to the poor people of the pueblas, largely ignorant of the world beyond the lake, Huntington's disease was, like their mean poverty or the sudden violent storms that swept across the lake, something they had come grudgingly to accept. That El Mal was an inherited disease affecting the villagers first became known to Venezuelan health authorities after World War II. And that finding might not have occurred if Americo Negrette, a young Venezuelan doctor, hadn't been sent to San Luis, a barrio south of Maracaibo, to serve a rural internship.

Fortunately, for the field of genetics, Negrette had a scientist's natural curiosity. He was alarmed by the number of people who walked San Luis's streets in what he first assumed was a state of drunkenness. Upon inquiry he found that they weren't vagrants, but were sick. Conducting an investigation similar to the one made by George Huntington a century earlier, Negrette learned that those who were sick often had brothers and sisters who also were ill, and that many of the children and grandchildren also had the disease.

Despite having been schooled in only rudimentary genetics, Negrette was able to construct a pedigree, from whose rough outlines he could see a path cut by the disease as it marched through a large San Luis family, uniformly sidestepping those not related to the affected group. From this map, Negrette found he could predict who among the people of San Luis were at risk and who were free of any chance of ever getting the disease. He concluded, quite correctly, that El Mal was inherited as a dominant gene, that it need only be inherited from one parent to carry out its inevitable derangement of the body's nervous system later in life. In a book he published in 1955, Negrette suggested, also quite correctly, that Huntington's disease could be wiped out by sterilizing certain very specific members of the pedigree.[2] Not surprisingly, the proposal stirred quite a controversy and was never carried out or even seriously contemplated by Venezuelan health authorities. But his report drew attention to the pueblas as a place where Venezuelan medical students could conduct genetics research and brought him students who were eager to pursue such studies.

One of these students was Ramon Avila Giron, whose black-and-white home movie of the local Huntington's victims captivated the attendees of the research workshop in Ohio in 1972. When Nancy Wexler finally received some federal funding to travel to Venezuela to search for a Huntington's homozygote—someone who had received two copies of the mutant gene, a copy from each parent— Negrette, Avila Giron, and another Negrette student, Ernesto Bonilla, agreed to serve as her guides. The Venezuelans, however, hadn't had the need to identify a homozygote and didn't know whether such a person existed.

Armed with Negrette's pedigree, Wexler and two research companions, Tom Chase and Ted Bird, began walking the hot and squalid streets of San Luis looking for a nuclear family in which both parents were affected. Chase and Bird were neurologists whose help would be needed to verify, first, that both parents indeed had Huntington's and not some other similar-appearing disease, and that the offspring of the two affected parents also truly had the illness.

San Luis is only a twenty-minute drive from Maracaibo, where the Americans were staying in a hotel, and they and the Venezuelan scientists would travel together to the barrio every day, spending the bulk of each afternoon tracking down members of Negrette's pedigree. After several days, the group moved the search south to Barranquitas, a three-and-a-half-hour drive inland from Maracaibo. The trip to Barranquitas was exhausting. Even though they began early in the morning, the trackers soon became frustrated by the few hours of searching they had during the day.

For the three Americans, the trip rapidly became daunting. Chase recalls that he and Bird were "hellishly uncomfortable." Says Chase, "It was an impossible place to conduct research. There was no scientific infrastructure from which to draw any kind of help, nor were there health records or personnel to guide us, other than Negrette and Avila Giron, who were as uncertain as we in figuring out what needed to be done. It was oppressively hot and I remember the food being disagreeable. I really never recovered from something that happened the first night in Maracaibo. I was trying to sleep when suddenly I heard a stamping noise as if someone was going to bolt into my room. I opened the door just enough to see a group of soldiers armed to the teeth. I'm not sure what they were doing in the hallway. It was a very strange place." (Chase stayed on, despite his discomfort, but within a few days, Bird suffered the effects of heat stroke and returned to the States.)

While Bird and Chase quickly became disoriented, Nancy recalls being "supercharged." She could barely sleep despite feeling exhausted. Yet, she, too, was becoming disheartened by their failure to find a homozygote. "It was late one afternoon and we were getting pretty desperate," says Wexler. "I sat down and thought the way we were going about this was pretty stupid. When I was a graduate student I learned in an anthropology class that when studying a different culture you needed to find an informant, someone of the group who could lead you to others. Here, we'd been pretending we were geneticists picking apart a pedigree, when what we really needed to do was act like anthropologists and investigate the people themselves. We really weren't sure what the best strategy was or what we really should be doing.

"So I went up to this fellow and I asked him if he knew of a family where both parents were ill. He said, 'Oh sure, in Laguneta there is such a family. The mother and the father are sick and they have thirteen or fourteen children.' I couldn't believe it."

At dawn the next day, Chase and Wexler hired a local fisherman to motor them across a large inlet in his chalana. The boat ride to Laguneta took over two hours. What they found when they finally arrived was a puebla de agua. "It was astonishing," says Wexler. Tin and wooden shacks fitted with small porches rose six to ten feet off the surface of the swampy lagoon. The heat was stifling. The air was dense with humidity, and all manner of mosquitoes and other bugs swarmed the boat. Pelicans and cormorants bobbed on the water's surface. Behind the shacks, rising up out of the shallows, was a thick low jungle crisscrossed by muddy canals of water. From it boomed a racket of indistinguishable animal sounds.

Chase recalls the trip across the inlet as being "dreamlike." He says, "I felt the further we went, the further back in time we were traveling. The landscape got progressively more primitive. All of a sudden we rounded this bend and before us was this startling village. There were these ramshackle stilt houses, and, on the porches scurried these half-naked kids, some with their stomachs protruding from malnourishment, their heads shaved to fight the heat and stave off lice. Yet, some of these kids wore T-shirts with logos from American colleges. I remember one of them wearing a T-shirt from Harvard. I never did find out where they got them."

As the chalana slid close to one of the outermost huts, the two Americans saw a woman in a brown print dress sitting quietly, cross-legged, on the porch. Another former student of Negrette's was accompanying them, and he asked the woman which of the houses was the home of Carmen and Eduardo, the couple who supposedly were both sick with Huntington's. The appearance of the Americans—Chase is exceptionally tall—apparently bothered the woman and she got up to go back inside her house. "Suddenly this choreic movement sort of unfolded from her," says Wexler. "Here was this setting that for me couldn't have been more exotic and more different than anything I'd seen in my life. And yet, here was this totally familiar disease. It was such a clash of total bizarreness and total familiarity. I was exhilarated and frightened, I felt connected and alienated. I was overcome."

A few minutes later, they found Eduardo and Carmen. The two were lying on hammocks stretched across their porch. From their movements it was immediately clear they both had the disease. And running about the small hut were a "passel" of children, recalls Wexler. Since there were fourteen children, and each had a one in four chance of inheriting two copies of the HD gene, it was likely that three or four of the kids were homozygotes. The Americans believed their research would require obtaining blood samples from both parents and an affected child, and they would need to examine both parents to verify that they had the disease.

Chase and Wexler, despite the heat, frantically began to assay their finding. They wanted to construct a family pedigree, and then take blood specimens back with them to the United States. "We didn't know when we'd be back, or if Carmen and Eduardo would still be alive whenever we were able to return," Wexler says.

But gathering the facts wasn't easy. Wexler encountered the first of what would be many cultural chasms she would have to bridge. Carmen gave Wexler the names of her children and their ages. Checking, then, with the children, Wexler was given different ages, and sometimes even different names. Carmen said one son was named Angel Eduardo, but he insisted his name was Anguito, a nickname he had long since taken for his given name. After much arguing, the mother and son agreed that the two names were indeed applicable to the same man.

Another problem was that none of the thirteen children they saw that day had yet developed Huntington's. Which of the children, if any, might be the homozygote? One older daughter was described as being sick, but she wouldn't let the Americans see her. Based on a description of her illness, it wasn't clear she had the disease. Also, Carmen had had several miscarriages. Could it be that the miscarried fetuses had received two copies of the gene, but that they were unable to survive a double dose of the disease? wondered Wexler. Perhaps, therefore, a homozygote for Huntington's didn't exist at all, she fretted.

Nonetheless, as the day in Laguneta progressed, Wexler's elation grew. "For days I'd felt unsure of what we were doing," she says. "But once we found the family I knew that some kind of study was not only feasible, but might even be critical. I felt if we studied these people, something about them would provide an important insight about HD." The next day, the Americans got on a plane and flew back to Washington with a suitcase full of blood samples. The trip, they felt, had been a success.

Wexler's exhilaration from the trip lasted weeks and was, in part, a result of the frustration she, Milton, and Alice were feeling about other research prospects. The discovery of the Venezuela project, as it became known henceforth, came for Nancy Wexler at a time when she was desperate for a research project she could participate in. She was trained as a clinical psychologist and her own research until then was in developing methods for fostering more thoughtful care of Huntington's disease victims and their families. While her research was having an impact on the care and counseling of those affected by the disease, it had no effect in predicting who might eventually develop the disease, or how to treat or cure it. That was a growing source of despair for Wexler. "I'm a doer freak, and [the Venezuela project] was something I could do," she says.

Within weeks of their return, Chase and Wexler were told that the funding under which Wexler had been working at NINDS would come up for review in October, when all federal funds are rebudgeted. She was told to organize a presentation to defend her continued funding and to prepare to make her arguments in late October.

Unbeknownst to Wexler or Chase, or anyone else at the institute, research similar to the Venezuela family project was being heatedly discussed by the small band of molecular biologists who were ambitiously advancing the riflip concept. By the time Wexler returned to the United States, Ray White and David Botstein were plotting a course designed to identify and map the location of riflip markers throughout the human genome. To undertake such a project, the two determined that they would need a series of extra-large families whose DNA they could investigate by securing samples of blood taken from family members. But there was the problem of who would find the families and who would convince these people to participate in such studies and provide their blood. Botstein and White were theoreticians and laboratory experimenters who left contact with human subjects to others. Indeed, for Botstein and White, the search for families had been a trivial matter. Mark Skolnick had suggested that the Mormon families in and around Salt Lake City might provide the needed DNA, and White began courting the richly funded Howard Hughes Medical Institute to back the research in Utah. White assumed, once funded, the matter of securing the DNA would be a minor issue.

Among the scientists Botstein and White were talking to about their concept was David Housman, whose lab was located near Botstein's at MIT. After Allan Tobin had spoken to Housman in California in early 1979, Tobin convinced Milton Wexler to invite Botstein, White, Housman, and several other molecular biologists to Milton's next scheduled science workshop, which was planned for October at a meeting hall at the National Institutes of Health. Tobin helped put together the meeting even before Nancy left for Venezuela. "I'm not sure I really knew what the MIT guys were up to," recalls Tobin. "Housman led me to believe it was important."

Botstein and White had no particular interest in Huntington's disease. But they were looking for opportunities to discuss their ideas. Since White was still scouting for grant money, they saw the workshop as a chance to talk up their concept at a meeting that, because it was based at NIH, would attract some of the people whose support they would need to get federal funds to back their research. The meeting turned out to be crucial, for it provided some of the country's foremost molecular biologists their first opportunity to hear about the new riflip concept, and it allowed Botstein, White, and others a chance to thrash out experimental strategies for systematically hunting down previously unidentified genes. The meeting also was punctuated by a surprisingly furious debate which, to this day, has left bruised feelings and, among some, outright animosity. The debate was set off by a question posed by Nancy Wexler.

The meeting started off innocently with presentations by Botstein and White. Wexler recalls, however, that it was straightforward enough for her to grasp that Botstein and White had found the technology she had long dreamed of. Several years later, she described her first understanding of the concept this way:

> The riflip technique was a method for finding undiscovered genes using the emerging tools of recombinant DNA biotechnology. The two biologists explained that riflips were small variations in DNA that could serve as signposts pointing the direction to specific genes. Their theory was that these signposts were scattered throughout the chromosomes and that some sat very nearby important genes. They are so close that when a gene is passed from parent to child, the signpost travels with it. Signposts, or as they also are called, DNA markers, are much easier to see in the laboratory than human genes. But by studying large families, one could track a marker as it passed from parent to child along with nearby genes. If every time a disease is passed from generation to generation a specific marker is also passed, then the marker would be considered close to the gene responsible for the disease.[3]

Wexler's mind began racing and she pelted the biologists with questions. "As I listened to the speculations at the workshop it occurred to me I knew just the right family," Wexler says. "There have been few times in my life when I felt convinced that something was right, times when my heart leapt into my throat, times when I couldn't sit still and wanted to race as fast as I could, laugh wildly or explode. Venezuela, which I had visited that July and where I had been planning to go for a much less ambitious project, suddenly opened a new dimension."[4]

Wexler asked if her Venezuela family would be useful. Botstein and White said that for their own purposes of finding riflip markers, the Venezuelans would indeed help. They would back her in her efforts to undertake future studies there. But the two dismissed the idea that the family could lead to the genetic culprit for Huntington's. That, they suggested, might take fifty years. Botstein and White said they had bigger fish in mind. Their goal was to find riflips on every chromosome, located so closely to one another that for the first time scientists would have a map of known DNA landmarks scattered throughout the genome. Then, when this riflip map was complete, scientists interested in looking for disease-causing genes, or any gene for that matter, would have landmarks from which to launch their searches. Without a full complement of riflips, the two said, a search for a disease gene would be a waste of time. Anyone embarking on such a search would be foolish, Botstein remembers thinking. And anyone who said they could pull off such a discovery was dreaming, he felt.

But Housman disagreed. Then what began as a simple disagreement soon turned into a major clash. Housman argued that the riflip technique should immediately be used to begin searching for genes that cause disease. Botstein dismissed the idea. "David can be very dogmatic," recalls Housman. "In a discussion with him, you basically do a lot of listening. He had his approach, and he was convinced it was right. I thought another way would be just as useful. I think he felt very proprietary about the whole thing."

Says Botstein: "Here we only had one riflip, the one found by Wyman and White, we didn't know yet if we'd find others on other chromosomes, and we didn't know yet if you could link the probes with families. And here's Housman saying, 'Let's cure a disease.' I thought you did people a real disservice when you start making those kinds of claims. To be honest, I thought he was taking advantage of [the Wexlers'] desperation."

Later, Botstein would say that he was convinced Housman really wanted to use the Venezuelan family to edge his way into riflip mapping. "I'm sure he knew how slim his chances were of really finding the gene," Botstein says.

The rift between the two was to become so wide that, while their laboratories stood side by side at MIT, they rarely spoke after the October meeting. "For me science flourishes with discourse," says Housman. "But we just stopped talking. It was one of the more famous non-discussions I've had."

Housman had strong personal motives for carving out his own route, and for fiercely sticking to it. Although still quite young, Housman had gone to MIT in 1975 as a budding star scientist expecting to make his mark in the emerging field of gene cloning. Yet, by 1979 he had made little progress. Housman had received his doctorate in 1971 at Brandeis University in Waltham and then worked for several years in Toronto at the Ontario Cancer Center, where he helped isolate chemical proteins in blood produced by genes that cause several rare inherited blood disorders. While he had helped find the proteins, the genes themselves still awaited discovery. When scientists first reported their ability to deduce the structure of a gene from the amino acid sequence of its protein, and then to clone the gene, Housman decided to tackle the cloning of the blood disorder genes at MIT. But soon after he arrived, the local government in Cambridge, where MIT and Harvard are located, passed an ordinance outlawing the synthetic manufacture of genes—cloning—until further research was done on the technique's safety. The Cambridge government had reacted to the fears, understandable at the time, that the gene-cloners at the universities might accidently concoct a genetically altered germ that might somehow be loosed upon the local populace. Until the restrictions were lifted in 1978, Housman was forced to pursue other work. At other labs, where cloning wasn't banned, scientists raced ahead and identified the blood disorder genes. A disappointed Housman could only look on while they received the accolades he had hoped would be his. "He felt like the rug had been pulled out from under him," says one of his colleagues. "He was very upset."

Throughout the fall of 1978, however, he became increasingly tantalized by the prospects created by the new riflip technique being discussed by Botstein. When he met Allan Tobin in February 1979, "things began to finally click," he says. Here in Huntington's disease, he felt, was a gene in which he could test the theory posed by Botstein. "Even though nothing was known of the gene's structure, what was known about its impact on people made it look like an ideal candidate," Housman recalls. For one, the gene was "highly penetrant," that is, everyone who gets the gene, gets the illness. Also, it was a dominant gene; victims needed to inherit only one copy from one parent for it to carry out its effect. These two characteristics would keep the gene search simple. Also, the disease itself was easily distinguishable from other illnesses, so that research wouldn't be clouded by studying people who didn't actually have the gene. Finally, families who have the disease could be identified and studied, since victims almost always know others in their family who have it.

Even before the October meeting, Housman learned of a way to get financial support for research into Huntington's disease, and he jumped at the chance. Under a recommendation made by Nancy Wexler's commission at NINDS, the institute was offering grants to university medical centers to develop the first federally funded Huntington's disease centers, where patients could be provided up-to-date treatment and where they could also be recruited for basic research studies. The grants were by far the biggest commitment the government had made to combat the disease. Having heard about the grants, Housman convinced Joseph Martin, chairman of neurology at Boston's Massachusetts General Hospital, directly across the Charles River from MIT, to apply jointly to become one of the national centers. In their application, the two proposed that Mass General would provide the patient care, while Housman's nearby lab would pursue the basic research.

Thus, Housman went to the October workshop having already set in motion plans to finance an assault on the Huntington's gene. But "I hadn't met Nancy and I didn't know about Venezuela," Housman says. When Wexler described the family she had met during her brief hunt for a homozygote, Housman says, "As far as I was concerned everything had fallen into place. We had a strategy, we were going to get funding, and now we had the family."

Despite objections by Botstein and White that taking on the gene was premature, Housman huddled with Wexler and told her she should abandon the search for the homozygote. Instead, he told her to return to Venezuela and find as large an extended family as possible with many living people affected by the disease. He told her she should then retrieve blood samples from them, from which his lab could begin DNA marker studies.

Wexler couldn't believe her good fortune, and neither could Housman. The two fueled each other's enthusiasm. Despite the fact that the Botstein strategy was as yet unproven, Housman told Wexler he was convinced it would work. He warned Wexler that, once she began funneling the blood samples to his MIT labs, it might take five to ten years or more to locate the Huntington's gene. It might take years more to discover the structure of the gene itself, and then years more until scientists could understand how the gene did its deadly work.

Wexler, for her part, required little convincing. After all the leads that headed nowhere, here was one she felt was on track, even if it would take years to reach its destination. She had gotten to the point where she wanted to believe that she truly would find an answer. Housman wanted to be as honest as possible about the slim chances of finding the gene soon. But he was absolutely convinced of the merits of the search; that it could be done given enough time and effort. As importantly, he told Wexler, he had the beginnings of a plan, a technique for finding riflips that differed from what Wyman and White had used. The DNA from the Venezuelans would provide the material he needed to prove he was right.

"The idea was exhilarating to me," Wexler says. "What I found so inviting was that, if you stuck with it long enough, you had to find the gene. There wasn't anything we'd come across that could make that promise. Sooner or later, [Housman] said you'd have to find the gene using this technique. Even if it took fifty years. So I figured we might as well get started right away because sooner was better than later."

She decided to commit herself completely to the riflip strategy. She took one further step, encouraging Housman to apply immediately to her father's foundation for a grant to get the initial research started right away.

Several weeks later, Housman traveled to Los Angeles, where the private Hereditary Disease Foundation's board of advisors was meeting to consider several grant requests. Housman had been sick with the flu, but flew across the continent anyway, popping one throat lozenge after another into his mouth to keep his voice going. Those present remember the young man's enthusiasm and excitement, most especially his self-confidence. Housman isn't a big man, but he has broad shoulders that make it hard to ignore his presence. He likes to look his listeners in the eye. He has a knack for expressing his ideas in simple, direct terms, and making even the simplest, most banal project appear as important as anything he's undertaken.

"No one else had ever made such a straightforward, confident presentation to us before," says Michael Conneally, the Indiana University geneticist who was a member of the advisory board. "David enthralled us and, of course, he won the grant."

Thus, practically assured of receiving one private grant and hopeful of receiving another from the federal government, Housman began lining up a molecular biologist to carry out the actual work of finding new riflips. Back at MIT, Housman telephoned James Gusella, a protégé who had worked in his laboratory in Toronto and had followed him to MIT. Under Housman's tutelage, Gusella had received his doctorate in molecular biology and was planning to head off to the California Institute of Technology to pursue additional postdoctoral work. "Once I explained what we had going for us, Jim decided to stay in Cambridge, taking a position he was offered at Harvard, rather than go to Caltech," says Housman. "It was a real fortunate decision. But he wouldn't have stayed if we hadn't gotten the funds from the Wexler Foundation to pay for his support staff."

Two months later Housman returned to Washington, and, during a conversation over dinner at a restaurant in Virginia, he and Wexler sealed their collaboration. By this time Wexler already was planning a new trip to Venezuela. After the workshop, she convinced her superiors at NINDS to back at least one additional trip—perhaps several trips—to the water villages along Lake Maracaibo by explaining the families' potential importance to the Botstein theory. Housman had come to Washington to discuss plans for the Venezuela expedition. It was at dinner that he explained to Nancy the power of finding a DNA marker—a riflip—close to the gene.

Housman explained that the marker itself could be a valuable diagnostic tool. Once a marker was found, its presence in someone could easily be determined by testing the DNA from blood cells, skin cells, or even from fetal cells in the amniotic fluid surrounding an unborn child. "I told her if the marker was close enough, right away it could be used in a diagnostic test to determine if someone has the gene, long before their symptoms appeared. It was the first time we'd really discussed the predictive power that the marker itself contained—that it could immediately be more than just a tool for finding the gene. Suddenly, it hit me. I was telling her that she could use a marker to tell if she was at risk, whether she had inherited the gene from her mother. I could see it had a profound impact on her."

The two also began to ruminate about some of the potentially troubling sides to the research. Housman said a tightly linked marker could be used in a prenatal test to see if a developing fetus had inherited the Huntington's gene. While sitting over dessert, the two discussed the very real possibility that for the first time in history, scientists would have a tool to forecast, with almost absolute certainty, the onset of an illness in a human being four or five decades before the disease's first symptoms appeared. Wexler had given some thought to this, having wondered whether at some distant time the *gene* could be used this way. But she hadn't fully appreciated the fact that an informative marker could also be used to diagnose the disease years before a gene was found. But without the gene in hand, there'd be no chance for a cure. The idea of diagnosis without any advance in treatment hit Wexler for the first time.

Says Housman: "Nancy asked, 'What will people do with that kind of knowledge? What would you do if you knew you were going to get the disease for certain or if you knew your yet unborn child would get it?' Nancy said most people were not prepared for these kinds of questions. Many people she knew who were at risk were able to live functioning lives by assuming they had dodged the Huntington's disease bullet. And here, years before the bullet arrived, you could tell them they were a target for sure.

"We both were really shook up. Here Nancy and her father had taken on this disease because of its personal effect on them, and now they could be involved in research—not just inspiring or funding it, but directly involved in research—that would have an immediate impact on themselves and people they knew. For Nancy, things were coming around full circle."

But to round the circle Wexler would have to return to the Venezuela jungle. In March 1981, she led the first month-long expedition back to Lake Maracaibo. With her that first year she took along a team that included several neurologists, a geneticist, a nurse, and a photographer.

The work in Venezuela soon became addictive. Aside from the scientific need to identify the large family, the expedition was an emotional tonic. (Indeed, she has returned for an annual dose.) The trips were an adventure that energized Wexler, and, at the same time, allowed her to channel her anxieties and obsessions.

Tom Chase says that the rigors of the first brief trip in 1979 discouraged him from joining the 1981 expedition. "I fled and never returned," he says. "But Nancy seems to have become intoxicated by the place."

In the years that followed, she became fluent in Spanish and developed a deep bond with many of the villagers. She survived nights asleep on dirt floors covered by cockroaches and rats, curfews and martial law during a month of protests over inflated food prices, threats on her life, and a sudden storm that nearly drowned her and her coworkers one night as they traveled by boat across the lake. But, bit by bit, she began to draw a picture of the Venezuelan family tree.

There was no precedent for piecing together a genealogy of a family that lived in such a harsh and alien environment, whose members were suspicious of the North Americans' motives, and for whom few health or other personal records existed. Wexler relied on her instincts and the advice of a few who had developed some techniques for hunting down Huntington's disease families in the United States.

Wexler decided to try a technique she had heard about from Michael Conneally, who had been collecting blood samples from Huntington's families in the Midwest. Using Negrette's pedigree as a guide, she once more traveled about San Luis with several local residents, finding parents, siblings, and cousins related to people who had the disease. As members of the pedigree were located, they were asked to come to a clinic the North Americans had set up at a San Luis schoolhouse. Each of those who agreed to participate was offered a small payment. At the clinic, the villagers would be examined by a neurologist and then interviewed to get the names, locations, and health status of as many other relatives as the villagers could identify. Up on a wall, the name of each new person interviewed was placed on a growing family tree, and the names of relatives, who would be sought at a later date, were also placed in their appropriate spot. Many in the family had identical names or names so similar that the researchers became confused as to whom they were interviewing. Wexler decided to take Polaroid snapshots of each member who came for an interview and place the photos on the tree under each person's name. When others in the family came in, they were told to find their place on the pedigree by finding the picture of a close relative and then asked to stand under the photo until a researcher was able to talk with them.

During that first month in 1981, the researchers filled in branches of the tree from San Luis, Barranquitas, Laguneta, and several other villages. Within weeks the researchers had covered several walls of a hotel suite in Maracaibo with yards of paper crowded with photographs, vertical lines connecting the generations, and horizontal lines linking siblings. The immense size of the emerging family astonished Wexler. "I thought the family would include about a hundred or so people, but we soon realized it numbered in the thousands," Wexler says. "Housman had said the bigger the family, the greater the chance we would succeed sooner rather than later. I became obsessed with finding every single member who was related to someone with the disease, no matter where he or she lived or whether they wanted to help us or not."

But the logistics of doing so were daunting. For instance, the researchers quickly realized they couldn't undertake the enormous expense of keeping blood specimens refrigerated or making daily air shipments to the United States. Instead, the team decided to have everyone who was part of the pedigree come back on a "draw day," a day the researchers would draw blood specimens from everyone who had been interviewed and examined during the previous week. The vials of blood were packed in small Styrofoam boxes together with ice packs, and then hand-carried to airplanes by researchers who traveled back and forth to Boston.

"Draw days were chaos," says Wexler. The researchers asked a few dozen people to come to the clinic. As an added incentive they sometimes held a party, serving Cokes and candies. And, as a result, the Venezuelans would come with dozens of their relatives in tow. The small schoolhouse room overflowed with crying babies and older children running to and fro. For hours the room would undulate in a mass of confusion.

"It was madness," recalls Anne Young, a University of Michigan neurologist recruited to the expedition by Wexler. "It was boiling hot. The room was packed. It was certainly the most exciting thing happening in town and lots more people than we wanted showed up. The noise was unbelievable. Some people wanted to know why their blood was being taken, some would resist at the sight of the needles, and others would complain because they hadn't been asked to give their blood, too. There was lots of yelling and, of course it was all in Spanish, which had to be translated."

The pandemonium upset some of the researchers, who argued with Wexler to impose greater organization. "Nancy's style looked very loose and haphazard, and, in fact, it was," says Young. "But the truth is you couldn't be organized, at least not in a North American style. You couldn't make appointments. People would say they'd come or that we should meet them someplace and then they wouldn't show up. The fishermen would disappear for weeks or then when we'd find them they might be drunk or belligerent. I'm convinced that you could have put a lot of people down there and nothing would have been accomplished."

One of the researchers' greatest problems was that some Venezuelans, especially some of the men, refused to give blood on draw days. Some were clearly frightened, while others were suspicious and would become angry and stubborn when pressed. The researchers would then make house visits, often traveling up into the hills, or to an isolated water village to coax the resisters to relent. Anne Young remembers one time the researchers were desperate to get the blood of an older man in order to have a complete DNA library of his immediate family. The family was what the researchers called "highly informative," since he had several brothers and sisters with the disease, but he had escaped. He was old enough that the researchers believed he didn't carry the gene and would never get the disease. The DNA studies required that researchers check the DNA of siblings with and without the disease. His siblings had already given their blood. The fellow, a heavyset gruff fisherman about sixty years old, had declined several requests to participate, even ignoring the pleas of his wife and children. Finally, the need for his blood became urgent, as the next day the last shipment of refrigerated specimens was being taken north.

"Ten of us descended on his home one weekend afternoon, piling out of three cars," says Young. "The group included blood drawers, pedigree-takers, and several physicians." Once inside they came upon the normal mayhem of the family's life. About twenty people lived in the three-room house and many of them were home or quickly came home, once they heard that the North Americans had come. Out back was a garden of sorts, crowded with hammocks and aswarm with goats, pigs, and chickens. Wexler and Fidela Gomez, a young nurse from Florida considered one of the expedition's "greatest persuaders," started talking to him. Says Young, "He didn't want anything to do with us." He was eating dinner at a table in the backyard when the Americans arrived and had been drinking beer.

"Nancy and Fidela went into the corner and started on him. The house was bedlam. As usual it was very hot and muggy and some music was playing on the radio, the mosquitoes were terrible, babies were crying. It was hard to have a private conversation. His sons teased him and his mood became foul. Nancy was trying everything. Finally, after an hour or so, Nancy told us all to leave. So we sat out in the cars. It was nighttime when Nancy and Fidela finally came out. Nancy sort of did this skip and held the tube of blood over her head, and we all cheered. Basically, she wore him down."

Wexler often used the fact that she was also at risk of getting the disease as a personal bridge to the villagers. "HD can be such a lonely disease, even for the Venezuelans who certainly saw so much more of it than we do," she says. "But there is something about being at risk that isolates you from the world. I felt it and I know many of the people down there felt the same. I told them that I came to Venezuela because my mother and her three brothers had died from the disease, but that only they could help provide a way of finding a cure. They thought that they were the only people with the disease, that they were freaks. Some of them couldn't believe a disease like this existed in the United States. 'You put a man on the moon and you can't cure this disease?' they'd say to me."

At one draw day, Fidela Gomez struck upon an idea the researchers later would use as a ritual to overcome reluctance. In the first few years, the researchers took skin samples from the villagers as well as blood, taking a small biopsy from the upper arm that would leave a distinctive round scar. Wexler had had a similar biopsy taken several years earlier. When one older woman drew back from a needle, she noticed Wexler's scar. "Aha," she said, "you have the mark." Gomez, a woman whose brashness closely paralleled Wexler's, stood up on a chair, and holding Wexler's upper arm up high, yelled to the crowd, "She has the mark, she has the mark," which soon was echoed through the room by the villagers. At some towns, Wexler and her mark would be passed around as a sort of physical letter of introduction.

Despite appearances, Wexler often had moments when her confidence faltered. "I did not always retain my conviction that this was the chosen path to the HD gene," she wrote in 1988.[5]

Sometimes I would be terrified that we were collecting endless mounds of data and vast quantities of blood and skin samples which would sit in a freezer somewhere and never lead to anything. I dreaded that I had convinced my colleagues, who very quickly became my best friends, to come on a wild-goose chase which would take them away from more important work at home.

The most grueling aspect of the work is watching patients we have come to know and care for worsen each year. It is seeing a vibrant, intelligent, loving and beautiful little girl with limp hair, freckles, a rare but stunning smile and expressive eyes, age 12, begin to stiffen and fall with the earliest signs of the disease, which killed her younger brother and now has two other brothers in thrall. Six children of nine affected by the disease.

Following the third March expedition in 1983, the researchers had collected 570 blood samples from people whose crucial places in the pedigree produced an "informative kindred." Back in Boston, at his labs at Mass General Hospital, Jim Gusella was feverishly developing a technique for isolating riflips. By this time, Housman had turned over his Huntington's disease research grants to Gusella, who had devel-

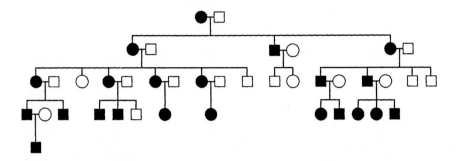

A portion of the Huntington's disease family pedigree from Venezuela showing the inheritance of a dominant gene. The black circles are affected females; the black squares are affected males.

oped five probes of his own and had an additional eight probes he had obtained from other molecular biologists, including Ray White. Gusella, at first, had no idea on which chromosomes these probes resided. It didn't matter, of course, since he had no clue at all as to where the defective gene was located. "We were picking up probes at random," Gusella says. "There was no rationale for beginning the search at one chromosome over another, so selecting probes at random seemed as good as any approach at the time."

Anxious to get his lab up and running, Gusella began testing the probes against the DNA extracted from the blood of a Huntington's family in Iowa that Michael Conneally had collected. Gusella expected little to happen. He had guessed that he would have to test at least three hundred probes before finding one linked with the gene. Perhaps, he thought, he might have to test hundreds, or thousands more. But just as importantly, he wanted to perfect his techniques for finding markers.

In 1982, while testing the Iowans' DNA with the third probe, called G8, the Harvard researchers in Gusella's lab produced a curious finding. In a highly consistent pattern, the probe was matching the DNA of Iowans who were sick, but not matching the DNA of those who were healthy. Gusella was suspicious that his technicians had made a mistake, but he sent the data from the experiment to Conneally, who had developed a computer program to analyze the results. The program was designed to calculate the odds that there was linkage between the gene and the marker. The program's accuracy increased significantly as more family members with and without the disease were tested. Conneally, therefore, was also skeptical when his program popped out a reading that indicated that the G8 probe was close to the gene. According to the program, there was only a one in sixty-three chance of no linkage. Conneally felt certain that his twenty-five-member family was really too small to expect any results. "It was just too preposterous to be right," Conneally says.

Nonetheless, as soon as the specimens from the March 1983 expedition were in his lab, Gusella's group began testing the entire batch of Venezuelan DNA with the G8 probe. The results were once more sent to Conneally's lab for computer analysis. The link was unmistakable, this time producing odds of 1,600 to one that the probe and gene were linked.

Wexler knew that Conneally's computers were digesting the data and that Conneally was away in Grand Canyon National Park in Arizona traveling with his family on the way to a scientific meeting. Eager to hear how the computer run was going, she called the lab from her office at NINDS and was read the results over the phone. "I just screamed and screamed and screamed," Wexler says. "I remember running upstairs and running up to some office people, jumping up and down and yelling, 'We found the gene, we found the gene.' I'm sure they thought I was crazy."

After hearing the news, Wexler called her father, and then her sister. "It was really hard to fathom," says Milton Wexler. "I think we felt we'd have the gene and then cure the disease in a matter of years."

Conneally's lab associates also were elated and wanted to share the news immediately with Conneally, who, they knew, was camping out somewhere in the Grand Canyon. Although it was late evening in Arizona, one of Conneally's associates, Peggy Wallace, put in a call to the park's rangers and cajoled one of the rangers to track down Conneally. Conneally recalls that "we'd just put out the campfire and I was sitting and having a beer, when suddenly I saw these flashing lights and I hear a voice saying, 'This must be it, these plates are from Indiana.' Out of the darkness stepped a ranger who asked me if my name was Conneally. He told me I had an urgent message to call my office. Now I knew that if they were trying to reach me that the Lod score [the odds of linkage] was high. I decided to have another beer; I think I might have had two. By then it was almost 1 A.M. in Indiana, so I decided to go to bed and call later in the morning. It turns out the folks back in Indiana were so excited they waited up all night for my call."

Gusella received the results calmly. "I really don't think I believed it," he says. "I wondered, 'What do I do now?'" Word spread throughout the molecular biology labs at Harvard and MIT. Incredibly, one of his first riflip probes had landed next to the Huntington's gene. Gusella was dubbed "Lucky Jim."

Indeed, the unaccountable luck even shocked the originators of the riflip concept. In Utah, Mark Skolnick says he was unwilling to believe it. "It was as if you wanted to hook one particular fish, but you had two dozen ponds to choose from," he says. "Not only did they pick the right pond, but, inexplicably, they hooked the right fish only the twelfth time they cast their rods."

(The researchers' astoundingly quick result, in fact, is near-mythic, and scientists familiar with it found a certain pleasure in outdoing each other with analogies to describe it. For instance, Roswell Eldridge, a top researcher at NINDS, wrote in an essay in 1985 that locating the marker was the result of "extraordinary good fortune."[6] He added: "The amount of luck involved in this discovery can be compared to knowing that an extremely long cookbook contains a single but serious misprint. Not knowing where in the book the misprint occurred, but having access to a correct edition, the researchers started to compare individual pages selected at random, and found the critical error on only the 12th page examined.")

David Botstein's reaction, like the man, was complex. In a conversation several years later he acknowledged that, of course, he was pleased and proud that an important discovery had its roots in a technique he helped create. Yes, he acknowledged, Gusella and Housman were the first to prove his theory correct, thereby validating it. But, perhaps because of understandable professional envy or sincere concern, Botstein made no effort to hide a certain disdain for the discoverers. He said icily that he felt the "lucky" discovery would breed unwarranted expectations. "I knew other [researchers] would look at this as something they could easily replicate in other diseases," he says. "I'm not sure the amount of real work needed to find most genes will be understood."

Unwarranted or not, expectations by researchers weren't merely raised by the finding, they went into the stratosphere.

In November 1983, Gusella, Wexler, Conneally, and their colleagues published their report in *Nature* saying that the genetic defect for Huntington's disease had been localized to chromosome 4.[7]

The discovery of a marker linked to the Huntington's disease gene makes it feasible to attempt the cloning and characterization of an abnormal gene on the basis of its map location.

Understanding the nature of the genetic defect may ultimately lead to the development of improved treatments. Furthermore, this study demonstrates the power of using linkage to DNA polymorphisms to approach genetic diseases for which other avenues of investigation have proved unsuccessful. It is likely that Huntington's disease is only the first of many hereditary ... diseases for which a DNA marker will provide the initial indication of chromosomal location of the gene defect.

(After the report was submitted to *Nature* but a few weeks before it was published, Gusella found that the G8 probe was landing on chromosome 4, thus mapping the HD gene to that chromosome.)

On November 9, an article on the front page of the *New York Times* reported the discovery as the "first genetic test to detect Huntington's disease."[8] The *Times* said that, while the article in *Nature* wasn't planned for publication until the following week, a news conference to announce the discovery "was moved ahead one week because scientists who learned about the findings [had] become so excited that they discussed it openly."

The discovery of a previously unknown gene by use of an entirely new technology captured the attention of hundreds of scientists and doctors worldwide. Suddenly, the use of DNA markers had been shown to be effective. Scientists searching for genes everywhere were riveted by the news. Janice Egeland, a sociologist at the University of Miami, for instance, remembers her excitement. Egeland had spent much of the previous two decades searching for a genetic cause to manic-depression. Over the years, she had collected data on a large family among the Amish farmers of Pennsylvania in which manic-depression had struck in unusually large numbers. But after years of searching for a specific gene or genes that might be at the heart of the family's problem, she was about ready to give up. Within hours of reading about the *Nature* report, she began placing phone calls to the scientists at MIT and Harvard, one of several dozen the researchers received within the first weeks of the *Nature* publication. Within a month, Egeland was collecting blood samples from members of her Amish family. Says Egeland, "It was a most wonderful breakthrough."

CHAPTER 5

Bruce Bryer and Reverse Genetics

While the headlines were hailing the mapping of the Huntington's disease gene, an unpublicized series of events was taking place in laboratories in Boston, London, and Toronto that would produce an even more impressive demonstration of the new riflip technology's power to solve the mysteries of genetic diseases. The target of these quiet efforts was muscular dystrophy, one of the most common of the three thousand inherited diseases.

In muscular dystrophy—indeed, in most of the major genetic diseases—researchers were faced with an almost insurmountable biochemical wall. This tragic disorder is a gradual, relentless wasting of the muscles that begins in early childhood and almost always ends fatally in early adulthood. This muscle destruction is the end result of a Niagara of chemical events triggered by a single defective gene buried somewhere in a single chromosome, the X chromosome. But all that is visible and measurable is the muscle destruction itself. For decades, researchers had little choice but to study the wasted muscles of muscular dystrophy victims, seeking a biochemical aberration, such as a missing or defective protein, that might possibly yield a clue to exactly what was going on. The search had been futile.

The advent of riflip technology offered a new strategy against genetic diseases, a strategy to become known as "reverse genetics." Instead of beginning with the disorder's consequences, as medical scientists had done for decades, the riflip technology offered the means to begin with the source of the disorder, the defective gene, and then work up through the chemical cascade to reveal the muscle-destroying mechanism. Reverse genetics is analogous to exploring the Mississippi River by starting with springs in Minnesota instead of the river mouth in Louisiana.

The key to solving muscular dystrophy was a sixteen-year-old boy from Spokane, Washington, Bruce Bryer. Bruce was born in Spokane in 1966 and immediately put up for adoption by his mother. At ten days of age he was placed with foster parents, Robert and Laurene Dixon, a generous Spokane couple who, over two decades, had cared for more than two hundred unwanted children.

From the start the child was sickly. "The boy's early months were filled with medical crises. Time and again, Mrs. Dixon had to take him to the hospital because of fevers, infections, and stomach ailments. During one hospitalization, when Bruce was nine months old, his pediatrician insisted that the Dixons bring in the chaplain to baptize him—his death seemed imminent."[1] As he would many times over the years, the boy hung on to life tenaciously and recovered.

When Bruce was three, a specialist at the University of Washington, Seymour Klebaroff, discovered the boy had been born with chronic granulomatous disease, a genetic disorder that crippled his immune system and rendered him susceptible to repeated infections. But Bruce also suffered muscle weakness which doctors, at first, attributed to his bouts of life-threatening illnesses. When he was nine, however, doctors realized the boy had Duchenne muscular dystrophy. By age twelve, Bruce depended on a wheelchair.

Muscular dystrophy had baffled medical researchers for more than a century. It was first recognized as a distinct disease in 1858 when a pioneer French neurologist, G.A.B. Duchenne, described seeing the muscle disorder in several young boys. In the 1950s, a German neurologist, P.E. Becker, identified a milder and less common form of the disorder.

By the time Bruce was diagnosed as suffering the Duchenne-type of muscular dystrophy, the disorder was recognized as one of the most common of the inherited diseases, striking one of every 3,500 male infants born—but never females. The infant boys seem more or less normal during their first few years of life. But sometime between ages three and seven those with the Duchenne muscular dystrophy begin to develop a strange form of waddling and a tendency to walk on their toes. They stumble and fall frequently, and have trouble climbing stairs. By age ten or twelve they are confined to a wheelchair. The weakening of the muscles is unremitting. Many of the victims die in their twenties as the muscles that control breathing wither. The Becker form of the disorder differs only in that it is milder and of later onset, and its victims live much longer.

It also was known by the early part of this century that muscular dystrophy, because it struck only males, stemmed from a defect in the X chromosome. Of the twenty-three pairs of human chromosomes, twenty-two pairs, known as the autosomes, are thought to have little or nothing to do with the determination of whether one is male or female. On the twenty-third chromosome pair, however, buried midst thousands of other genes, are the genes that determine sex. The larger and longer of the sex-determining chromosomes is known as the X chromosome while the second, shorter, almost fragmentary sex chromosome is designated the Y chromosome.

In females an X chromosome inherited from the mother is paired with a second X chromosome inherited from the father. But in males

How X-Linked Inheritance Works

The mother has a defective gene on one of her two X chromosomes but she is protected from its effects by the normal version of the gene on her other X chromosome. The father has normal X and Y chromosomes. Each male child has a 50 percent chance of inheriting the mother's defective X chromosome and suffering the effects of its defective gene. Each female child also has a 50 percent chance of inheriting the mother's defective X chromosome and becoming a carrier of the defective gene like her mother and a 50 percent chance of inheriting the mother's normal X chromosome.

the X chromosome, inherited from the mother, is paired with the tiny Y chromosome, inherited from the father.

This sex difference in the chromosomes—XX for females and XY for males—has its own peculiar consequences in inherited disorders like muscular dystrophy. Females are resistant to such disorders since their second X chromosome can partially or totally compensate for the defects in the first X chromosome. Males, on the other hand, don't have such a backup set of genes for their single X chromosome since it is paired with the tiny Y chromosome. Thus, males are highly vulnerable to damage and defects on the X chromosome.

A second consequence of the XX/XY distinction between females and males is that females can become unknowing carriers of an X-linked inherited disease. Unlike males, in whom an X-linked disease is apparent usually early in life, females, with their double X, can go through life without any apparent signs of an X chromosome defect. A healthy woman with a normal X and a defective X chromosome has a fifty-fifty chance of passing the defective chromosome on to each of her children. If, by the toss of the genetic coin, the child is a boy, he will suffer the consequences of his mother's hidden genetic disorder. (If the child is a girl, she won't suffer such a fate, since her mother's defective X chromosome will be offset by her copy of the father's normal X chromosome—but the daughter, like her mother, becomes a carrier of the X defect and risks passing it on to her children.)

Since the earliest days of their discipline, geneticists had been able to identify an X-linked disorder simply from the fact that over the generations the disorder shows up exclusively in males in an afflicted family. As early as 1911 color blindness was identified as an X-linked disorder on such evidence. Hemophilia, muscular dystrophy, and a host of other, much rarer disorders were similarly traced to the X chromosome.

(Defects in the Y chromosome also would show up solely in males but only one such consequence, which affects the development of the testes, is known. On the other hand, a woman theoretically could suffer an X-chromosome disorder if her mother's defective but hidden X chromosome happened to have the same defect as the father's single defective X, an extremely rare possibility.)

Except for its mode of inheritance, muscular dystrophy was as much a puzzle to medical scientists of the 1970s as it had been to Duchenne more than a century earlier. The disease had defied all attempts to determine why the children's muscles suffered such a slow, steady breakdown. There are, by one guess, an estimated ten thousand proteins involved in muscle development and function, and, hence, ten thousand genes, which, if defective, might underlie muscular dystrophy. The vast majority of these muscle-related proteins were—and still are—unknown. Try as they might, biochemists had been unable to spot a protein abnormality in muscular dystrophy patients that would explain why their muscles began to deteriorate a year or two after birth.

Fortunately for the future, Hans Ochs, a University of Washington immunologist, had taken an interest in Bruce's immune disorder, chronic granulomatous disease, as early as 1970. The disease, like muscular dystrophy, is due to a defect on the X chromosome. Ochs, suspecting that Bruce's X chromosome might hold clues to other X-linked disorders, had cultured some of the boy's white blood cells, thus preserving his chromosomes for future study.

On top of his other problems, Bruce also suffered failing eyesight. In 1981, the boy was found to have retinitis pigmentosa, a degeneration of the retina that leads to blindness. Roberta Pagon, a medical geneticist interested in genetic eye diseases, confirmed that the boy suffered a rare form of retinitis pigmentosa caused by a defect on the X chromosome, making Bruce a tragic victim of three X-linked diseases.

Pagon speculated there was a gross structural defect at a single spot on Bruce's X chromosome. Ochs asked Christine Disteche to examine the boy's X chromosome. Disteche was a cytogeneticist, a scientist who looks for chromosome damage by studying the arrangements of light and dark bands that appear on chromosomes when they are stained by a special dye (see photo section).

Disteche thought she could see hints of a deletion, or a bit of missing DNA, in the short "arm" of the chromosome. (The X chromosome, like all the chromosomes, is pinched at a spot somewhat off center, dividing the chromosome into a short arm and a long arm.) The hints of an error on the short arm, however, were too ambiguous for any definite conclusions. Ochs thought that the uncertainty might be resolved by Uta Francke, an expert in chromosome mutations at Yale University in New Haven, Connecticut. When the experienced Francke looked at the banding patterns on Bruce's X chromosome, she thought she, too, could see an error on the short arm, possibly the absence of a fairly large chunk of DNA in the chromosome.

The cytogeneticists, however, needed more samples of cells to confirm their suspicions that a large fragment of the chromosome somehow had been deleted. They also needed to compare Bruce's X chromosome with that of his mother and, if possible, his maternal grandparents. But the boy's young mother, who had left him for adoption ten days after his birth, couldn't be located.

While the researchers puzzled over Bruce's X chromosome, another pivotal development was taking place in the cytogenetics laboratory of the Hospital for Sick Children in Toronto. The laboratory was headed by a Canadian, thirty-eight-year-old Ronald G. Worton. Worton brought an unusual background to molecular genetics. He had earned a bachelor's degree in physics and a graduate degree in biophysics, had done his postgraduate work at Yale in biochemistry and genetics, and finally in 1971 had arrived in Toronto to do research in cytogenetics, the study of the chromosomes.

In 1979, Worton's laboratory was host to a young cytogeneticist from Belgium. Christine Verellen-Dumoulin brought with her details of an astonishing discovery her colleagues in Brussels had made several months previously. A fifteen-year-old Belgian girl, identified only as Anne, had been diagnosed with muscular dystrophy, one of the first females ever known to have the disorder.

When Anne was only two years old, her parents had noticed that she had trouble climbing stairs and was prone to falling.[2] By the time she was five, the calf muscles of her legs were enlarged, a typical feature of muscular dystrophy. By age thirteen, she walked with a waddling gait and suffered muscular weakness in the arms, legs and neck. Only then did her doctors begin to suspect the young girl might have muscular dystrophy, as incredible as it might seem.

In Anne's case the diagnosis of muscular dystrophy had to be made solely on the basis of the girl's symptoms. There was no chemical or genetic test that could verify that the young girl's dystrophy was caused by the X-chromosome defect that produced the disorder in boys.

Nevertheless, when Verellen and her Belgium colleagues had examined Anne's chromosomes in 1976, they had spotted a suspicious aberration in one of the X chromosomes. (The error was on the short arm of the chromosome in the same area where Disteche and Francke, in 1985, thought they saw a deletion in Bruce Bryer's X chromosome.)

The error in Anne's case was a translocation, a mistake that sometimes occurs during a process known as recombination. During formation of a parent's germ cells—the father's sperm cells or the mother's egg cells—there is a rearrangement of the parent's paired set of chromosomes to produce a set of twenty-three single chromosomes. During this rearrangement—or recombination—there are occasional errors. In Anne's case it appeared that a piece of the short arm of the X chromosome had been tossed out and replaced by a piece from one of the twenty-two autosomes (non-sex chromosomes). The translocation appeared to be a spontaneous mutation, since the mistake couldn't be seen in the chromosomes of either the father or the mother, neither of whom showed any hint of muscular dystrophy.

Anne's jumbled X chromosome suddenly offered hope of a solution to the muscular dystrophy enigma. It was possible that the translocation had occurred at the site of the Duchenne muscular dystrophy gene, the DMD gene. If the translocation was actually the cause of her muscular dystrophy, then the DMD gene lay near the breakpoint where the X chromosome ended and the inserted piece of the autosome began.

But such a translocation had never been seen in the X chromosomes of boys afflicted with muscular dystrophy or in their mothers who carried the X chromosome defect that caused muscular dystrophy. It was entirely possible that the translocation was unrelated to Anne's muscular dystrophy, that, instead, her disorder was due to an invisible mutation somewhere else on the X chromosome.

There was a second mystery surrounding Anne's muscular dystrophy. Thousands of women who had borne boys with muscular dystrophy were obvious carriers of the mutant DMD gene. Yet, unlike Anne, they didn't suffer muscular dystrophy, themselves; they were protected from the disorder by their normal second X chromosome. In such women, nature apparently had a way of turning off the damaged X chromosome and letting the normal X function. Anne, herself, had a second X chromosome that presumably was normal, but it failed to protect her against the devastating effects of the muscular dystrophy mutation. Somehow, nature had erred and turned off the wrong X chromosome.

In short, the translocation on Anne's X chromosome was a tantalizing lead, but it also could be a false trail. Geneticists and biologists could easily spend years of tedious study of the translocation without coming up with a single clue to the disease. Nevertheless, given the frustrating lack of progress in muscular dystrophy, it seemed a chance worth taking.

Verellen arrived in Worton's laboratory with cultures of Anne's cells and a desire to study the nature of the translocation in the X chromosome. She quickly pinpointed the spot where the normal X chromosome had broken off and the autosome fragment had been inserted. This breakpoint was near the dark bands designated bands number 21, on the short arm of the X chromosome. The short arm of any chromosome is designated the "p" arm (and the long arm, the "q" arm). The location of the breakpoint, hence, was designated Xp21.

Just as Verellen·began trying to uncover the Xp21 breakpoint, welcome news came in from opposite sides of the globe. "About this time we heard of two other girls with muscular dystrophy, one in England and one in Hawaii," Worton recalls. In both cases, there was a translocation on the X chromosome. Moreover, the breakpoints for both translocations were in the same region, the Xp21 band. "It was clear this wasn't a coincidence," Worton says. The Xp21 translocation, as suspected, was directly involved in all three girls' muscular dystro-

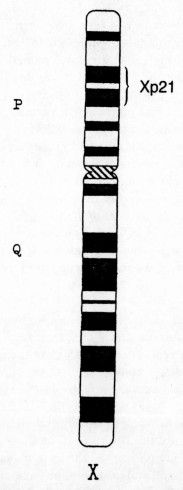

Normal X chromosome showing the Xp21 bands where the breakpoint appeared in Anne's X chromosome.

phy. Hopes suddenly rose that the defective DMD gene lay somewhere near the breakpoint at Xp21.

By 1981 muscular dystrophy researchers sensed that the foundations for breakthrough were falling into place but exactly how and where it would occur were frustratingly unclear. In Toronto, Verellen and Worton suspected the DMD gene lay near the Xp21 bands where the breakpoint in the muscular dystrophic girls was seen. At Yale in New Haven, Francke and others studying Bruce Bryer's chromosomes also suspected that the boy's muscular dystrophy had some link with the aberration on the short or "p" arm of the boy's X chromosome.

But suspecting the gene lay on the short arm of the X chromosome and actually confirming it were two different matters. While geneticists had tracked down the approximate locations of many genes on the X chromosome, no one had ever zeroed in on the exact position of a gene, much less plucked the gene out to examine it.

Meanwhile, the concept of using riflips as markers for locating genes was spreading. Botstein, White, and Skolnick had been talking widely and openly for two years about their idea for a riflip map. Their paper proposing the idea appeared in the *American Journal of Human Genetics* in 1980, followed a few months later by publication in the *Proceedings of the National Academy of Sciences* of the Wyman-White paper describing how to locate a riflip. Some molecular biologists, nagged by the geneticists, already had begun searching for riflips that might be linked to defective genes.

Among these riflip seekers were two young researchers working under molecular biologist Robert Williamson at St. Mary's Hospital School of Medicine in London. Kay E. Davies and J.M. Murray in 1981 began a search for riflips on the X chromosome that might be linked to the DMD gene.

In a matter of months the young British scientists hit paydirt. They found a sequence of bases along the short arm of the X chromosome that was polymorphic: slight variations in the sequence from individual to individual would produce different lengths of DNA fragments when this section of the chromosome was cut with a restriction enzyme. In other words, the sequence of bases in this spot was a riflip, a restriction-fragment-length polymorphism. This unique sequence could be pinpointed and plucked out by DNA probes that homed in on the sequence and hybridized with it, the same kind of riflip-homing DNA probes that Wyman and White used in their keystone experiments in Worcester in 1980.

Davies, Murray, and their clinical colleagues in London immediately applied their riflip probes to the DNA of families in which one or more boys with muscular dystrophy had been born. Four out of five times the riflips inherited by the stricken boys were the same lengths as the riflips carried on their mothers' X chromosomes, but of a different length than the riflips inherited by the boys' healthy brothers and other unaffected family members. The riflips, in other words, could distinguish muscular dystrophy boys and their carrier mothers and sisters from their unaffected relatives. The riflip and the DMD gene were linked, passed along together, 80 percent of the time.

In the remaining 20 percent of cases, the scientists figured, the link between the riflip and the DMD gene had been severed, the separation occurring during the recombination of chromosomes when the germ cells are formed.

This 80–20 percent ratio of linkage-unlinkage for the riflip and the DMD gene was a key clue. The closer a marker is to the target gene, the less likely the two segments of DNA will be separated or unlinked during the recombination process. The rough rule of thumb was that each 1 percent chance of separation represented a distance between the gene and the marker of about one million DNA bases. The fact that the riflip and the DMD gene separated 20 percent of the time hinted that the two must lie about twenty million DNA bases apart on the short arm. Unfortunately, this didn't pinpoint the exact location of the gene. The gene could lie twenty million bases "above" the riflip or twenty million bases "below" it. And even that calculation was crude and uncertain. Nevertheless, it confirmed that the DMD gene was in the neighborhood of the Xp21 band.

Discovery of the riflip marker coincided with other hopeful news. Researchers looking for dystrophic girls found two more, bringing the worldwide total to five. All five, counting Anne, had translocations near the Xp21 region.

A small meeting to pool information on the girls already had been scheduled by the Muscular Dystrophy Association for the summer of 1982 in a hotel in Scottsdale, Arizona. With news from London that a riflip linked to the DMD gene had been found, the Scottsdale meeting quickly turned into a brainstorming session on whether the means were at hand to find the DMD gene. The answer was: Possibly. The dozen or so researchers agreed to reassemble the next summer at the Cold Spring Harbor laboratory on Long Island. Each would bring a "wish list" of what he or she felt was needed to find the DMD gene.

Before the Cold Spring Harbor gathering, a second piece of electrifying news broke from across the Atlantic. Davies, using a probe from Peter Pearson's laboratory in Leyden, the Netherlands, pin-pointed a second riflip on the X chromosome, closer to the chromosome's pinched center than their first riflip. Tests on the DNA of muscular dystrophy families indicated it, too, was within twenty million bases from the DMD gene.

More important, about 60 percent of the muscular dystrophy boys inherited both riflips from their mothers, that is, the two riflips were linked 60 percent of the time. In the remaining 40 percent the two riflips had been separated, or unlinked. The two riflips must be about forty million bases from each other, yet each was only twenty million bases from the DMD gene. Obviously, the DMD gene must lie somewhere between the two riflips.

"Davies literally put the DMD gene on the map," says Donald S. Wood, research director of the Muscular Dystrophy Association in New York. The map, to be sure, was crude, somewhat analogous to saying that the location of Manhattan Island, known to be on the northeast coast of the United States, had been narrowed down to somewhere between Cape Cod and Cape May. But it was the first firm evidence of where the DMD gene lay on the X chromosome, between the two riflips and in or near the Xp21 band.

"Everyone thought that finding the DMD gene was just around the corner," Wood recalls.

None were more elated at the news from London than the executives and directors of the Muscular Dystrophy Association. For years, the volunteer fund-raising agency had been tugging millions of dollars from the pocketbooks of sympathetic Americans by means of an annual twenty-four-hour-long "telethon" hosted by movie comedian Jerry Lewis. In between acts by famous entertainers, Lewis, his black bow tie hanging loose and his voice hoarse from hours of talking, would appear on camera with a crippled child in his arms and beg for a few more dollars that could lead to a cure for the child and others like him. It was—and still is—a highly effective fund-raising technique, imitated but rarely matched by other disease-oriented funding agencies. Many Americans, who tend to confuse muscular dystrophy with multiple sclerosis, cerebral palsy, or cystic fibrosis, instantly recognize muscular dystrophy when they're told it's the disorder suffered by "Jerry's kids."

The news from Davies was the kind of advance the association needed to fulfill Lewis's promises.

The association's grantees gathered, as planned, at Cold Spring Harbor in the summer of 1983. They were determined that the association would foster what appeared to be an imminent breakthrough. It was decided that the association would back an all-out effort to find the DMD gene. "They estimated it would take five million dollars and over five years to get the gene," Wood recalls.

Ordinarily, the association, like all private and government funding agencies, would wait for various scientists to propose particular projects. The association's scientific advisors would then pick the most promising proposals to fund. But it was the association that proposed a crash effort. "We decided to find out who had the strongest motivation to go after the gene and to give them anything they needed," Wood says. The one million dollars a year earmarked for the effort would go to three different laboratories, each pursuing a different avenue to the gene.

The decision, however, raised a delicate question. The three chosen scientists, young and highly motivated, were almost certain to compete with each other to be the first to find the gene. Science may progress by the sharing of information but scientists progress by beating each other to a discovery. After all, the cliché goes, who remembers the second person to write $E = mc^2$? The competition to be first often can become bitter and even personal. In such situations, it's an unwritten rule that you never tip off the competition of the progress you're making until you're sure it's too late for them to catch up. Discoveries, even if it's the discovery of a blind alley, might remain under wraps for months as the discoverers try to insure their lead or hope their competitors will become sidetracked on experiments that will be futile and time-consuming.

Such secrecy could seriously delay the finding of the DMD gene, the Cold Spring Harbor attendees realized. Thus, it was agreed that each of the three scientists would keep his or her two competitors abreast of every development, even if it permitted one researcher to leap ahead of the others. It was an agreement that was faithfully followed. "It turned out to be science at its best, friendly competition," says Wood.

Two of the choices were obvious: Davies in London, who had found the riflip markers for the gene, and Worton in Toronto, who was homing in on the Xp21 breakpoint on Anne's chromosome. Worton already was receiving support from the Canadian Muscular Dystrophy Association, and his Canadian grant would now be supplemented by the U.S. association.

Not so obvious was the choice of thirty-three-year-old Louis Kunkel of the genetics division at Children's Hospital in Boston and the pediatrics department at Harvard Medical School. At the time, Kunkel probably was known more for his ancestry than for his research. He was the third generation of a budding dynasty of scientists. His grandfather, Louis O. Kunkel, had been a noted botanist at Rockefeller University while his father, Henry, was an eminent immunologist at Rockefeller. Growing up in a New York City suburb midst such scientific talent had imbued him with a strong curiosity and a deep respect for nature. "I always thought I would be a scientist," he says. As a teenager his summers had been spent in biology laboratories at the U.S. Department of Agriculture's big research and quarantine facility at Plum Island, New York (off the northern fork of Long Island), and at Cornell University Medical Center in New York City. It was at the urging of a Cornell biologist that he did his graduate studies in human genetics at Johns Hopkins Medical Institutions in Baltimore.

The young scientist already was a grantee of the Muscular Dystrophy Association. In 1980, after a postdoctoral stint at the University of California, San Francisco, he had gone to Children's Hospital in Boston to launch an effort to discover how one of the two X chromosomes in females is inactivated. Since this was a key question in the case of girls like Anne, where the erroneous inactivation of the normal X chromosome left them suffering muscular dystrophy, the association had been funding him. Kunkel also had been one of a dozen or so researchers at the Scottsdale meeting and was thoroughly up-to-date on the state of the search for the DMD gene.

Kunkel also was well versed in the use of riflips and DNA probes. While in San Francisco, his mentors in molecular biology had been the first to pluck a gene—the insulin gene—out of human DNA and induce bacteria to reproduce it, that is, clone it. Morever, he was married at that time to a young researcher who was working in Y.W. Kan's laboratory when Kan and Dozy discovered the first riflip in the sickle-cell anemia gene. Thus, he was intimately familiar with the riflip concept.

"Kunkel wasn't widely known but he had a solid background in molecular genetics," Wood says. In sum, Kunkel fit all the criteria set out at Cold Spring Harbor for the candidates who would pursue the DMD gene: He was young and enthusiastic, and he had the talent and experience to carry out a gene hunt. He also had a plan.

Kunkel's plan was built around Bruce Bryer's cells, or the B.B. cells as they were called to protect the boy's privacy. He was fully aware of Francke's work with the B.B. cells at Yale. "I had known Uta for years," Kunkel says. "When I was a graduate student at Johns

Hopkins I met Uta at [genetics] meetings we would go to." He knew of Francke's suspicion that the aberration on the short arm of Bruce's chromosome was a deletion of a large chunk of DNA and that the deletion probably had something to do with the boy's muscular dystrophy.

The young Boston molecular geneticist had compiled a library of DNA probes that clung randomly to parts of the X chromosome. These X-chromosome probes suddenly took on a new importance when word came that Francke at Yale had found that Bruce's X chromosome did, indeed, suffer a deletion.

"I was quite excited because it confirmed that the library we built here was well worth screening" for probes that could detect deletions in Bruce's X chromosome, he explains.

Kunkel's strategy, in a nutshell, was to find out what genes were missing from Bruce's X chromosome and then to see if any of the same genes also were missing from boys with muscular dystrophy. Whatever pieces of DNA they lacked in common likely contained part or all of the DMD gene.

Kunkel needed cultures of the B.B. cells themselves to launch his search for the DMD gene, but Ochs at the University of Washington already had given Francke first claim on the cells. Kunkel's department head, Sam Latt, approached Ochs and worked out an agreement that Kunkel could obtain the B.B. cells after Francke had had sufficient time to exploit them for her purposes. "After about a year Ochs released the B.B. cells to us," Kunkel says.

His next step sounds far simpler in the telling than it was in practice. He took normal X-chromosome DNA, broke it into fragments, and tossed the fragments into a flask. He then took copies of Bruce's DNA, cut them with a restriction enzyme, and tossed those fragments into the same flask.[3] The fragments of normal DNA and of Bruce DNA that complemented each other in their base sequences hybridized, base A in one fragment clinging to base T in the other, base G to base C, and so on. These hybridized fragments, representing normal DNA that Bruce possessed, were then put aside.

Among the fragments of normal DNA that failed to find matching pieces in Bruce's DNA were eight fragments that, it was hoped, contained some small bits of the genes that Bruce was missing. There was even a chance that one of the fragments of normal DNA might contain a piece of the DMD gene or at least of a piece of DNA that lay close to the gene. There was one way to find out: Use the eight fragments as DNA probes and apply them to the DNA of other boys

suffering muscular dystrophy. If these fragments of normal DNA failed
to hybridize, it meant the boys were missing the same bits of DNA
as Bruce.

Anthony P. Monaco, one of Kunkel's young associates, headed the
next step.[4] Monaco approached several clinics for samples of DNA of
both muscular dystrophy boys and normal men and women. He and
Kunkel then settled in for what were certain to be tedious and often
frustrating months of DNA probing.

The probes were applied to the DNA of fifty-seven muscular dystrophy boys and sixty-one normal persons. Seven of the probes struck
out; they hybridized with DNA from both the boys and the normal
persons. But, in an almost totally unexpected stroke of fortune, an
eighth probe, labeled pERT87, failed to match up with pieces of DNA
from five of the boys with muscular dystrophy. The five boys and
Bruce, then, had two things in common: They were missing the same
short stretch of DNA, defined by the pERT87 probe, and they had
muscular dystrophy. Since none of the boys suffered Bruce's other
disorders, this missing stretch of DNA must be related to muscular
dystrophy.

From experience, molecular geneticists knew that loss of a gene
or bit of a gene, that is, a deletion, would be found in only 5 to 10
percent of the patients suffering a genetic disorder; the other 90 to 95
percent of patients with the same disorder would have other mutations
in which only a single DNA base had been altered. The fact that five
of the fifty-seven boys, or 9 percent, suffered the same deletion "was
right in the ballpark of what one would have expected for a genetic
disorder," Kunkel says. "So we were really ecstatic about it."

There was another reason for ecstasy. "It meant we were damn
close [to the DMD gene]," Kunkel explains. "The probe was potentially within the gene itself."

The pERT87 probe told the scientists that Bruce and the other
boys suffered a deletion, but it didn't tell them how much of the DNA
had been knocked out. What the biologists now needed were more
precise probes that would tell them where the deletion began and
where it ended. Monaco began another tedious procedure, called
"chromosome walking," that has become a keystone of locating unknown genes.[5] By laying short, overlapping probes up and down the
spot where the pERT87 probe had landed, Monaco gradually built
up a long fragment of normal DNA that was sure to bracket the deletion. The long fragment was then cut into pieces or subprobes, each
of which was tested on the DNA of the five muscular dystrophic boys.

By early summer of 1985, the Boston researchers were almost certain their pERT87 probes were bracketing around or even on the damaged DMD gene. The size of the deletion, Monaco and Kunkel estimated, must range between 250,000 and 600,000 bases long, again in the ballpark of known disease-causing genetic deletions.

At this point, Monaco and Kunkel uncovered what became one of the first practical applications of the hunt for the DMD gene. Some of the pERT87 subprobes clung to a stretch of intact DNA that lay near the deleted segment. The base sequences in these intact stretches varied from boy to boy; they were polymorphic, and produced different lengths of DNA when they were cut with the same restriction enzyme. In other words, they were riflips. Even though Kunkel and Monaco hadn't yet found the DMD gene itself, the riflip probes were close enough to it to be used for diagnosis. When these riflip probes were used on the five boys' mothers, it was clear that each mother carried the same riflip as her afflicted son. In one family, two brothers, both of whom had muscular dystrophy, had the same riflip as their mother— and their grandmother. The riflip was tightly linked to the DMD gene.

"It appeared that these probes [for the riflips] could be used for carrier detection," Wood of the Muscular Dystrophy Association recalls. The probes would be invaluable in determining whether sisters of muscular dystrophy boys had also inherited the DMD gene and were at risk of passing it to their future sons.

But passing out the subprobes to muscular dystrophy clinics for carrier detection posed a delicate problem. Almost any competent molecular geneticist could take the probes and start his own hunt for the DMD gene. He might even beat Kunkel to the gene. While the Boston researchers would receive the proper credit—their names would appear as coauthors on any paper describing the discovery of the DMD gene—the discoverer's name would be first. From then on, references in the scientific literature to the discovery of the DMD gene would read "Smith, Joe, et al." Smith, or whoever, could become the scientist remembered for the discovery of the DMD gene at Kunkel's expense.

Despite this risk, "Lou agreed on the spot" to release the probes to the clinics, Wood says. But, he adds, there was a kind of gentlemen's agreement among the recipients; none would use the probes to go after the DMD gene.

Within months, copies of the probes for the DMD-linked riflips were in the hands of genetic diagnostic laboratories. In addition to detecting female carriers of the DMD gene, the probes could tell a carrier who was pregnant whether the fetus she carried also had inherited the gene. This precise prenatal diagnosis could allow the woman to abort an affected fetus—but also could prevent the abortion of a normal fetus. Until that time, a prenatal diagnosis could tell a woman known to carry the DMD gene only whether the fetus was male or female. If the fetus was male there was a fifty-fifty chance it would have muscular dystrophy. Many women who already had borne one boy with the disorder declined taking a fifty-fifty chance and chose abortion of a male fetus. Presumably, half of these aborted male fetuses were normal, although there wasn't any way to tell. The riflip probes, however, provided a direct test of a male fetus, revealing with high accuracy whether it was normal or affected. As the riflip probes became available, the number of abortions for muscular dystrophy began dropping dramatically.

Worton in Toronto also was closing in on the DMD gene but had run into a time-consuming problem. Worton was pursuing a strategy that initially seemed less risky than Kunkel's. Kunkel easily could have been wrong in his guess that Bruce Bryer's muscular dystrophy was due to the deletions picked up by the probes he was using. Worton, on the other hand, was using Anne's fractured X chromosome to track down the gene. There was general agreement that the translocations seen in all the girls like Anne had caused their muscular dystrophy. Thus, it was highly likely that at least part of the DMD gene was at the breakpoint. Indeed, some of the gene might still be lying at the edge of the breakpoint. What Worton planned to do was to find the exact spot where Anne's X chromosome had broken off and where the piece of inserted autosome began.

There was a way to find the junction. Verellen had confirmed that the piece of autosome wedged into band 21 on Anne's X chromosome had by sheer and confusing coincidence, come from *autosome* 21. Fortunately, in 1982, Roy D. Schmickel and his colleagues at the University of Pennsylvania had pinpointed a set of genes on autosome 21 known as rRNA genes that are involved in the cell's protein-making machinery. Moreover, the Penn scientists had probes for these rRNA genes. A collaboration was struck up, and by the beginning of 1984, Worton and Schmickel had good news.[6] The chunk from autosome 21 had broken off its parent right in the middle of the rRNA genes. In Anne's X chromosome, then, pieces of the rRNA genes must lie next door to what was left of the DMD gene. A probe for the rRNA genes might well land next to, or even overlap, the DMD gene.

Worton's young protégé, Peter N. Ray, headed the attempt to use the probes for the rRNA genes to find the junction. But he and his mentor suddenly ran into a major technical snag. In some of their early experiments Worton and Schmickel had had to fuse some of Anne's cells with mouse cells. The pieces of Anne's X chromosome they were now working with were still contaminated with bits of mouse DNA. Because mice and humans share many genes, Ray's probes couldn't distinguish between Anne's DNA and the contaminating mouse DNA. It took Ray more than four months to develop probes that detected human DNA but not mouse DNA. It would prove to be a costly delay in the race for the DMD gene.

By early fall of 1985, the Canadians had a probe that spanned the junction where the piece of the inserted autosome 21 joined the X chromosome.[7] The probe was dubbed the XJ probe. It contained, they fervently hoped, a piece of the damaged DMD gene.

When the XJ probe was applied to the DNA of muscular dystrophy patients, it detected deletions in six of the patients, including the five boys who had deletions detected by the probes from Kunkel's laboratory. The Canadian XJ probe and the Boston pERT87 probes must be landing on or near the same spots and detecting the same deletions. Moreover, the probes were landing within the region marked out by Davies's riflips. There seemed to be no question now, the probes must be hitting on or near the DMD gene.

The Canadians, like the Kunkel team, also "walked" the chromosome around the deletion and found a riflip that could be used for carrier and prenatal diagnoses.

The Canadians' elation was tempered, however. It wasn't until late September and early October 1985 that Ray, Worton, and their collaborators had convincing evidence that their XJ probe was on or near the DMD gene, six months after Kunkel's group had confirmed that their pERT87 probe was linked to the gene.

Like the metal detectors of treasure hunters, the DNA probes had picked up unmistakable signals of the DMD gene. The next task was to dig through layers of DNA to find the DMD gene itself.

CHAPTER 6

Dystrophin

by the fall of 1985, everything was in place for the final push to find the Duchenne muscular dystrophy gene. There was general agreement that the early suspicions were correct. The DMD gene was a normal gene, involved somehow in muscle function or development, which had been damaged and was malfunctioning, causing the destruction of the muscles. The nature of the damage was the deletion of part of the gene. Davies's riflips had narrowed down the location of the gene to the Xp21 band. Within this same area Worton's and Kunkel's probes were detecting pieces missing from the DNA of 5 to 9 percent of muscular dystrophy boys. Since the probes were composed of normal DNA, they presumably contained at least some of the pieces that had been deleted in the muscular dystrophy patients. Some of these missing pieces, then, must be pieces of the DMD gene. The goal now was to find those missing pieces.

But there was a pitfall. A gene doesn't exist as a continuous, unbroken length of DNA. Instead, a gene is fragmented into short, functioning sections of DNA with long, meaningless segments of DNA inserted between the gene fragments. The probes were too short to cover the entire gene, which presumably was sprinkled along several hundred thousand bases of DNA. There was no way of knowing for certain that the deletions the probes were detecting actually lay inside a gene fragment. The deletions very easily could be in one of the meaningless non-gene sections of DNA and, in fact, could be quite a long way from any part of the gene. In short, if they focused their gene hunt on the wrong deletion, there was a chance the researchers could miss the DMD gene entirely.

During the 1985 annual meeting of the American Society for Human Genetics, held in Salt Lake City, an impromptu strategy conference was organized. "Virtually every major researcher and clinician who had anything to do with Duchenne muscular dystrophy got together in a hotel room," Wood recalls. "There must have been twenty or so people representing every major laboratory."

The first step, it was decided, would be to construct a map of all the deletions detected in Bruce Bryer's damaged X chromosome. If the deletions tended to congregate around one particular spot, that would be the most likely place to look for a piece of the gene.

Kunkel and his young associates returned to Boston and began making up a new set of subprobes to pin down, more precisely, the deletions in Bruce's X chromosome. Three of these subprobes were picked for an unusual exercise in international cooperation. A general appeal went out to almost every medical center in the world that dealt with muscular dystrophy. Within weeks, twenty-five different centers, from Helsinki and Sydney to Iowa City and Cardiff, had obtained samples of DNA from 1,346 patients. (In the resulting report, published in *Nature*, Kunkel had more than seventy coauthors, something of a record for a biology paper.)[1]

Many of the DNA samples were forwarded to Boston, where Kunkel and his colleagues began using their subprobes to locate deletions. They detected deletions in eighty-eight, or 6.5 percent, of the muscular dystrophy patients, again "right in the ballpark" of what one would expect for a deletion-type mutation in a genetic disease.

Kunkel decided to take a closer look at the deletions in fifty-five of the patients.[2] The deletions in twenty-five of these patients overlapped each other. Here, then, was the spot most likely to be within the DMD gene.

But the exercise also uncovered a worrisome dilemma. Taken as a whole, the deletions were spread up and down an astonishingly long segment of DNA, far longer than expected for even a large gene. Moreover, the probes encompassed riflips that the Boston and Toronto teams had found. If the probes were within the gene, then the riflips presumably were within the gene also. Yet, when the mothers and other relatives of the muscular dystrophy patients were checked for linkage between the riflips and the DMD gene, it appeared that some of the riflips were a few million bases away from the gene. Obviously, the riflips couldn't be both in the gene and millions of bases away from the gene.

There were three possible explanations for this disconcerting situation, Kunkel says. "It said that either a mutation [deletion] at one point was affecting a gene a long ways away, perhaps two million bases away, or there was more than one gene involved." If the former was true, then the probes were not even close to the gene and the gene hunt would have start over again. If the latter were true, that more than one gene had to be damaged to cause muscular dystrophy, it would confound all that was known about inherited disorders in humans. There was no known human genetic disorder that obeyed the laws of inheritance as faithfully as muscular dystrophy yet required mutations in more than one gene.

The third possibility, Kunkel explains, was that the DMD gene was far larger than anyone had guessed, "a big gene dotted over a large area of DNA."

There was little choice but to proceed on the assumption that the probes, composed of normal DNA, contained a piece or pieces of the undamaged DMD gene. By this time, Worton in Toronto was finding deletions with his XJ probe in the same region as the Boston group were finding with their pERT87 subprobes. And the Canadians were finding the same disconcerting evidence that the probes may be landing outside of the gene.

Kunkel and Worton now faced a new problem. Even if a deleted piece of DNA was identified, how would anyone know whether it was part of the DMD gene? After all, no one knew what the normal gene's function was, what protein it produced. Without the protein available there wasn't the slightest clue as to the sequence of DNA bases that comprised the gene.

Worton decided to start flipping through the sequence of DNA bases in the XJ probes, looking for telltale sequences that denoted the beginning or ending of a gene fragment. It would be a tiresome and time-consuming task.

In Boston, Kunkel and Monaco decided on a different tack. If the gene was vital to human muscle, it presumably was vital to the muscles of other creatures, they reasoned. Nature had a habit of saving or conserving vital genes as species evolved from one another. Therefore the DMD gene, in its undamaged state, should exist in other animals— and should be identical to the human DMD gene. Kunkel and Monaco decided to test their pERT87 subprobes on a DNA "zoo."[3] Monaco gathered DNA from monkeys, cattle, mice, hamsters, and chickens and tested the animal samples with the subprobes. To his delight, two subprobes matched up—hybridized—with DNA in all the animals. The subprobes were, presumably, part of a DNA segment that had been conserved throughout millions of years of evolution. Because nature conserves only the genes, the segment was possibly a gene or a part of a gene.

Monaco turned quickly to the mouse to see what DNA base sequences in the segment were common to man and mouse. Two DNA bits in the segment were identical in both species and there were hints in the sequences that both bits might be parts of a gene or genes. One of these bits of mouse DNA was then tested to see if the genetic machinery in a human muscle cell could "read" the genetic message. It could. A human muscle cell would be capable of transcribing the message into a protein. It must be a gene or a piece of a gene shared by humans and mice. The bit of DNA also was one of the pieces that had been deleted from the DNA of some muscular dystrophy patients.

The elated Boston team was now convinced they had plucked out a part of the undamaged DMD gene. They quickly prepared a paper for publication and submitted it to *Nature* in July 1986.[4]

"Word got out [about the Boston feat] and everybody made a crash effort to get the gene also," Wood recalls. "Kay [Davies] and Ron [Worton] really pushed hard."

"Lou phoned me right away and told me they had the gene," Worton recalls. "Some of the people in my lab were a little depressed about it but I was glad," he adds.

A few months later, the Canadians, following much the same strategy with their XJ probes, also pinned down a piece of the DMD gene. The Canadians, however, were too late. The Monaco-Kunkel paper appeared in *Nature* in mid-October and few readers doubted that Kunkel's laboratory had won the race to the DMD gene.[5] Worton's report of finding a piece of the gene appeared in *Nature* the following July.[6]

"Jerry's Telethons Pay Off," declared *USA Today* in a page one story following a December 22, 1986, announcement by the Muscular Dystrophy Association that the DMD gene had been found.

By the beginning of 1987, a total of eight pieces of the DMD gene had been plucked out with the various probes. These tiny bits of gene averaged only about 150 bases each, but they were scattered along 130,000 bases of DNA. At that, the pieces appeared to constitute only a sixteenth of the entire gene. To the startled molecular geneticists, this hinted that the full gene was spread along one million to two million bases of DNA on the X chromosome. By June, enough of the gene had been identified by Kunkel's lab and several others to confirm that the DMD gene was dotted along at least two million bases of DNA, perhaps more.

"It was the biggest gene anyone had ever seen," Kunkel says. It comprised almost a thousandth of the entire human genome.

In the next step, Michel Koenig in Kunkel's laboratory pulled out a huge fragment of normal X-chromosome DNA, one long enough to encompass the entire DMD gene, and induced bacteria to copy, or "clone" it.[7] Using pieces of the huge gene as probes, he tested the DNA of 104 boys with muscular dystrophy. Half the boys had deletions in the gene. At the same time, Arthur H. M. Burghes in Worton's laboratory confirmed that the translocation suffered by Anne had broken into the same gene and unquestionably was the cause of her muscular dystrophy.[8]

If there were any doubts that the DMD gene had been found, they were now dispelled.

It was, all in all, a grand piece of science. "There were no false leads and no one published anything that they had to retract later," Wood notes.

But finding the DMD gene, as exciting as it was to the scientists, was only the penultimate step. The Boston and Toronto teams had launched their search for the DMD gene with one specific goal in mind, to find the protein. It had been obvious for decades that muscular dystrophy was the result of a defective protein produced by a defective gene. Years of meticulous searching through the labyrinth of unidentified human proteins had failed to find even the slightest hint as to what the defective protein was or what it did to destroy muscles. Kunkel, Worton, and Davies, by finding the DMD gene, were attempting for the first time to go through the back door to find an unknown protein. This was the new "reverse genetics" strategy, to find the gene in order to find the protein, instead of what had been usual in genetics, going from protein to gene.

The key to the protein was the message spelled out by the bases of the DMD gene. Each three-base "word" in the DNA, of course, designated the position of an amino acid in the final protein (the "word" ACG designating the amino acid cysteine, for example). Thus, with pieces of the gene in hand, both the Boston and Toronto groups could deduce, from the sequence of the bases in the gene fragments, what the sequence of amino acids was likely to be in fragments of the mysterious protein. They also could infer the approximate size of the protein.

If the enormous length of the gene was to be believed, the protein would be far bigger than any protein then known. To some, the protein's putative size was literally incredible. Wood of the Muscular Dystrophy Association recalls that a month before the Monaco-Kunkel paper reporting discovery of the gene was published, Monaco described the discovery orally at a meeting in Genoa, Italy. "Tony was up [on the stage] describing the message that the gene made and presenting figures on the size of the message," Wood recalls. "One enormously respected molecular geneticist in the audience said flatly that no protein was that big."

Kunkel put another of his young protégés, Eric P. Hoffman, in charge of the effort to find the protein by decoding the DMD gene. Until now, the researchers had been dealing with a piece of DNA that was "recognized" by both mouse and human muscle cells and, hence, was a gene that functioned in both species. But this didn't necessarily mean that the protein the gene produced performed the same job or that it was identical in both species. The human cells might be able to read the mouse gene and produce a protein but it might be a peculiarly mouse protein, different from its human counterpart. In short, Hoffman now had to make sure that the protein the gene produced in the mouse was the same as the protein produced in the human.

Hoffman began the monotonous task of determining the sequence of the DNA bases in both the mouse and the human gene, base by base.[9] He sequenced about a fourth of the gene and found the base sequences were 90 percent identical in mice and humans. It was sufficient to assure that the sequence of amino acids in the mouse and human proteins must also be 90 percent identical, close enough to believe that the proteins were one and the same and served the same function. The foundation was laid for using the mouse protein as a stand-in for the human DMD protein.

Hoffman then turned to a tool long used by immunologists, antibodies. Whenever a "foreign" protein enters the body of a human, a mouse, or any other mammal, the immune system produces antibodies that attack the intruder. The antibodies are highly specific, homing in on the intruding protein—and only that protein—and signaling the rest of the immune system to destroy it. This unerring specificity of antibodies makes them ideal for determining whether two unknown proteins are the same.

Hoffman used this antibody tool. He first plucked out the segment of mouse DNA containing parts of the DMD gene. The gene fragments were then transplanted to bacteria, which promptly began churning out pieces of a protein, the products of the DMD gene fragments. Several rabbits and two sheep were then inoculated with the protein fragments. Since the fragments were from mice and foreign to these animals, they immediately raised antibodies against the fragments.

Armed now with a supply of antibodies that reacted only to the DMD protein, Hoffman turned to muscle tissues from both normal humans and muscular dystrophy patients.[10] The antibodies reacted with normal human muscle tissue, indicating that human muscle normally contains exactly the same protein as that produced by the undamaged DMD gene. But the antibodies failed to react with muscle from two muscular dystrophy patients, strong evidence that the protein produced by the DMD gene was missing in the muscular dystrophy patients.

There no longer was any question. The DMD protein found in normal human and mouse muscle was lacking in muscular dystrophy patients who suffered deletions in their DMD genes. This was the protein that held the secret of muscular dystrophy.

Biochemists over the years had frequently run across mysterious proteins in human tissues. Even though they mightn't have the slightest inkling as to a protein's purpose, they nevertheless would record its discovery. But a search of the records showed that no one before had ever encountered the huge protein turned out by the DMD gene.

Like explorers, the Boston researchers had the right to name their discovery. "Since we know of no previously reported protein that shares the abundance, sequence, or size characteristics of the DMD protein, and since this protein was identified by molecular genetic studies of patients affected with Duchenne muscular dystrophy, we have named this protein dystrophin," Kunkel and his colleagues proclaimed in the journal *Cell* in December 1987, a year after the discovery of the DMD gene.[11]

Within weeks, appeals were sent out to muscular dystrophy clinics.[12] The clinical researchers reached into their stores of frozen tissues and sent back samples of muscle tissues taken during diagnostic biopsies of 103 patients, including forty with a variety of neuromuscular disorders other than muscular dystrophy. Hoffman began testing the muscle tissues with his anti-dystrophin antibodies.

Among the thirty-eight patients who had been diagnosed with the severe Duchenne type of muscular dystrophy, thirty-five had either no dystrophin or only minuscule amounts. Among the eighteen patients diagnosed with the milder Becker form of muscular dystrophy, twelve had an abnormal type of dystrophin, the protein being either longer or shorter than normal dystrophin. Among the forty patients with other types of neuromuscular disorders, thirty-eight had normal amounts of normal dystrophin.

What was now clear was that, if the damage to the DMD gene was severe enough, disabling the gene entirely, it would cause the severe Duchenne type of muscular dystrophy. If the damage was less severe, the gene might still function but it would turn out a crippled form of dystrophin that led to the milder Becker type of muscular dystrophy.

The findings also hinted that occasionally muscular dystrophy is misdiagnosed. "A 21-year-old man with a diagnosis of Duchenne's dystrophy had a dystrophin phenotype indistinguishable from normal (that is, one that indicates neither Duchenne's nor Becker's dystrophy)," stated the report subsequently published in the *New England Journal of Medicine*.[13] "Although he has been confined to a wheelchair since the age of 12 years, he is currently not near death," the report noted cryptically. Ordinarily, few victims of Duchenne's dystrophy survive to age twenty-one. The patient, in other words, probably was afflicted with some muscle disorder other than muscular dystrophy. Similar suspicious anomalies were noted in several patients whose dystrophin levels were inconsistent with a diagnosis of Becker's dystrophy, hinting they also might be suffering some muscular disorder other than muscular dystrophy.

"This discovery is the culmination of research that can be called a triumph of 'reverse' genetics," enthused Lewis P. Rowland of the Neurological Institute at the Columbia-Presbyterian Medical Center in New York in an editorial accompanying the report published in the *New England Journal of Medicine* in May 1988.[14]

"The work of Kunkel and his associates establishes without any doubt the power of the theory of Botstein et al. that provided the basis for reverse genetics," Rowland continued. "Botstein et al. postulated that it would be possible to find the gene for a disease with an unknown gene product by using nucleic acid markers for the gene—restriction-fragment-length polymorphisms (RFLPs)," the neurologist explained, adding parenthetically, "Who would have thought that an acronym devoid of vowels would become a conversational word?"

"When the paper of Botstein et al. appeared in 1980, many were

George Huntington, the Long Island, New York, physician who, in 1872, was the first to describe the inherited nature of the "choreic" disease later named after him. Huntington, like most physicians of his day, was unaware that Gregor Mendel had discovered the laws of inheritance in 1866, yet the Long Island country doctor accurately described the inherited nature of the dominant-gene disease as it passed through the generations of several Long Island families.

(Photo courtesy of the National Library of Medicine)

80 are you kidding?
he can't be
don't believe it
SAVE THE DATE
MILTON WEXLER'S 80TH BIRTHDAY &
THE HEREDITARY DISEASE FOUNDATION'S
20TH BIRTHDAY SEPT. 24TH SATURDAY NIGHT.

Milton Wexler as pictured on a 1989 invitation to a celebration of his eightieth birthday and the twentieth anniversary of the Hereditary Disease Foundation he established. In 1968, Wexler, a Los Angeles psychoanalyst, discovered that his ex-wife, Leonore, suffered from Huntington's disease and that there was a fifty-fifty chance that his two daughters, Nancy and Alice, had inherited the fatal Huntington's disease gene. His and Nancy's crusade to find a cure led to the discovery in 1983 of a diagnostic test for the disease

(*Photo courtesy of the Hereditary Disease Foundation*)

Nancy Wexler (*right*) and a Venezuelan woman stricken with Huntington's disease. The woman was one of the hundreds of Venezuelans living along the shores of Lake Maracaibo whose tragically high incidence of Huntington's disease helped lead to the mapping of the Huntington's disease gene and the diagnostic test.

(*Photo by Nick Kelsh, Kelsh Marr Studios, Philadelphia*)

A Venezuelan "water village" on Lake Maracaibo where Nancy Wexler and her colleagues compiled the huge pedigrees of families afflicted with Huntington's disease.

(*Photo by Steve Uzzell, Vienna, Virginia*)

(*Botstein photo by Donna Coveney, courtesy of MIT. Davis photo courtesy of Stanford University News Service*)

David Botstein (*left*) at Massachusetts Institute of Technology and Ronald W. Davis of Stanford University. In
1978, while attending a retreat of geneticists at Alta
Lodge, a ski resort near Salt Lake City, Botstein and
Davis (*right*) first articulated the notion that there
were unique segments of DNA scattered throughout
the human genome that could be used as markers for
locating genes. This concept of restriction-fragment-
length polymorphisms (RFLPs, or riflips) became the
basis for mapping the human genome. Botstein is now
at Stanford University.

Ray White shown in his laboratory at the University of Utah preparing samples of DNA for riflip testing. In 1978, when he was at the University of Massachusetts, White was recruited by Botstein to work out the laboratory techniques for finding riflips that Botstein believed to be scattered throughout the human genome.

(*Photo by University of Utah Medical Illustrations and Photography*)

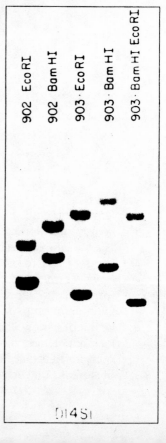

01451

Mark Skolnick (*right*) and Randall Burt, the University of Utah population geneticists who are tracking down the genes that predispose individuals to cancer. The pedigree of a large Mormon family lies on the table. Skolnick, excited by Botstein and Davis's impromptu 1978 lecture at the Alta ski lodge, worked out the population genetics for the landmark 1980 Botstein-White-Skolnick-Davis paper proposing that riflips could be used as markers to map genes.

(Photo by Brad Nelson, University of Utah Medical Illustrations and Photography)

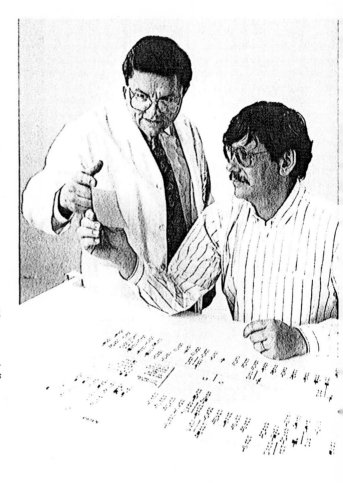

OPPOSITE The first riflip (restriction-fragment-length polymorphism) found in 1980 by Arlene Wyman and Ray White. Radioactive tracers attached to fragments of DNA from two individuals (labeled 902 and 903) produced the black smudges on X-ray film. The DNA fragments had been chemically cut out of the individuals' DNA by two different restriction enzymes (EcoRI and BamHI). The vertical distance between the smudges shows that the fragments cut from the DNA of individual 902 are of different lengths than the fragments cut from the DNA of individual 903. This demonstrates that this tiny portion of human DNA varies among individuals and thus can be used as a marker to trace the inheritance of genes that lie nearby. The technique for revealing such length differences among DNA fragments is known as a Southern blot.

(Photo courtesy of Ray White, University of Utah)

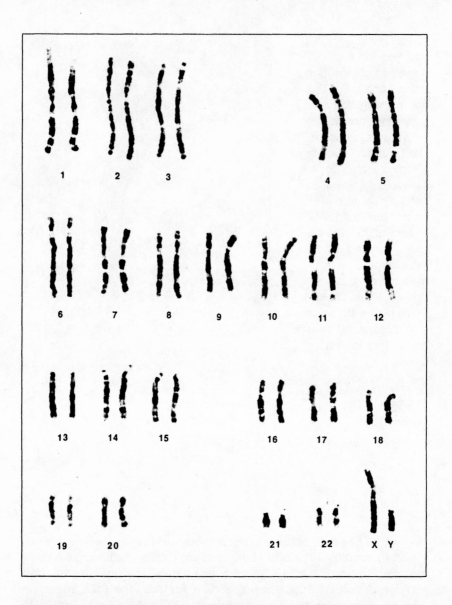

The twenty-three pairs of human chromosomes, arranged and numbered according to length. The last set, in the lower right corner, contains an X and Y chromosome, showing that these chromosomes came from a male (females have two X chromosomes, instead of an X and a Y). The light and dark bands are the result of staining of the chromosomes with a fluorescent dye. Deviations from the normal banding pattern help scientists determine whether there has been gross damage to a chromosome, producing one or more gene defects.

(Photo courtesy of Jorge J Yunis, Hahnemann University, Philadelphia)

David Housman, the Massachusetts Institute of Technology molecular biologist who decided to prove the validity of *Botstein et al.* by using a riflip to find the gene that causes Huntington's disease. It was Housman who convinced Nancy Wexler to collect DNA from the Venezuelan families.

(Photo courtesy of David Housman)

James Gusella, a Harvard Medical School molecular biologist, was recruited by Housman to search for the Huntington's disease gene. In 1983 Gusella led the team that found a marker for the disease gene using DNA from one Venezuelan family. Gusella is shown here examining an X ray of a sequence of DNA from a family member. The marker for the Huntington's disease gene was the first discovered by use of the new riflip technology.

(*Photo courtesy of Massachusetts General Hospital, Boston*)

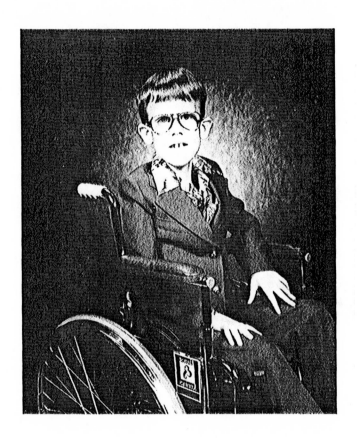

Bruce Bryer, the Spokane, Washington, boy who was born with three disorders caused by damage to his X chromosome. Experiments with the DNA from Bruce's X chromosome led to the discovery of the Duchenne muscular dystrophy (DMD) gene. Bruce died in 1983, three years before the DMD gene was found.

(Photo from the Washington State Historical Society, courtesy of Mrs Robert E Dixon, Sr, Bruce's foster mother)

The Boston Children's Hospital team that found the Duchenne muscular dystrophy (DMD) gene. *Left to right*: Eric Hoffman, Michel Koenig, Chris Feener, Marybeth McAfee, Corlee Bertelson, and team leader Louis Kunkel.

(*Photo courtesy of the Muscular Dystrophy Association, New York*)

OPPOSITE **B**ert Vogelstein (*left*) and Eric Fearon of Johns Hopkins University School of Medicine examine the Southern blots of riflips in DNA from a colon cancer patient. Vogelstein used Knudson's "two-hit" theory and the riflip technology to discover that defects in several "tumor suppressor" genes cause colon cancer. They subsequently found suppressor genes in other common cancers.

(*Photo by J Pat Carter, Baltimore, Maryland*)

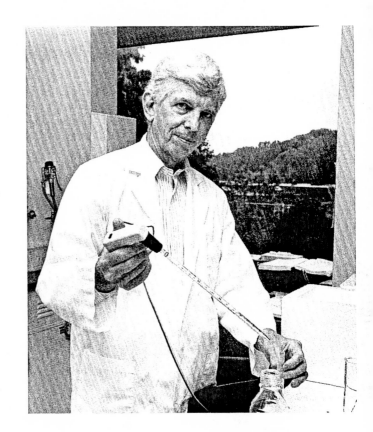

Alfred Knudson, the physician-biochemist-geneticist whose "two-hit" theory of cancer was proven by the discovery of the genetic damage that causes retinoblastoma, an inherited eye cancer.
(*Photo by Paul Cohen, courtesy of Fox Chase Cancer Center, Philadelphia*)

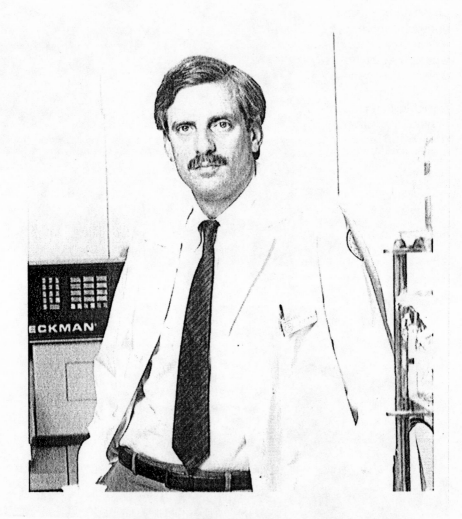

Jan Breslow, the Rockefeller University molecular geneticist who is tracking down the genetic defects that underlie atherosclerosis, the clogging of the arteries with cholesterol-rich deposits that can lead to a heart attack.

(*Photo by Ingbert Gruttner, courtesy of Rockefeller University*)

Victor McKusick, the pioneer Johns Hopkins medical geneticist, shown with all eight editions of his encyclopedia of inherited diseases, *Mendelian Inheritance in Man*. The first edition in 1966 listed 1,487 known genes, but only 68 of them had been mapped to, or located, somewhere on the X chromosome. The eighth edition in 1988 listed 4,600 known genes, of which 1,500 had been mapped to specific chromosomes.

(Photo courtesy of Johns Hopkins Medical Institutions)

James Watson, shown with a model of the double-helix structure of DNA, which he and British physicist Francis Crick discovered in 1953. Watson, director of the Cold Spring Harbor (New York) Laboratory, heads the Office of Human Genome Research at the National Institutes of Health, which funds and coordinates the U.S. effort to map the human genome.

(Photo by Susan Lauter, Cold Spring Harbor Laboratory)

Lap-Chee Tsui, the co-discoverer of the gene that causes cystic fibrosis, the most common of all inherited diseases. The development of the riflip technique inspired Tsui to begin searching for the gene when he joined the molecular biology labs at Toronto's Hospital for Sick Children in 1981. In August 1989, Tsui and Francis Collins of the University of Michigan finally reported isolating the gene. Unexpectedly, the discovery set off a controversy over whether the millions of "silent" carriers of the gene should be identified through widespread population screening.

(Photo by Rob Teteruck, courtesy of The Hospital for Sick Children, Toronto)

skeptical. It was thought that the markers would take years to find, that investigators would not know when they had found the gene, and that even if they did find the gene, it would not be possible to identify the protein encoded by that gene," the enthusiastic Rowland declared. There no longer was any basis for skepticism, he added, reviewing, step-by-step, the discovery of the DMD gene and its product, dystrophin.

Even as Rowland was writing his glowing editorial, Worton's group in Toronto and Kunkel's in Boston were tracking down the role of dystrophin in normal muscle. The Boston researchers, when they had seen the gene's base sequence, suspected the protein was a structural component of the muscle cell rather than an enzyme that triggers a chemical reaction. Both groups now traced the protein to the thin sheath of tissue, the so-called sarcolemma, that envelops muscle fibers. The Toronto researchers differ with the Bostonians on exactly where in the sarcolemma the protein lies. But they both speculate on how the muscle damage might occur: Perhaps dystrophin helps provide structural or mechanical strength to the membranes of muscle fibers. Without dystrophin, the membranes would be weak and highly susceptible to eventual tearing after months or years of contractions. Membrane tearing could lead to the death of a muscle cell. Such spreading cell death as the months passed would explain the progressive muscle weakness seen in muscular dystrophy patients.

In the summer of 1988, Kunkel sat down with a group of reporters during a meeting of geneticists at the Jackson Laboratory in Bar Harbor, Maine, and contemplated whether it would be possible to prevent or stop the deadly muscle deterioration of muscular dystrophy, when and if the secrets of dystrophin are ferreted out. Perhaps, he suggested, it might be possible to boost the supply of dystrophin in children diagnosed with the defective DMD gene. It might also be possible to find a protein that would function in place of dystrophin; after all, the muscles seem to function normally in the first several months of life, hinting that in fetal and newborn stages other proteins may be at work.

The answer, once again, may lie with the mouse. In 1984, after Davies had pinpointed the approximate location of the DMD gene with her second riflip, a geneticist in London, Graham Bulfield, found a strain of inbred mice that suffered a mutation on the X chromosome. The mutation was in about the same spot where Davies's riflips had pinpointed the DMD gene in humans. At the time, hopes were raised that the mutant mouse might be an ideal animal model for human muscular dystrophy. But there was a problem. Although there was some evidence that the muscles of these little rodents were dying off, the mice themselves didn't seem to suffer muscular dystrophy. Nevertheless, because the mutation was X-linked and it affected the muscles, this inbred strain of mice was dubbed the *mdx* mouse.

When Hoffman in Kunkel's laboratory was hot on the trail of dystrophin, he used his anti-dystrophin antibodies on the muscle tissues of the *mdx* mice. The mouse lacked dystrophin. It was, indeed, a true animal model of the human disease.

In mid-1988, Hoffman met pathologist Terry Partridge of the Charing Cross and Westminster Medical School in London. For several years Partridge had been experimenting with injecting normal muscle tissue into the muscles of mice such as the *mdx* mice to see if the normal muscle cells prevented the deterioration of the rodents' muscles. Partridge struck up a collaboration with the Boston group.

Partridge took dystrophin-producing muscle cells from newborn mice, supplied from Boston, and injected these normal cells into the muscles of the dystrophin-deficient *mdx* mice.[15] In many of the mice the injected normal cells fused with the recipient's defective muscle cells. The fused muscle cells shortly began producing quantities of dystrophin.

The experiment, published in early 1989, shows that it well may be possible to stave off the disastrous effects of muscular dystrophy by bringing in an outside source of dystrophin. Whether the mouse experiment can be transferred directly to humans, however, is another question. For one thing, there is the problem of obtaining large amounts of normal fetal or newborn human muscle tissues that would be required for treating the thousands of muscular dystrophy patients. The patients also may be prone to rejecting such grafts of muscle tissue in the same way they would reject organ transplants. And lastly, Partridge and his colleagues warned, there is the formidable problem that the injected muscle cells supply dystrophin to only a small surrounding area of muscle tissues; any patient treated in this manner might have to undergo innumerable injections of muscle cells spaced every fraction of an inch before such therapy could have any significant effect.

In early 1990, less than a year after Partridge's experiment with the *mdx* mice, researchers at the University of Tennessee in Memphis began implanting healthy but immature human myoblast muscle cells into the crippled muscles of eleven young boys with muscular dystrophy. These eleven were the first of forty-three young boys afflicted with the disorder who would receive the experimental treatment at four U.S. and Canadian medical centers in 1990 in an attempt to determine the efficacy of the myoblast transfer procedure and to establish protocols for future experiments.

In the Tennessee experiment, the boys, ranging in age from five to eleven years, received two injections of the healthy myoblasts in the foot muscle that moves the big toe. This small muscle is one of the earliest to be affected by muscular dystrophy and in all eleven boys it was 25 percent weaker than normal. The cells, taken from either the brother or father of each boy, were implanted in only one foot, while sham injections were made in the other foot to serve as a control.

The point of the experiment is to see if the implanted muscle cells will thrive and produce dystrophin, stopping or even reversing the steady destruction of the boys' muscles.

By mid-year, similar experiments were about to begin at Montreal's Neurological Institute and Hospital, Children's Hospital in San Francisco, and Tufts University-New England Medical Center in Boston.

Thus, six years after Davies's riflips had pinpointed the location of the DMD gene on the short arm of the X chromosome, there was talk of a cure for muscular dystrophy. As Rowland wrote of the discovery of dystrophin: "All this is wonderful and an unbelievable change from the state of ignorance of only a few years ago."[16]

Bruce Bryer, however, didn't live to see the breakthrough. On Christmas night, 1983, he and his foster parents, the Dixons, were returning home from a holiday celebration when a car hit their van. Although Bruce was uninjured, the stress of the accident was too much for his fragile body and he died the next day. The breakthrough was made using Bruce's DNA, which had been preserved in bacteria.[17]

CHAPTER 7

Two Hits

In 1985, doctors at Karolinska Hospital in Stockholm extracted a small vial of blood from the umbilical cord of a newborn baby boy and rushed it by air-express to a team of waiting scientists in Cincinnati. Led by a young researcher named Webster Cavenee, the scientists at the University of Cincinnati sifted through genetic material from the blood cells and found that the child had inherited a mutant gene from its mother. The gene was responsible for a rare and potentially lethal eye cancer called retinoblastoma that had struck the mother when she was barely two months old, costing her both eyes and her sight.

Within weeks Cavenee relayed word back to Stockholm that the baby had inherited his mother's mutant gene. The Swedish doctors decided to undertake a perilous procedure. They put the boy, just five weeks old, under general anesthesia. For more than an hour, they closely examined his retinas. In both eyes the doctors found evidence of cells already massing into minute but dangerous malignant clumps. While the child was still unconscious, the doctors directed a narrow burst of radiation at the retinas, vaporizing the malignant cells while leaving the rest of the eye unharmed. Periodic examinations over the next sixteen months found no hint of the cancer. Several years later, the cancer still hadn't returned.

News of the feat, reported in the *New England Journal of Medicine* in May 1986,[1] reverberated through the halls of medicine. Never before had doctors used a genetic test to predict the emergence of a cancer accurately. It was a turning point in genetics as well as cancer research.

Detection of the gene was made possible by a riflip probe discovered by Cavenee while working with Ray White at the University of Utah. The discovery of the Huntington's gene in 1983 had provided rock-solid proof of the power of the new riflip probes. But the probes' use in 1986, in alerting doctors to the presence of a cancer before it could ravage its victim, was something else again. The riflip could help predict the future.

It's not surprising, therefore, that the experiment had a riveting effect on medicine and on the scientists carrying out gene-hunting experiments throughout the world. In an editorial accompanying the report in the *New England Journal,* Fred Gilbert, a molecular biologist at Mt. Sinai School of Medicine in New York, proclaimed that the experiment was "the first demonstration of the feasibility of predicting, either prenatally or post-natally, whether a person at risk for an . . . inherited cancer is likely to have the disease."[2]

The implications caught the attention of many cancer specialists studying the most basic aspects of cancer. While it was widely accepted that people inherited a tendency to develop certain diseases, such as diabetes, and that some families seemed to have higher-than-normal occurrences of certain cancers, prior to the Cavenee report there was little direct evidence that a proneness to cancer could be detected at or before birth. Now there was.

Moreover, there was proof that the riflip concept put forward by Botstein et al. in 1980 could help diagnose illness before symptoms appeared and predict a person's susceptibility to future cancer. "Cavenee's report showed convincingly what we knew would be true," Ray White recalls. "Its publication convinced everyone else, too."

The retinoblastoma report confirmed a little-heeded notion about cancer that was quite different from anything previously postulated. Heretofore, cancer researchers had thought that a mutant cancer gene had to "turn on" and actively cause a cell to become malignant. In the case of retinoblastoma, however, a gene's failure to function apparently produced the eye tumor. Scientists now had proof that the loss of a certain gene, or the loss of its normal function through mutation, could *allow* human cells to spin wildly into a state of malignancy.

The report sparked new speculations about the causes of cancer. Researchers began suggesting that the inheritance of similar nonfunctioning or missing genes might explain a host of rare childhood cancers that afflict certain families in unfathomably large numbers. Others went even further, noting that while retinoblastoma itself was rare, striking only about two hundred children a year in the United States, the mechanism underlying its genesis might also be at work in the more common cancers that claim the lives of a half-million Americans annually. The report inspired several laboratories around the world to begin using riflips to plumb tissues, extracted from a widening range of common cancers, for evidence of the nonfunctioning or missing genes. Several scientists began calling them "anti-cancer" genes.

The cascade of gene discoveries made in the wake of the retinoblastoma episode prompted Samuel Broder, director of the National Cancer Institute, to remark in 1989, "It turns out that studying a very tragic but uncommon tumor (retinoblastoma) made possible some fundamental insights about the most basic workings of cancer. All of this may not yet be obvious to the public, which is more concerned about advances in treatment, but I'm convinced the genetic work will eventually begin showing results there, too."

The 1986 retinoblastoma report involving the Swedish baby boy also placed Web Cavenee firmly in the vortex of the emerging cancer gene research. He was propelled there by a series of events that stretched back through Ray White to a Houston pediatrician obsessed with a theory of cancer. Five years earlier Cavenee had gone to work as a postdoctoral researcher with Ray White in Utah.

"Back in 1981 we really were just trying to prove White and Botstein were right, that [riflip] probes could be found throughout the genome and that a map of these riflips could be produced," says Cavenee.

Cavenee was hired by White to help construct the riflip map, although the strategy for finding these DNA polymorphisms was uncertain. "We were looking for ways to find particular chromosomes and then to find where on a chromosome particular [polymorphic] fragments of DNA were located," says Cavenee. "Once a known piece of a chromosome was found, we'd go after restriction-fragment-length polymorphisms in particular areas. If we found a riflip, we then had a marker for a very specific spot on a chromosome, and that piece of chromosome would be 'mapped.' Ray's goal, of course, was to find riflips on all the chromosomes, providing the landmarks for a map of the entire genome."

In those early days, Cavenee was concentrating on riflip mapping without any particular gene or disease in mind. He was scouting around for a place in the genome to start pinning down riflips when Mark Skolnick returned from a trip to Houston with a bit of intriguing news. Skolnick had gone to a cancer meeting at M.D. Anderson Hospital and Tumor Institute in Houston, where he met Louise Strong, a cancer researcher who, along with pediatrician Alfred G. Knudson, had been studying the childhood eye cancer called retinoblastoma. Strong and Knudson had discovered that in malignant retinal cells from a few young patients who developed retinoblastoma, a piece of chromosome 13 seemed to be missing. Under a microscope, they could see that chromosome 13 was shorter than normal. (See illustration of chromosome 13.)

Strong told Skolnick that she and Knudson had uncovered an especially strange mutation in one family. It seemed that, in some members of this family, a piece of DNA had been chucked out of chromosome 13 and moved into the middle of chromosome 3. Indeed, only those members of the family who had the eye cancer had this translocation. Skolnick immediately saw that this event could be a marker that White might use in tracking down the retinoblastoma gene.

So Cavenee called Strong. "Louise suggested we take a whack at chromosome 13," Cavenee recalls. "She and Knudson were trying to find a gene involved in retinoblastoma, and they believed it was in the deleted area of chromosome 13, but they had no way of getting at it. At that time, of course, there wasn't any way to do it except, perhaps, by using a [riflip] probe," says Cavenee. "I decided 13 would be as good a place as any to begin mapping riflips, and Ray agreed."

The pieces of DNA missing from the "malignant" chromosome 13 were isolated from a normal chromosome 13 through DNA hybridization. Cavenee then took the unmatched strands of "normal" DNA and had a bacterium clone copies. He began scouring the copies for riflips.

"It was an extraordinarily fortunate place to begin," Cavenee says. "It really wouldn't have happened if it weren't for Louise and Al. Nowadays, it's well known that Knudson is one of the great guys in science. But originally his predictions about retinoblastoma and cancer, in general, were very hard to swallow. I wasn't very certain about what he was hypothesizing. I really was more interested in his shortened chromosomes."

Knudson's cancer theories had been more or less ignored because he was delving into a malignancy that most cancer researchers regarded as a quirk. In their judgment, study of retinoblastoma had little relevance to the more common cancers, which, they believed, were triggered by exposure to some environmental carcinogen such as a chemical, nuclear radiation, or excess in the diet. The tragic eye cancer had caught Knudson's attention in 1969 and he was convinced that it represented a significant piece of the unsolved puzzle of cancer.

For Knudson, solving the retinoblastoma mystery was the kind of intellectual mountain climb he had sought his entire scientific career. Tall, square-shouldered, but quiet and modest, Knudson all too often throughout his career had closeted his ambition to be a research scientist, accepting, instead, administrative jobs that rewarded his scholarliness and calm leadership skills. But he had quietly been itching for an opportunity to tackle a problem in basic science. When he first came across the eye cancer in 1969, his principal responsibility was building a program of medical genetics at M.D. Anderson. That left him little time for creating hypotheses and testing them out in experiments.

Knudson had been attracted to genetics early in his training. As an undergraduate student at the California Institute of Technology during the early years of World War II, Knudson learned about genetics from the elderly Thomas Hunt Morgan, the biologist whose study of the chromosomes of fruit flies early in the century was the foundation for modern research into genetics. "I was truly inspired," Knudson says. "I was convinced that studying the genesis of traits was to be my life calling, that studying genetics would allow me to study human life at its most basic level."

Confident and persistent, he set about constructing a career plan. His idea was to get a doctorate in embryology—the study of human prenatal development—and genetics, thereby tackling heredity by studying how genes biochemically manifest themselves at the dawning moments of a human life.

World War II knocked the determined Knudson off course. He enlisted in the Navy because he was told he could defer active duty for several years if he attended medical school. "The war effort had little need for embryologists," Knudson says. He completed his medical studies shortly after the war ended, choosing to specialize in pediatrics, the medical discipline closest to embryology. While in residency training at New York Hospital on Manhattan's East Side, Knudson says he "got sucked into" cancer work after treating several youngsters across the street at the world-famous Memorial Sloan-Kettering Cancer Center.

"There was something about tumors I intuitively felt was interesting," says Knudson. "I couldn't put my finger on what it was exactly, but I knew it was something I should pursue. Some of us were beginning to think that cancer research would provide important clues into understanding normal human development. By looking at the aberrations in life—a cancer—we might better understand the normal processes of life."

He had plans to pursue pediatric oncology but once more his plans were interrupted, this time by the Korean War. He was obliged to serve, since he owed the military time. "I thought I'd never get back to research," he says. "I figured by the time I got out of the army I'd be too old to start off as a junior researcher, doing postdoctoral work in somebody's lab."

Nonetheless, once out of uniform he returned to Caltech, where he finally got his doctorate, not in genetics, but in biochemistry. "I thought learning the basics of biology would be good for me," he says. Because he wanted to continue treating patients, Knudson accepted a job heading the pediatrics department at City of Hope Medical Center in Duarte, California, just outside Los Angeles. Once more, he found himself ruminating about the strangeness and tragedy of childhood cancers. Most cancer researchers ignored deep investigations into the causes of the childhood cancers, believing that they were rare freaks of nature, tragic examples of where the growth that had formed the fetus from the embryo had failed to stop in specific organs. This view implied that childhood cancers were unrelated to the biological systems underlying the common cancers of adulthood, those of the lung, breast, and colon.

But Knudson clung to a stubborn conviction that hidden within the mysterious causes of childhood cancers were secrets that, if unveiled, would illuminate concepts that could be applied to all cancers.

He was especially puzzled by the rapidity with which a childhood cancer evolved. Most other tumors, those that arose in adults, appeared to take years or even decades to develop. In the case of the childhood cancers, Knudson believed that something perverted cells while still in the embryo. He thought this made sense because these cells were rapidly dividing and were especially vulnerable to some kind of outside interference.

But, he was convinced, a cancer was a cancer, and whatever triggered the early onset of the disease in the embryo probably did so by acting through the same unknown biochemical pathways involved in the more common, late-occurring forms of the disease. Early-onset disease, he figured, might be more easily studied because the disease's course was telescoped over months, instead of decades.

In 1962, Knudson finally began tackling the problem full-time when his supervisors accepted his request to conduct research while serving as chairman of the medical center's biology department. He first investigated childhood leukemia, a cancer of the blood-forming cells, and pursued, without success, the idea that some type of virus initiated the youngsters' cancer. The idea wasn't farfetched, since it was well known that viruses caused leukemias in mice, chickens, and other animals.

It was during this unsuccessful foray into leukemia that he happened upon a case that brought him full circle back to genetics. A young child was admitted to City of Hope with symptoms of an inherited disorder called neurofibromatosis, a curious illness characterized by light-brown spots on the skin, and nonmalignant tumor growths that formed just below the skin. In some cases, the lumping masses grotesquely distort a victim's face, leading laymen to mistake it for the disorder that afflicted Britain's famous Elephant Man at the end of the nineteenth century. There was ample evidence in neurofibromatosis that the skin growths resulted from a dominant gene, meaning that one of the child's parents also had to have the disease. But, oddly, the parents in this case didn't have the illness. The child also had another, nastier problem, a childhood cancer of the nerve cells called neuroblastoma, which was lethal. There was no evidence that neuroblastoma was inherited.

Knudson made a reasonable assumption based on his earlier thinking: both disorders must have resulted from damage to a region of the child's DNA. The damage mimicked the gene defect that normally was inherited in neurofibromatosis. But that same injury also inflicted other damage that gave rise to the nerve cancer. Knudson reasoned that the grievous blow to the genes occurred during early cell division in the embryo.

"The case reinforced my earlier ideas," Knudson recalls. "Cancers that looked alike might arise from two types of destruction to the same bit of genetic material. I began to think that sometimes the damage was inherited as a mutated gene. But sometimes it was the result of mutation by some external force or, perhaps, the result of a natural error during cell replication."

At about the same time, researchers were beginning to uncover examples of "cancer families" haunted by cancers of the breast and colon. It was a widely held belief that these familial syndromes where cancer struck many family members were the result of genetic aberrations that differed from the causes of more common types of cancers. But Knudson believed these familial cancers were caused by the same genetic mutations as the more common cancers, except that the mutation was inherited, instead of being inflicted by an environmental carcinogen. It was an extraordinarily brash bit of thinking for the day.

Once more, however, Knudson "wandered away" from the research, mostly keeping his notions to himself. In 1966 he accepted an offer to organize a medical school for the State University of New York at Stony Brook on Long Island. But three years later, when he was offered a similar job at the M.D. Anderson Cancer Center, he jumped at the opportunity. "I'll finally get a chance to devote my time to cancer," he thought.

In Houston Knudson grew intrigued with retinoblastoma. It was a childhood cancer and some cases clearly were inherited, the stricken child having one parent who had suffered the same eye cancer in early childhood. The eye cancer, while sometimes studied as a curiosity, was largely ignored as a freak occurrence. In fact, it wasn't until the middle of the last century that it was possible for children to inherit the cancer. If the eye tumor is left untreated, it eventually invades the brain and is fatal. Prior to 1866, most children afflicted with it died before they were ten years old. But the invention in 1866 of the ophthalmoscope made it possible for doctors to look into the eye and detect a tumor when it was still a small mass at the rear of the eyeball. Thus diagnosed, the cancer could be eradicated by surgical removal of the affected eyes. Although blinded, children began to survive into adulthood and to have their own families.

Gradually, physicians began reporting cases of adults who had been blinded by the disease having children victimized by the same cancer. Presumably, a gene was being passed from parent to child. If it was a gene, however, it surfaced in patterns that defied the Mendelian laws of inheritance. Sometimes the disease would strike half the children of a previously affected adult, acting like a single, dominant gene. Sometimes none of an affected individual's children would have the disease, suggesting the gene was recessive. But then the disease might reappear in the grandchildren, striking in the pattern of a dominant gene.

Moreover, the disease manifested itself oddly, sometimes causing the growth of several tumors in both eyes, sometimes causing tumors in one eye, and sometimes producing only one tumor in one eye. The disease sometimes hit very early, exploding into a full-blown tumor mass by the time an infant was six or ten months old. But other children weren't stricken until they were five or six years old.

By the time Knudson began looking at the disease, its pattern of inheritance was so complicated that most researchers despaired of finding its cause. Ophthalmologists concentrated on detecting and treating the cancer at the earliest possible moment. Caught when still microscopic in size, the retinal tumors could be effectively treated by radiation, thus saving the eye. This, however, required that all children of an affected parent be hospitalized and placed under general anesthesia every two months or so for the first three or four years of life. While the examination was life-saving for infants who inherited the disease, it was an unnecessary trauma for the 50 percent or more who didn't.

And, there were additional complications. About 60 percent of the retinoblastoma cases weren't inherited at all. Instead, they resulted from a spontaneous genetic accident, one that must have occurred in the developing embryo itself. For these children, the disease had no antecedents in earlier generations.

"I believed that, inherited or not, it was the exact same disease," Knudson says.

In a flash of insight, Knudson decided that the cancer must be the result of more than one mutation, or "hit," upon the genetic machinery in the retinal cells known as retinoblasts. If the retinoblastoma gene were inherited, he reasoned, it would have been passed on in either the father's sperm or the mother's egg. Therefore, the mutant gene must be in every cell in the body. There should be tumors everywhere. Even if the inherited gene were expressed only in the retinoblasts, every retinoblast in the eye should be malignant. Yet only a tiny proportion of the retinoblasts turned malignant. Apparently, something more than just the inheritance of the mutant gene had to occur before the cells turned malignant. The inheritance of the mutant gene was the first "hit" but the malignancy didn't develop until a retinal cell received at least one additional "hit."

By early 1970, Knudson was convinced his multiple hit theory was right. But he needed to know how many additional "hits" against the cell were needed to trigger malignancy. So he set up a statistical analysis by studying sixty-six retinoblastoma cases treated at M.D. Anderson Hospital and elsewhere in previous years. By adding up all the tumors in all the victims, he found that the mean number of tumors per case was three. That number was important. His mathematics told him that three malignant cells could develop out of the millions of fast-dividing retinoblasts if there were as few as two separate mutating "hits."

But the evidence that just two mutations were enough, even though based on a complex bit of mathematics, wasn't very solid. And Knudson knew how shaky it was. But he found great pleasure in the "two-hit" theory. It neatly explained the inherited form of retinoblastoma. It could even explain the 60 percent of the victims who developed the cancer spontaneously. A child who was born without the mutant gene might get the first "hit" in one of the highly sensitive retinal cells from exposure to an environmental mutagen such as cosmic or ultraviolet rays, or a random error during chromosomal replication. The likelihood of the same cell being slammed twice in such a way is what made the disease rare—but it sometimes happened.

Knudson liked the idea so much that he began extrapolating it to other childhood tumors. "My idea was this: Some people are born with a genetic predisposition, that is, under certain circumstances they are more likely than other people to develop a cancer," says Knudson. "Those people who don't inherit the gene defect can still, of course, develop the cancer. Also, the first step toward initiation of a nonhereditary cancer must involve a mutation to the same DNA contained in the inherited gene. I felt I had shown some evidence that hereditary and nonhereditary forms of a specific cancer involved the same genetic mechanisms, except that the inherited form was more likely to occur, and to occur earlier in life, and in a more pronounced way."

Knudson published his two-hit hypothesis in 1971 in the *Proceedings of the National Academy of Sciences*.[3] It was titled "Mutation and Cancer: Statistical Study of Retinoblastoma." But no one pounced on the notion, or tried to extend it to other cancers. The "two-hit theory" languished in the backwaters of cancer research. Knudson wasn't surprised. He hadn't identified the gene, where it might be located, or what its normal role was. Moreover, he had no idea of what exactly produced the second hit, or where in the young retinal cells the second hit had to land. "I was kind of stuck," he says. "I didn't know how to go about proving all this."

Over the next several years, Knudson and his associate, Louise Strong, reproduced the same statistical findings in two other childhood cancers, Wilms' tumor of the kidney, and neuroblastoma, the nerve tissue cancer. Then in 1976 Knudson married Anna Meadows, a physician at Children's Hospital in Philadelphia. The previous year Meadows had treated a two-year-old girl with retinoblastoma who also was mentally retarded. The combination of the two anomalies was very unusual. For geneticists the appearance of two otherwise unrelated illnesses is a clue not to be ignored. Both illnesses might be due to a mutation of DNA that affects two separate but physically close genes on the same chromosome.

A quick review of other cases of combined mental retardation and retinoblastoma similar to that seen in Meadows's two-year-old patient showed that most of these retarded children suffered damage to chromosome 13. The fact that the mentally retarded little girl also suffered retinoblastoma hinted that the retinoblastoma gene might be on chromosome 13, too. As far as he was concerned, Knudson had located the first inherited hit, a mutant gene on chromosome 13.[4]

Knudson rarely shows emotion or describes his own delight or satisfaction with anything more evocative than "interesting." But Louise Strong says, "Al was excited, or about as excited as he ever gets." In a paper published in the *New England Journal of Medicine* in 1976, Knudson proudly declared that the discovery showed that he and his colleagues had found the "only instance in animal or man in which a specific chromosomal abnormality can . . . consistently predispose to a specific tumor."[5]

Knudson speculated further. He suggested that the first hit involved the loss of a gene on the copy of chromosome 13 that a child inherits from the parent who has retinoblastoma. The second hit, he guessed, might involve loss or mutation of the same gene on the copy of chromosome 13 inherited from the other parent. How the loss of the second copy might occur, however, was a mystery.

"Although this very interesting situation cannot be resolved at present we look forward to further . . . data . . . from chromosomal-deletion cases of retinoblastoma," he wrote at the end of the 1976 paper.[6]

And further evidence started coming in. Researchers elsewhere began conducting their own examinations of the chromosomes of retinoblastoma families. And they too noticed material had been deleted from chromosome 13. Indeed, the deleted material could be seen in cells taken from tissue anywhere in the body, a sure sign that the shortened chromosome was inherited. Even when they looked at the chromosomes from affected children with no prior family history of retinoblastoma, researchers sometimes found a shortened chromosome 13. The difference was that those with the inherited form of the cancer had a shortened chromosome 13 in cells throughout the body, while those with the spontaneous form of the cancer had the shortened chromosome only in the tumor cells.

At least two aspects of Knudson's theory were proving true. Inherited or not, the disease involved damage to the same stretch of DNA. Knudson's guess that this first mutational hit landed on DNA specifically located on chromosome 13 was holding up, too.

But the target of the second hit was still unproven. Knudson was sure that it involved the second copy of the still-unidentified gene on

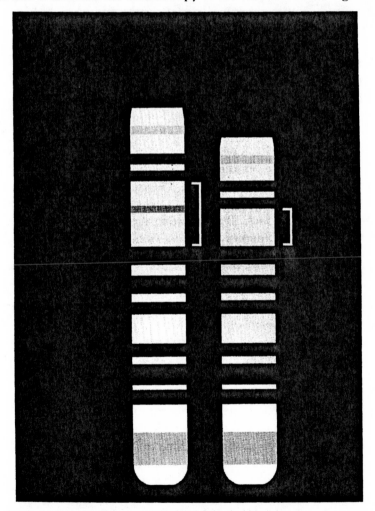

Credit: Jorge Yunis, Hahnemann Hospital School of Medicine

Depiction of a normal copy of chromosome 13 (left), and a copy of a chromosome 13, missing a bit of DNA. The deleted DNA contains the RB gene.

chromosome 13 inherited from the other parent. "By then I was convinced that we were on the trail of just one gene that caused cancer," Knudson recalls. "Lose one of the gene's two copies and you are predisposed, lose the second and you get cancer. I knew of no other gene ever found anywhere that was such a potent carcinogen. The idea was hard for most people to accept. Nobody had ever found a gene that, by its absence, could cause cancer."

In 1980, a team of geneticists and eye doctors associated with the University of California and Children's Hospital in Los Angeles set out to test Knudson's hypothesis.

Robert Sparkes, a UCLA geneticist, was looking for a gene that could serve as a marker for retinoblastoma. When he checked the catalogue of genes whose location had already been mapped, he found that only one gene was known to lie along chromosome 13, a gene for an enzyme of unknown purpose called esterase D. He then conducted a study of retinoblastoma patients, each with a shortened chromosome 13, and, to his delight, found that esterase D was produced in lower than normal amounts, hinting that one of the two copies of the esterase D gene had been lost or damaged. This suggested that the esterase D gene and the retinoblastoma gene might lie close to each other, an extraordinary bit of luck. By measuring the esterase D levels in different generations, Sparkes hoped he would have a variable marker to track the passage of the unknown retinoblastoma gene through families.[7]

Indeed, esterase D also was polymorphic, that is, it was produced in two distinct versions. If each parent carried a different version of the enzyme, the researchers could use those two versions as an easily traceable proxy for the retinoblastoma gene. Sparkes and his colleagues found cases where a father had the eye cancer and version F of the enzyme whereas the mother had version M of the enzyme. If a child had the eye cancer and version F of the enzyme, it would be strong evidence that father's retinoblastoma gene and his esterase D gene had been passed along in tandem to his child. Tracking the esterase D versions in three generations of three different retinoblastoma families, Sparkes discovered that the two genes were in such close proximity to one another that they were always passed along together, that is, the two genes were tightly linked.

Once they knew the esterase D gene was a good proxy for the retinoblastoma gene, Sparkes and his colleagues wanted to use it to uncover Knudson's second hit. If the loss of one copy of the esterase gene resulted in a person producing only half the normal level of esterase D, then the loss of both copies of the gene should result in zero production of the enzyme. Thus, complete absence of the enzyme would mean that the second copy of the esterase gene had been knocked out—and the second copy of the retinoblastoma gene, being tightly linked to it, would have gone with it. In other words, total lack of the enzyme would be a marker for Knudson's long-sought second hit.

It was this reasoning that stirred excitement in the California team when they came upon a three-year-old girl with a unique combination of traits. She had retinoblastoma and her esterase levels throughout the cells in her body were half of the normal level. But when they tested tissue removed from the girl's eye cancer, they found no enzyme activity at all. Within the tumor, therefore, both copies of the gene for the enzyme had been lost. Given the esterase gene's tight linkage with the retinoblastoma gene, both copies of the retinoblastoma gene also had been lost. The little girl had suffered both of Knudson's hits.

In a report published in *Science* magazine in February 1983, the California researchers' excitement leapt off the page. "We maintain that there is only one retinoblastoma gene and that this gene represents a prototype of a class of human cancer genes characterized by a loss of genetic information at the . . . tumor level," they wrote. Furthermore, they declared that the "potential importance of the patient we describe in helping us to understand the nature of human cancer cannot be overemphasized . . . ," noting that it was the first evidence that a cancer was a recessive illness, caused by the "loss" of two copies of one gene.[8] Another piece of Knudson's hypothesis was vindicated.

But Sparkes's confirmation involved what some scientists called a "clever trick." The discovery had been made by hunting down a marker, a sort of genetic hitchhiker, rather than the retinoblastoma gene itself.

The retinoblastoma gene itself had yet to be identified. It would take a report by Web Cavenee nine months later in *Nature*,[9] the British scientific journal, for scientists to begin to accept the notion that the loss of a gene might serve as a template for other cancers.

Cavenee, working with White in Utah, understood why. "For most of 1983, everyone's attention was focused on a very different gene story," he recalls. "Everyone was talking about oncogenes."

Oncogenes are a class of apparently normal genes that regulate cell growth. In the early 1980s, through completely different research, it was found that malignancy couldn't occur unless the growth-regulating genes were present and active. A prevailing theory was that somehow, perhaps by mutation, these growth-regulating genes had been switched on at the wrong time or in the wrong places and they were actively but erroneously promoting the growth of a cancer. Scientists speculated that over the course of a lifetime any number of mutations could convert the normal genes into oncogenes. But it was evident to the researchers that the transformation of a healthy gene into an oncogene was just one step in the development of a cancer. However, oncogenes were created by spontaneous mutation of normal genes in somatic cells, not the germ cells, and therefore oncogenes weren't inherited.

Still, even if not inherited, the discovery that cancer arises from a special set of genes was a liberating notion. Suddenly, it was apparent that "the roots of cancer lie in our genes," wrote Robert Weinberg, a molecular biologist at the Massachusetts Institute of Technology, in an article in *Scientific American* in 1988.[10] Weinberg's own discoveries concerning oncogenes dominated the cancer research news in 1982 and 1983. Increasingly, cancer specialists and researchers were turning their attention to the study of genes as the linchpin for cancer.

Initially, Cavenee and White weren't interested in cancer, and paid only peripheral attention to the doings of the oncogene researchers. Instead, in the wake of the riflip discoveries, a race was on among several laboratories seeking to be the first to construct a complete map of riflip landmarks. White was determined to be first and was focusing his attention on developing techniques for finding the riflips. In fact, Cavenee spent more than a year picking apart DNA from retinoblastoma tumor cells before he even thought to see a patient with the disease.

When Cavenee first met White in Massachusetts back in 1980, Cavenee was studying how genes control the production of cholesterol while he was doing postgraduate work at MIT. He had received a grant from the National Institutes of Health to study at Cambridge University in England, a highly coveted award. But before the grant took effect, Ronald Reagan was elected president. The federal institutes grew nervous after the election, telling Cavenee and other grantees that they were worried the Reagan budget-cutters would be upset about overseas grants. Cavenee's award was rescinded.

"I was looking for a project and talked to Botstein, who was becoming a very big deal, of course," Cavenee recalls. "He told me to talk to White about this thing they had started. I wasn't sure what they were doing but I talked to White and he offered me a job. Everyone I knew except Botstein told me I was crazy to go out to

Utah, that it was out there in the middle of nowhere and that I'd just disappear in the mountains," says Cavenee. "But Botstein was very persuasive. He said I would be getting in at the ground floor of an entirely new field of research, that if I didn't go I'd miss the boat, that I would lose out if I played it safe. He's a very forceful guy. So I thought for a while and then I left for Utah."

White had told Cavenee that the Hughes Institute was building a new genetics facility at the University of Utah, but when Cavenee arrived, White's staff was jammed into temporary laboratory space. "Ray's vision, to make the map, was clear, but his methodology wasn't," recalls Cavenee. "So we spent a lot of time talking, just batting around ideas. Ray is full of ideas, constantly posing possibilities, problems, and solutions. It was an invigorating time because we felt like we were cutting a path others could follow. So, of course, everyone worked like crazy. We had this feeling we were part of something that was going to explode."

On the floor below their laboratories, the researchers had a Ping-Pong table. "We were all very young and we'd work all night, take a break to bash the crap out of the Ping-Pong ball, and then go back to work again," Cavenee recalls. "I remember being exhausted and having a lot of fun."

Cavenee spent about a year searching for and finding riflips on chromosome 13, the effort that brought him into contact with Louise Strong from Houston, and through her, with Knudson's "two-hit" theory. "One day I realized that I'd become as intrigued as Knudson by the mechanics of the cancer," he recalls. Suddenly, he wanted to know if some minuscule change in a chromosome, as proposed by Knudson and Sparkes, could actually produce a tumor. "Cancer was thought to be a result of some really nasty shock to the DNA, not some subtle, almost inconsequential event," he says. "I wanted to see if it was possible, because it would really change the way scientists thought about the things that make a cell go crazy. The riflips gave me a way to look."

Within months, Cavenee isolated seven riflips in the area of chromosome 13 where Knudson and Strong had found major chromosome deletions. The riflips were found, of course, in a normal chromosome 13. Cavenee assumed that at least one of the riflips was close enough to the gene so that it would always be present when the retinoblastoma gene was present, and gone when the gene was gone.

The power of the riflip probes was that they could differentiate between the two copies of chromosome 13 that a child inherited. The

DNA probe for one of the riflips thought to be near the gene would land on the same spot on both the father's and the mother's chromosome 13. This probe-demarked section of both chromosomes would then be plucked and cut with a restriction enzyme. The restriction-cut fragment from the parent with the normal chromosome 13 would be one length; the fragment cut from the chromosome with the defective retinoblastoma gene would be a different length. This, of course, is the nature of a riflip, a restriction-fragment-length polymorphism.

In one of the first retinoblastoma patients studied by Cavenee, the young scientist encountered an astonishing result. The fragments cut out by the restriction enzyme from the child's chromosome 13 were all the same length, the length of the father's chromosome 13 fragment. None of the fragments, it appeared, had come from the mother. If the riflips were to be believed, the patient had inherited two damaged chromosome 13s from the father and none from the mother, an impossibility as far as genetics was concerned.

Cavenee was fascinated. Well, he thought, this patient certainly is an oddity. There was a possible explanation. The probe for the father's chromosome was landing on or near the spot where the chromosome deletion had produced the mutant retinoblastoma gene. Somehow, in this patient, this damaged section of the father's chromosome had been duplicated and substituted for the normal chromosome inherited from the mother. During this mystifying duplication, the normal section of the mother's chromosome had been cast out. As far as the riflip probe was concerned the paternal and maternal chromosomes were identical, both damaged. Was this duplication Knudson's postulated second hit? "I figured that was one way to explain the cancer, but I assumed I'd stumbled across a real unusual case. I mean, it just made sense that the second hit would involve some damage to the otherwise healthy DNA, but not a duplication of the same bad one."

So Cavenee repeated the experiment in another retinoblastoma patient—and got the same result. The riflip probe was so sensitive that it could detect whenever the gene was present in a damaged form or if it was missing. Seeing this change in one gene was expected, but seeing exactly the same alteration in the other chromosome was stunning. "It was the first time anyone had ever seen a mutant gene duplicated like that and now I had seen it twice," Cavenee recalls. He did the same experiment on thirty-three patients, and each time the result was the same. As long as there was only one mutant copy of the gene and a healthy copy, the cancer was held in check. But when the healthy copy was cast out and replaced by a duplicate of the mutant gene, "the cancer was unmasked," Cavenee explains.

What was causing retinoblastoma, Cavenee concluded, was a phenomenon known as chromosome "crossover." Whenever a cell divides, it has to make a copy of its entire set of paired chromosomes in order to provide the daughter cells with a full complement of paired chromosomes. During this duplication process there sometimes are crossovers where a segment of, say, the paternal chromosome crosses over into and replaces the same segment in the maternal chromosome. As a result, the daughter cells end up with paired copies of the paternal segment. Ordinarily, this crossover is of little consequence since the paternal genes that crossed over are just as healthy as the maternal genes they replaced.

But Cavenee's experiments proved that, if a chromosome carries a retinoblastoma abnormality, this shuffling is a disaster. The segment that crosses over is missing a vital gene, a gene that was in the cast-out healthy segment. The daughter cells now have two chromosome 13s lacking the gene. The result is cancer. The crossover is Knudson's second hit.

Knudson's two-hit theory was validated. The experiment, published in November 1983 in *Nature*, made Cavenee famous.[11] "Knudson had proposed this wonderful idea, but of course, it was just that, an unproven idea, and there are lots of those out there," says Ray White. "Web proved it was right. In doing so, he provided absolute molecular proof, from an exquisite experiment, that a human cancer, whether it was inherited or arose sporadically, evolved from a defect in a bit of chromosome so small that it had to involve a specific gene. Moreover, he uncovered proof of a system that had to be working in other cancers, perhaps in many other cancers. An important, but hidden mechanism for cancer had been revealed."

Says Knudson, who by then had moved to Philadelphia's Fox Chase Cancer Research Center, "I was very pleased of course. I was a bit amazed how quickly someone was able to prove it all."

Cavenee's instant celebrity pulled him out of White's shadow and shoved him into the fledgling field of cancer genetics. His work had provided the final proof that a whole new type of cancer gene existed. The oncogenes that had been discovered several years earlier promoted uncontrolled cancer growth by their presence. But the new class of gene was its reverse, allowing cancer to flourish by its absence. Knudson had originally dubbed the gene an anti-oncogene, but later he and others began calling it by another term: suppressor gene. Like the cop on a beat, the gene's presence deterred cells from behaving like outlaws.

Cavenee presented results of his experiment at several cancer seminars in the spring of 1983, and within weeks he was fielding job offers from several research centers. The University of Cincinnati wanted him to move there, head his own team, and search other tumors for suppressor genes.

At Cincinnati, Cavenee used the riflip probes like an archeologist's pick, digging back through the layers of a family's genetic history as recorded in their DNA. By 1985 he and his new team were able to report a case of a young girl with retinoblastoma that left no question that the crossover phenomenon was the second hit in the eye cancer. "These results suggest a new approach for identifying . . . mutant genes that lead to cancer and a conceptual basis for accurate predictions of cancer predisposition," he and his colleagues declared in their report in *Science*.[12] Within weeks of the report's publication, Cavenee carried out the first such predictive test on the young boy born at the Karolinska Hospital in Sweden.

Cavenee's discovery, however, now pushed another mystery of retinoblastoma forward. What was the function of the gene which, when present, held cancer in check, but when absent, let the malignancy proceed so lethally? A race broke out among several teams to identify the gene. In California one team was headed by Sparkes, who was using his esterase D gene marker to find the gene. Another team launching a search was headed by Thaddeus Dryja, a research ophthalmologist at Massachusetts Eye and Ear Infirmary, a facility allied with Harvard University School of Medicine. Dryja had been interested in retinoblastoma for years, troubled by his inability to catch the tumor early in his young patients. Having heard of Cavenee's interest in finding the gene back in 1982, he collaborated with the Utah researchers by providing them tumor tissue from his patients.

By 1986, however, Dryja was working with molecular biologists at MIT's Whitehead Institute. Following months of experiments with riflip probes, they pinpointed the exact spot on chromosome 13 where the suppressor gene lay and plucked it out of a piece of normal DNA.[13] The discovery was widely acclaimed, for it immediately meant that scientists could begin piecing together the protein produced by the gene. Clearly, the protein of the normal gene is what slowed or turned off cellular growth.

Knudson was further elated. In a conversation following the discovery he said that he expected research into the normal gene's protein product to herald a new class of cancer therapy. "Instead of trying to kill tumor cells with toxic medicines that do all sorts of damage to healthy tissues, we could make malignant cells behave like normal ones. It's a stunning development."

Knudson was feted increasingly as the father of an entirely new theory of cancer genesis. In autumn 1988, he received the prestigious Charles S. Mott Prize for Cancer Research from the General Motors Cancer Research Foundation. Knudson's thesis was called "original and clearly visionary," an insight that had made genetic research "one of the most active and promising areas of cancer research."

Meanwhile in Boston, Dryja began using the suppressor gene as a probe to diagnose retinoblastoma in some of his patients. The test, which he first used in late 1987, has its limitations. It can detect only whether a child has inherited the first hit; it can't predict the explosion of a sporadic case of the disease. Still, "We get a lot of people who had the disease as youngsters and want to be tested to see if they carry the gene or whether the tumor was the result of a spontaneous set of mutations," Dryja says. "For others who know the disease was inherited, the test can be an early warning or it can relieve them of any concern."

For Bill and Bonnie Quinlan of Dedham, Massachusetts, a suburb of Boston, Dryja's new gene test was a godsend. In 1987 they were unaware of the test and spent Bonnie's pregnancy anxious about the health of the fetus she carried. Bill, thirty-four years old, had developed a retinoblastoma when he was two months old. By the time the disease was detected, his eye already was overrun by the ghastly white tumor mass, and both the tumor and eye were immediately removed.

Of his three siblings, only his younger sister had had the disease, the other two having won the fifty-fifty genetic toss. Bill and his sister inherited the first hit from their mother, who was struck by the cancer as a child. "When I first was diagnosed, the family was very upset," says Bill Quinlan. "But it also sort of relieved my grandmother. She had always thought my mother's eye problem was her fault, that maybe she'd done something to her as a baby. When my cancer was diagnosed, she finally realized something beyond her control was going on."

Quinlan told his wife about the potential cancer while they were dating. "It came up right away because of my plastic eye," he says. "My eye doctor told me when I was about eighteen or nineteen that I shouldn't have kids; he said that I shouldn't risk giving my child cancer." But once married, Quinlan and his wife decided to take the chance. His younger sister had recently given birth to a little girl who was diagnosed soon after birth with a tumor. The little girl's tumor was eradicated without the loss of her eye.

"My niece's problem upset us but we wanted a child," Quinlan says. Adds Bonnie, "When I saw the trouble my niece went through, it hit me for the first time how serious the disease was. My doctor talked to me about an abortion. It was a very hard time, but I was just convinced the baby would be born alright."

In early 1988, the Quinlans delivered a baby son, William Francis. Two months later, little Will was taken to Mass Eye and Ear for what the Quinlans were told would be the first of frequent eye examinations. "It was frightening," says Bill Quinlan, a hard-talking Irishman who runs his own janitorial maintenance business. "They had to put him under, and they tell you first all that can go wrong, like, maybe he won't wake up. I had to be in the operating room with him, and watch him be put to sleep. I wasn't prepared for how scared I got."

A few months later, however, the Quinlans' eye doctor told them that Dryja had begun offering an experimental test that could detect whether young Will had been born with his father's disabled gene. At a meeting with Dryja at Mass Eye and Ear, the Quinlans were told that the test could tell them merely whether Will needed more examinations. The test would cost eight hundred dollars and Quinlan wasn't certain his insurance would cover the cost. "That's a lot of money for us, but I figured it was worth the gamble," he said.

Just before Christmas, researchers at Mass Eye and Ear collected blood samples from the Quinlans, and from Bill Quinlan's mother and siblings. By using riflip probes that revealed the presence of the gene Dryja had isolated, the researchers could tell whether young Will had inherited the disabled chromosome that had traveled down from his grandmother to his father, aunt, and cousin—or whether he had lucked out and inherited his father's healthy copy of the chromosome.

Just after the New Year, the Quinlans were given the verdict. Will's DNA was free of the suspect gene. He had inherited a copy of Bill's good chromosome 13 along with one from Bonnie.

"Only after I got the news did I realize how worried I was," says Quinlan. "The world was lifted off my shoulders."

Two weeks later, Quinlan's mother, who had been ill for months, died of a type of cancer called small cell carcinoma. Within a year of the elder Mrs. Quinlan's death, researchers found that the retinoblastoma gene often was missing or defective in tissue samples removed from this type of cancer, an indication that the gene may be involved in more than just the rare eye cancer.

"It was a real emotional roller coaster for us," Quinlan says. "We were lucky she stayed alive long enough for her to give her blood. When she heard that Will didn't have the gene, she was very, very happy."

CHAPTER 8

Cancer Unleashed

On September 1, 1988, two scientific reports surfaced in the pages of the *New England Journal of Medicine* which, taken together, provided the firmest evidence yet that possessing a distinct group of genes sharply increased the odds of developing a deadly and common cancer.[1, 2]

The experiments outlined in the reports showed that cancer of the colon, the second leading cancer killer in the United States (after lung cancer in men and breast cancer in women), struck only in the presence of a specific set of defective genes. Both reports said the gene discoveries represented the first extension of the Knudson theory to a disease that was a widespread public health menace. But, unlike retinoblastoma, which required two hits against just one gene, the new research suggested that colon cancer required a two-hit derangement of each of several genes, perhaps as many as four.

That colon cancer is caused by genetic aberrations often shared by several members of certain families has long been suspected. The malignant tumors often arise from polyps, benign mushroomlike clumps of tissue in the bowel. Under certain conditions the polyps turn pernicious, puncture the lining of the intestinal wall, and drop malignant cells into the bloodstream, where they are carried to other organs. Doctors have observed that a tendency to develop potentially dangerous polyps is seen more often in some families than others. They also know that frequently colon cancer occurs in brothers and sisters. They weren't surprised when Ronald Reagan and his brother, Neil, both were diagnosed with mild forms of the cancer only ten days apart in 1985. Scientists believed that some family members share a genetic soil in which a colon cancer is more likely to grow, but the specific genes involved, if any, were unknown.

It was assumed that for most of the cancer's victims, the transformation of the polyps to cancers arises as a result of some random series of events, possibly the consumption of a high-fat diet or some other exogenous factors. That siblings sometimes developed the same cancer, doctors believed, might well be due to their shared exposure to food and other carcinogenic substances.

The two new reports didn't name specific genes—that was to come later. Instead, the researchers produced data that, like a shadow on the wall, provided convincing, but nonetheless circumstantial, evidence of the genes' existence. Even so, the reports' release was a watershed in cancer research. And the dual publication wasn't serendipitous. Weeks earlier the *New England Journal*'s editors had scheduled them to appear side by side. By simultaneously publishing the studies in a highly visible manner as the lead two reports in that week's issue, the editors were signaling to America's doctors that a scientific sea-change was in the making.

"An outline is beginning to emerge of the series of genetic events involved in a fully developed human cancer and of the underlying mechanisms," wrote Peter Nowell, a Philadelphia cancer specialist, in an editorial comment enlisted by *NEJM* to accompany the two reports.[3]

The *New England Journal*'s editors surely knew the issue would catch many of the magazine's readers by surprise. But, in truth, the reports were merely the loudly proclaimed mid-course results of a previously unpublicized scientific footrace. For almost two years, scientific teams in the United States and Great Britain had been waging an unusually intense search for the precise genetic changes that turn a normal colon cell cancerous.

The scientists themselves were aware of the stakes. "Most colon cancer is detected only when the disease already has progressed dangerously," said Mark Skolnick, one of the principal researchers of one of the studies. Skolnick and associates Lisa Cannon-Albright and Randall Burt at the University of Utah reported how, by studying numerous large families, they had found that one gene or perhaps several genes, inherited by as many as one-third of all Americans, could produce a biochemical environment in which a cancer can flourish. "If we can find the genes, or a marker for them, we could use a simple blood test to screen the entire U.S. population. Those with the gene or genes would know they are carrying within them a potentially dangerous genetic defect. They would be warned to get regular checkups, and avoid the kinds of high-fat foods thought to trigger the cancer. All that would be a measure of prevention far better than anything we have today."

The other study, led by Bert Vogelstein at Johns Hopkins University School of Medicine in Baltimore, identified four specific chromosome areas where genetic defects occur as a polyp progresses from a benign to cancerous state. His study proposed that, when accumulated in the genetic machinery of a cell, the combined defects set off the kind of uncontrolled cell division that characterizes cancerous growth. Tests to detect the presence of these four defects could flag whether a cell's normal programming was disturbed, allowing for the same kind of widespread predictive screening proposed in the Skolnick report.

"It's a super-exciting set of discoveries," Vogelstein said a few weeks after the reports appeared. "Only a decade ago cancer was thought of as a black box in which nothing could be seen. Now we know that, within the cell, certain genes must go awry. The next step is to identify the genes involved. That's no small feat, but at least, now, for the first time, we know what we are looking for, and how to look."

The experiments in Salt Lake City and in Baltimore were the result of two distinctly different avenues of research. Yet, both investigations were direct descendents of Knudson's now-proven two-hit theory and riflip technology. Vogelstein, Skolnick, and about a half-dozen other researchers had become convinced by the mid-1980s that tools were available to crack the most basic riddles of human cancer. Great glory would come to the scientist who did it first.

Even before the 1978 meeting at the Alta mountain resort high above Salt Lake City, Mark Skolnick had set a single-minded course to unmask genes that presaged a person's risk of cancer. The cancer rates within certain Utah families were much too high to ascribe to chance, Skolnick maintained. Skolnick also dismissed the possibility that carcinogenic agents in the environment alone could account for the familial cancers, because people living near the affected families often were cancer-free. Skolnick decided, before he could search for a gene or genes embedded within the shared family DNA, he would first have to prove through classic genetic segregation studies that, indeed, a cancer trait was being transmitted down through the generations of numerous family pedigrees. In other words, his first job was to determine whether certain cancers really did occur frequently in some families and not in others, a phenomenon that was widely reported through anecdotal accounts by doctors throughout the world, but which had yet to be proven.

Skolnick is tall and thick-shouldered, but his presence is muted by a soft chubbiness, owlish glasses, a wispy brown mustache, and a youthful face. He is unpretentious to a fault, often attending important scientific meetings in the same worn corduroy pants and battered blue sneakers that he regularly wears to his ground-floor office in a quiet building at the edge of the University of Utah campus. But his manner and tendency to set off into intellectual musings belie a fiercely stubborn and competitive streak. Skolnick had spent much of the four years prior to the Alta meeting in 1978 producing a series of population studies among the Mormons that convinced him he was seeing a phenomenon of unusual importance.

"The Utah families, most of whom are Mormons, provided us with a living laboratory of unbelievably fertile possibilities," Skolnick recalls. Family data was being collected regularly and stored by the Genealogical Society of Utah, which was amassing pedigrees going back five and six and more generations in a huge library controlled by the Mormon church. At the same time, the state of Utah was collecting its own registry of cancer cases. By meshing the two bodies of information in a series of computerized searches, Skolnick produced a series of studies revealing that cancers of the breast, colon, prostate, lung, cervix, and stomach were frequently found in unusually large clusterings in some of the Utah families.

These clusterings were rarely apparent when Skolnick and his colleagues looked at specific nuclear families or even closely related kin. But the Mormon-based populace kept detailed logs of their relatives, linking them to cousins in family branches that sometimes numbered into the thousands. Often hundreds of members of these far-flung branches were alive, living in towns north and south of Salt Lake City along the rugged Wasatch Mountains, which cut Utah down the middle.

Skolnick was entranced by what he called the "power of the family pedigrees" to reveal previously hidden traits. He frequently stumbled across clans in which a half-dozen to a dozen siblings had married and each had their own families of almost an equal size.[4, 5]

The combination of factors often made Skolnick wonder "if the omnipotent force that brought these people here, might also have in mind the needs of later generations of scientists."

Skolnick, of course, wasn't the first to notice the Mormons' potential for study. In 1947, Eldon Gardner, a professor at the University of Utah, came across a large family he subsequently called Kindred 107, which contained a significant number of cases of breast cancer.[6] Gardner's research and studies of other pedigrees over the next few years produced the first data in the United States suggesting that "hereditary components were associated with breast cancer." The studies found that sisters of breast cancer victims were twice as likely to also develop breast cancer as were sisters of victims of other cancers.

Gardner first found Kindred 107 when a student of his in a genetics class described how two aunts had died with breast cancer within two years of each other in the 1920s, and that they died at almost the exact same age, one at forty-five, the other at forty-six. Gardner was piqued by the apparent coincidence and arranged through the student to meet with living descendents of the two women's parents and to collect medical histories. By the time Skolnick met Gardner in 1975, Gardner had documented thirty cases of breast cancer in the family, an aberrantly high rate. The disease was of an especially virulent nature; five women had cancer before age forty, and two men in the family also had breast cancer.[7]

Gardner suspected that the women inherited a gene that made them vulnerable to the cancer, but he had no way of proving it. When he constructed a classic Mendelian pedigree chart, the disease failed to show up in succeeding generations in a pattern that was either recessive or dominant. Skolnick quickly became convinced the high-powered computer programs he was using to organize the Mormon genealogies would help Gardner.

Over the next few years, Skolnick, Tim Bishop, and Lisa Cannon-Albright, collected eight other Utah families with similarly large clusterings of breast cancer cases. One technique was to quiz patients who came to the university hospital's breast cancer clinic. One family, for instance, was discovered when a patient at the clinic said she had a sister and a niece with breast cancer. By sifting through the computerized Mormon genealogy, the researchers found several hundred living relatives. Another pedigree was found at the clinic when a patient told researchers that she thought there were four generations of breast cancer cases in her family. A genealogical search located every descendent of the woman's great-grandmother who had breast cancer. In that family, called Kindred 1001, the researchers found fourteen cases of breast cancer among 346 relatives.[8] The researchers even restudied Gardner's original Kindred 107, meeting with 118 family members at a family reunion in 1980 in a small rural town north of Salt Lake City.

Skolnick and his colleagues were certain that, when the families were combined in a computerized analysis, the frequency of these cases would prove the presence of a powerful gene. The suggestion of inheritance was there, but not the proof. The researchers were unable to find a clear detectable pattern of inheritance. Skolnick was stumped. Since 1975, he had bet his professional career that, by combining the power of the pedigrees, the cancer registry, and his computers he would unmask clearly defined inheritance patterns in breast cancer. Eventually, he expected to use similar tactics to expose Mendelian patterns in other common cancers. Using the riflips described by Botstein at Alta, he was certain he'd find the genes. But he first had to show that at least one of these common cancers segregated in an identifiable Mendelian manner, and neither he nor his colleagues were able to do that.

Then in 1981, Skolnick held a fortuitous conversation with Randall Burt, a gastroenterologist at Salt Lake City's University Hospital, who specialized in treating cancers of the bowel. Skolnick's interest in collecting cancer clusters was by this time becoming well known among local doctors, and every few months some doctor came calling or invited him over to discuss another example of the clustering phenomenon. Earlier that year, a Salt Lake City physician reported to Skolnick that he was treating a family in which a brother and sister, as well as their nephew, had colon cancer. Skolnick regularly tracked down these leads, punching the family names into his genealogy data base to see what popped out. The system scanned the connected family branches, searching for the oldest available common relative. The tactic often produced a number of related clans. Those families were then matched with the computerized registry of cancer cases in the state to see if, indeed, the family had a high incidence of cancers.

The genealogy search in this case found that the common relative was the siblings' great-grandfather, and that he had a twin brother who had married twice. Indeed, the extended family was found to have over five thousand members spanning six generations. When four branches of the pedigree were combined with the computerized cancer registry, Skolnick found fifteen cases of colon cancer, a frequency that was several times higher than expected in a family of that size, but no greater than that seen in other families with higher-than-normal cancer clustering. When all this data was fed into a computer, no pattern of inheritance emerged.

During a casual conversation with Burt, Skolnick explained his dilemma. "I was really frustrated and I was talking to everybody about the problem," says Skolnick. "Randy immediately suggested we look for polyps as well as full-blown cases of cancer." Burt told Skolnick that, if he was seeking a gene that predisposed people to colon cancer, he ought to study polyps, because colon cancers almost always arise from polyps.

"In my mind, polyps were precursors to the cancer, so why not include polyps in the segregation studies," Burt recalls. "Anyone who was inheriting a likelihood of getting colon cancer also had to inherit a likelihood of growing polyps."

Burt had always assumed that a predisposition to growing polyps was inherited, because it was rare to see the condition confined to just one member of a family that he treated. But until he spoke with Skolnick, he hadn't known there was a way to prove a genetic link. Skolnick, in turn, saw a chance finally to prove his cancer-inheritance theory. If he couldn't immediately show that a specific gene was causing cancer, at least he could prove that a tendency to develop polyps, which might later turn cancerous, was inherited.

Skolnick put aside the breast cancer research and threw himself into the study of colon cancer. Burt quickly showed him that studying polyps as they advanced to cancer would provide the statistical power his earlier studies lacked. Surgeons could see the polyp in the bowel and remove it, by use of a reedlike catheter that contained a fiber-optic lens at its tip. By combining the polyp and cancer cases, the researchers hoped they would accumulate a large enough base of data to meet the rigorous statistical requirements for segregation analysis. Skolnick and Burt started inviting into Burt's clinic as many of the family members as possible, whether they were sick or not, and examined them for the presence of previously undetected polyps.

Over a period of two years, from 1982 to 1984, Burt examined 191 relatives of the family known as Kindred 1002 and examined them with the fiber-optic device. "One important key to the study that just can't be understated was the phenomenal cooperation we got from the family," says Burt. "It is basically an uncomfortable exam, and we were asking people who were young and otherwise very healthy to come into a clinic. There was nothing in it for them except helping us. It's one thing to ask people to give blood for DNA testing, and quite another to ask them to undergo a colon exam. Some of them had to do it several times."

Even Burt, who suspected he would find a slew of asymptomatic polyps in healthy people, was taken aback by the findings. Adenomatous polyps, the type that can progress to cancer, were found in 21 percent of the kindred members. "Based on my experience, I assumed there'd be a lot of quiet polyp growth going on, especially in a family with a history of colon cancer," Burt says. "But I was surprised that one of five people had potentially dangerous tissue growth already taking place. It was a bit alarming to us, because it meant that in the presence of what we assumed was a gene, the rate of risk was very high."

Separately, Burt and his colleagues also searched for the presence of polyps in 132 spouses of the family members, a group whose bloodlines were unrelated to the kindred and who would be used as a comparative control group. Burt found that 9 percent of the spouses had polyps, which was believed to be the same as the rate that the disease randomly occurs in the population. The different rate of polyp growth between the blood relatives and the spouse group strongly suggested that the tendency to develop polyps was a dominantly inherited trait. It was the first time such a finding had been reported.[9]

By 1985, when the initial results of Kindred 1002 were published in the *New England Journal of Medicine,* Skolnick was becoming increasingly anxious to produce a high-profile breakthrough. Other research labs running with the riflip concept already had generated big-time splashes. Moreover, Ray White, whose labs were backed by funding from the cash-rich Hughes Medical Institute, was becoming world-famous. He and Cavenee had found a marker for retinoblastoma, and were hard on the trail of genes involved in other cancers, too. White's fame was especially galling because Skolnick had suggested that White come to Utah in the first place. The two had collaborated for the first year after White arrived, but they soon went their separate ways, as each sensed the other wanted to spearhead his own investigations. White was very focused on developing the riflip map, while Skolnick was principally interested in using riflips to pick out genes he was convinced predisposed people to cancer.

"I was after what I thought would be the biggest game, cancer, the disease I thought would benefit most from the new technology," Skolnick says. "It turns out the entire process was just going to take a lot longer than I thought. Quite probably, I was naive or just overly optimistic at first. In retrospect maybe I should have gone after something less complicated. But I felt that, if there was any place in the world where the genetics of cancer would be resolved, it was in Utah, in studies based on the Mormon family population. To be honest, there were plenty of times I wanted to pack it all in, especially as others began reporting important findings."

Indeed, in September 1985, Skolnick packed his wife and two boys off to New York, where he planned to spend a yearlong sabbatical studying at Memorial Sloan-Kettering Cancer Center. "I needed to get away, and get my bearings, to reinvigorate myself and my commitment," Skolnick says. When he returned to Salt Lake City in 1986, he sat down with Burt, Bishop, and Cannon-Albright and decided to reconfirm the original findings in Kindred 1002. Skolnick decided to mount a two-pronged attack. The first was to show that the original data was not an anomaly distinct to Kindred 1002. They decided to go after more families. Next, they would begin doing riflip studies on the accumulated families' DNA, the final step to pinpointing the location of a predisposing gene.

Fortunately, while Skolnick was in New York, Burt had continued to recruit families. By matching Skolnick's data base of families and hospital records, the scientists searched for families where two siblings had colon cancer, an indication of a shared predisposition to the disease. Almost every person contacted this way led to the eventual recruitment of large families. By 1987, the researchers had thirty-five families with an average of about one hundred members each under study. By 1990, they had almost twice the number of families under study.

In Kindred 1002 alone, the family Skolnick had first heard about in 1981, the scientists had examined 350 members. "And that's just a fraction of the descendents of an original pioneer," says Skolnick, noting that the pioneer's kin probably number about ten thousand. "The [family] founder had eighteen children and one of the children had four wives," he says. "If he passed along a predisposing cancer gene, we'll be able to track it."

In further studies, the research team found that polyp growth continued to occur in the kindreds at almost double the rate of the spouse families. Just as important, a computerized analysis found that the pattern in which the trait popped up within the families was similar to that first found in Kindred 1002 in 1985. It appeared to be the result of a dominant gene because only one parent needed to have the trait for 50 percent of their children to be similarly affected.

Moreover, the numerous polyps the researchers found and removed from the families indicated that the gene is a common one, carried by about one-third of the total population. It was this finding that was carried in the September 1, 1988, *New England Journal* issue.[10] For the first time, a research team was claiming evidence that a predisposition to a common cancer was passed along from parent to child. Skolnick's patience had been rewarded. He had yet to begin riflip studies, but, at least, "now we knew there definitely existed a gene worth going after," he says. "I felt vindicated, and pleased that I'd stayed with the cancer research."

Skolnick was especially gratified by the study's implications for public health policy in the United States. "It means one-third of Americans stand some risk of developing colon cancer, while the other two-thirds probably aren't at risk at all," Skolnick says. If the predisposing gene or a marker can be identified, researchers would be able to tell people which group they are in. Also, by studying who among those with the gene develop cancers, researchers will hold a powerful tool for identifying what about these people's lives—a type of diet, perhaps—promotes polyp growth and turns these polyps malignant.

But even with the marker technique, rummaging through all twenty-three pairs of chromosomes may take years. Armed with the experience of fifteen years of studies, Skolnick braced for a long-term assault on the predisposing colon gene. Meantime, Skolnick had linked up with two other physicians who were willing to use a similar strategy to look for predisposing genes for breast cancer and melanoma, the cancer of the skin. By the time the colon cancer results were being reported, Laurence Meyer, a research dermatologist at Salt Lake City Veterans Administration Hospital, was collecting pedigrees in order to mount a predisposing gene study for the skin cancer.

"You know there must be a reason some people get lung cancer and others don't," Skolnick says, adding that he believes there are physical traits caused by predisposing genes for most all cancers that are yet to be uncovered. "It's going to be hard finding the traits, but at least now we have an incentive for looking."

Across the continent in Baltimore, Bert Vogelstein was taking a different tack. Unlike Skolnick, Vogelstein wasn't a geneticist by training, but like Ray White, Web Cavenee, and David Botstein, was a molecular biologist, interested in the inner workings of the nucleus of the cell.

Vogelstein had come to study cancer as part of a classic scientific approach to understanding a molcular system. That is, by observing a system under stress, when it is behaving aberrantly or in an exaggerated fashion, scientists pick up clues to normal functioning. Quiet, well-behaved genes rarely call attention to themselves. When they act up, like an unruly child, scientists can see the products of their destructive or outrageous behavior. The cancerous cell is the most common classroom for observing misbehaving DNA. If cancer is normal behavior in a hyperactive state, the scientists reasoned, then, conversely, normal functioning is merely a well-mannered version of a malignancy.

In 1982, Vogelstein, thirty-three years old, jumped into the race to find Knudson's two-hit genes in the pediatric cancers. By identifying which chromosomes misbehave, Vogelstein reasoned he would find which snippets of DNA, or genes, were involved in normal aspects of cellular growth, and other regulatory activities. He, like other molecular biologists, wanted to know which genes tell a cell when to stop growing, when to divide, and when to differentiate, a process in which a cell matures into a different kind of cell to form a new type of tissue. "Cancer has always been viewed by the molecular biologist as a doorway to understanding normal cellular functioning. As far as I was concerned, [riflips] were the tools you could use to pry open the door," Vogelstein says.

Like Cavenee and White, Vogelstein began dissecting tumor tissue with riflips he and his colleagues were cloning. "I wasn't aware at first of what the guys in Utah were doing with retinoblastoma," Vogelstein says. "But, by 1982, it didn't require some special insight to know that using riflips to detect changes in chromosomes was the right way to go. [The riflip] idea had been described, was being used and developed, mostly by White. Also, it was becoming more and more certain, based on the work confirming Knudson, that a cancer could involve a loss of some bit of chromosome. People were looking at cells taken from tumors and they saw these losses under a microscope. I guess I thought it was obvious that the [riflips] provided a precise way of finding exactly what gene or genes were lost."

But unlike White and Cavenee, who confined their focus, at first, to chromosome 13, Vogelstein began combing through chromosome 11. Researchers elsewhere had reported seeing under a microscope losses of chunks of chromosome 11 in some cases of Wilms' tumor, the inherited childhood cancer of the kidney. At first, Vogelstein wasn't looking for gross chromosome losses, but rather for presence of an oncogene called *ras*, which was located on chromosome 11 and was known to promote cancer growth. He didn't find the oncogene, but he did find that a specific site in the chromosome was missing in half the cases he looked at. A riflip that was closely linked to that spot was present in chromosomes taken from healthy tissue, but absent in the tumor. The discovery galvanized Vogelstein, for it provided the first proof that a Knudson-like genetic loss transcended retinoblastoma.

"It might be universal," he thought. Vogelstein was hooked. The discovery itself was exhilarating. But, equally important, he felt a great satisfaction at latching onto a hypothesis that, conceivably, he could spend a lifetime trying to prove.

Vogelstein and his colleague, Eric Fearon, a young graduate student, published the Wilms' tumor discovery in 1984 in *Nature* magazine.[11] But their satisfaction was short-lived. The *Nature* issue also carried the exact same results discovered almost simultaneously at three other labs.[12-14] Following Cavenee's report the year earlier, in which he had used riflips to pinpoint chromosomal losses in retinoblastoma, several research teams obviously undertook experiments to search for similar findings in Wilms', another pediatric cancer.

Realizing the search for the Wilms' gene already had attracted a crowd, Vogelstein headed off in another direction. If sharing the Wilms' finding with others disappointed Vogelstein, it also fueled a competitive fire. The Wilms' tumor discovery convinced him that the Knudson hypothesis was correct, and that the riflip technique worked. He decided to use the technique to find out if gene losses like those in retinoblastoma and Wilms' also were involved in other commonly occurring cancers. Vogelstein was pleased because he now had the kind of focused objective he'd been searching for since entering research eight years earlier. To his surprise, he now realized that he was burning to arrive at the objective first.

Vogelstein wasn't just a biologist; he was also a trained physician. Although he majored in mathematics as an undergraduate, "I always had it in the back of my mind, since I was a teenager, to do medical research," he says. "I can't tell you why, it just seemed that it would be a reasonable way to spend one's life." While at Johns Hopkins School of Medicine, Vogelstein found that he was especially attracted to preventive care. The specialty that incorporated such an outlook most, he found, was pediatrics.

Unsure yet how to translate that interest into a field of research, Vogelstein enlisted for two years of service at the National Institutes of Health, where he eventually wound up working in the laboratory of Robert Gallo, an up-and-coming scientist trying to prove a link between viruses and cancers. (Gallo, of course, would make his name a few years later, when he, along with French researchers, was credited with discovering the virus that causes AIDS.)

At the same time, Vogelstein began treating children with cancer. "Treating the kids made a big impression on me," he says. "I realized it was a field of medicine that needed a great deal of work. I felt whatever advances might occur in cancer would come through an understanding of how to prevent the disease. I went back to Hopkins, to its oncology center, and for a few years began working on projects that looked at the nuclear structure of normal cells and tumors. I was in the right place, thinking about the right things, when the riflip technology came along. Suddenly, there was this technology that would make it possible to identify genes involved in the cancers I had seen as a resident."

Vogelstein, along with Eric Fearon, decided to extend the chromosomal loss search to bladder cancer, a fairly common adult cancer. They chose it because it would involve studying tissue similar to tissue in Wilms'. Once more, Vogelstein's riflips detected the disappearance of DNA on chromosome 11. He wasn't surprised. By now, he was convinced that Knudson's ideas were "visionary," he recalls. But he was a long way from finding a gene like the one that caused retinoblastoma. Many things yet needed to be proven. The DNA subtracted in the Wilms' and bladder tumors could well have been caused by cancer, not the initiating step. In other words, something else that ignited the cancers also might be knocking out or mutating the chromosome 11 DNA. Also, by 1985 Vogelstein was looking beyond bladder cancer.

"I wanted to find one gene that was involved in several cancers," he says. Discovering such a gene or genes, he felt, would be extraordinarily satisfying, and a discovery of great significance. Such a gene might be the target area in DNA for carcinogens. Perhaps scientists could figure how to protect these genes from the onslaught of chemicals and other environmental agents that triggered a cancer. It might lead to broad screening programs. Tests using a riflip to detect a specific gene loss might help to discover those people who were at risk of getting several types of cancer, or were already in the very earliest stages of the disease. And, finally, these genes probably would prove to be essential, when healthy or in place, to the smooth running of a cell's natural life cycle, thus providing a major addition to the understanding of some basic bit of human life.

Vogelstein and Fearon almost immediately began work on colon cancer tumors, mostly because tumor samples were readily available. Also, the disease developed in clearly defined stages. In its earliest precancerous stage, the disease was merely a polyp. Over years, a benign polyp can begin to grow and become progressively more dangerous, eventually exploding into fast-dividing carcinoma. Vogelstein figured that the loss of different genes might trigger these different stages. Vogelstein and Fearon first tested their chromosome 11 riflips on chromosomes from colon cancer cells but they failed to detect any loss of chromosome material. They tested Cavenee's retinoblastoma riflips for chromosome 13, but they weren't missing either. The deleted DNA had to be elsewhere. Stumped, and unsure where to try next, the researchers realized they had yet to take an elementary step. The two researchers trooped over to Hopkins's medical library and spent hours poring through the literature of medical research studies, looking for any report of cell scientists who had seen microscopic DNA losses in any kind of colon tumors. In an issue of an arcane genetics journal, which had been just published, they found what they were looking for.[15] A team of French scientists reported a colon cancer case in which, quite strangely, the tumor tissue contained a chromosome 17 with two long arms fused together. It somehow was missing its normal shorter arm. Moreover, an entire copy of chromosome 18 was gone. Something very destructive had mangled chromosomes 17 and 18. The French team guessed, and Vogelstein agreed, that the mutations had to be associated with the cancer. Did they cause the cancer or were they the result of an event associated with the disease? Vogelstein didn't know. But he felt he'd just found the lead he needed.

Vogelstein was greatly relieved by the discovery of these mutations. Without any chromosome candidate, he would have had to start searching for losses using riflips on chromosomes from all over the genome. It would have required a daunting "brute force" search, pure luck, and long hours.

One by one, Vogelstein and Fearon tested chromosome 17 riflips against DNA taken from colon cancer tumors. Immediately, they knew they had a problem. As they hoped, in some of the chromosomes they detected losses, but not nearly often enough to make a scientific claim. Vogelstein believed the DNA he was testing was contaminated with healthy connective tissue and blood-vessel cells that also existed within the tumors. They turned to another colleague, Stanley Hamilton, a pathologist who helped them purify the samples. Experiments on these cancerous cells were "electrifying," Vogelstein recalls. In twenty-two of thirty tumors, one particular spot in chromosome 17 was consistently missing, although it regularly showed up in healthy tissue examined from the cancer patients.

The researchers then probed adenomatous polyps, looking for the same DNA losses. Like Skolnick and Burt in Utah, Vogelstein looked at the adenomatous polyps because they are the type known to progress to cancer. The probes showed no DNA loss in every case, meaning there was no chromosomal deficit in the precancerous cells. But the experiment was important. Somewhere between being premalignant and malignant, the colon cells lost genetic material. It seemed that, as in retinoblastoma, the loss of a gene in colon cells let loose the forces of cancer. But unlike the eye cancer gene, genes involved in colon cancer were lost in cells already formed in the body. These colon cancer genes somehow were deleted during the course of a person's life.

In a paper published in October 1987,[16] Vogelstein and Fearon reported that they were on the track of a gene that probably was involved in turning an adenomatous polyp from a harmless nest of tissue into a potentially deadly malignancy.

The language of a research paper is, by tradition, restrained. Scientists generally choose their prose with moderation, and than bury it in long paragraphs of unvarnished type. The Vogelstein, Fearon, and Hamilton report ran just four pages in the middle of *Science* magazine and received almost no attention outside the scientific community. At the end of the paper, the authors wrote:

> On the basis of the data presented here and elsewhere, one can begin to formulate a hypothesis for the development of colorectal neoplasia [cancer]. Early in the neoplastic process, one colonic cell appears to outgrow its companions to form a small, benign neoplasm. The step promoting this outgrowth is currently unknown, but the cells that form these benign neoplasms have undergone a generalized alteration in their genomes.... [Later] a loss of tumor suppressor genes on chromosome 17 or, less frequently, on other chromosomes, may be associated with the progression from adenoma to carcinoma. This genetic model of neoplastic development, although rudimentary, is consistent with the multistep nature of human malignancy and can be tested in a variety of experimental systems.[17]

For fellow scientists, Vogelstein's claim was illuminating. The concept of a gene whose presence blocks unrestrained cell growth had been proposed and proven in retinoblastoma and in Wilms' tumor.

Already, several scientists, Cavenee among them, were predicting that defects in these growth-blocking genes would be found in all common cancers. But no one had yet proven that. Here, the Hopkins team was making the first claim that such a gene was implicated in a common lethal cancer. But Vogelstein had yet to find the gene. Moreover, it was becoming clear that DNA was lost in more than one chromosome in colon cancer. That meant, unlike retinoblastoma, more than one gene was involved. He'd need to find those genes, too, in what order they disappeared, and what were the forces that drove them away. Only then could Vogelstein claim that he had uncovered the full genetic landscape of a cancer.

By this time, Ray White was becoming a riflip guru. *Science* magazine reported that "White was providing a sort of mail order service for the genetics community, sending probes to anyone who requested them."[18] As a result, White's name appeared as a collaborator in a lengthening list of papers pinpointing the location of genes, most of which were for rare disorders, such as neurofibromatosis.

Vogelstein had used several of White's riflip probes in his search for a tumor suppressor gene, and noted his appreciation at the end of his 1987 report in *Science* magazine.[19] Afterward, he phoned White. "I hadn't met him, I only knew him by reputation," Vogelstein says. To his surprise, Vogelstein learned that White, too, was studying colon cancer. Moreover, the Utah researcher was uncharacteristically worried. He was locked in a feverish race of his own and he feared he was running second. It wasn't Vogelstein he was chasing.

Several years earlier, White had come across several large Utah families whose members developed colon cancer at exceptionally high rates, much more frequently than in the families studied by Mark Skolnick. As teenagers, the affected relatives began growing dozens of polyps in their large bowel. Often, these would advance into cancers. Treatment usually involved the removal of a large section of the intestines. The condition was called familial adenomatous polyposis, or FAP, and was thought to account for about 1 percent of all colon cancers worldwide. Geneticists already had shown that the condition struck family members in a Mendelian pattern similar to a dominant gene, but by 1986, no one had found the gene, or knew where in the genome it lay.

White decided the FAP gene would be a good one to tackle. He believed that whatever gene was involved in FAP certainly must also be involved in the more common version of colon cancer. For FAP families, the gene probably was inherited in a defective form, he guessed, while in the other common cancers, the gene might be mutated by a carcinogen later in life, or inherited in only a partially damaged form. White, too, was daunted, however, by the idea of shooting his armory of riflips at DNA from the FAP families without knowing in which chromosomes to begin.

A report by Avery Sanberg at the University of Arizona and Lemuel Herrera at Roswell Park Memorial Institute in Buffalo pointed the way. A mentally retarded boy was admitted to Roswell Park[20] suffering from what appeared to be FAP. His case immediately drew the attention of doctors because nobody else in his family had FAP. It was, once more, the anomaly geneticists needed to stumble across. The coincidental occurrence of two unexpected disorders often was a signal that some gross damage to the DNA wounded numerous nearby genes, causing more than one illness. Tissue sent to Arizona was analyzed under the microscope. Not unexpectedly, the researchers found that a slice of DNA was missing, taking with it genes that protected against colon cancer and mental retardation. It was in the middle of chromosome 5.

White immediately focused his search on that chromosome. But the report also tipped off Walter Bodmer, an especially astute molecular biologist at Imperial Cancer Research Fund, the prestigious genetics laboratory in London. Sir Walter had joined the international gene hunt with the publication of the first riflip discoveries, and his lab in London was tracking several of the same genes being sought by White and his associates at the Hughes Institute. By early 1987 Bodmer had collected tissue samples from 124 relatives in thirteen families with FAP from medical centers in Great Britain, the Netherlands, Canada, and Israel.

When the report of the Buffalo boy appeared, the Britons dropped their other projects and threw all their resources at the FAP gene. By spring, only several months after beginning experiments, they found that one probe on chromosome 5 was uniformly lost in FAP relatives, while healthy members retained the riflip. Bodmer had heard rumors that White was also sifting through chromosome 5, so he and his colleague, Ellen Solomon, rushed their finding into print. It appeared as two reports in August in *Nature* magazine,[21, 22] and was hailed by the worldwide press. *Nature*'s editors called it "a remarkable discovery." The *New York Times* carried the report on its front page with the headline, "Cancer of colon is believed linked to defect in gene."[23]

In the report, Bodmer's team wrote that it had located a gene that turns off the production of growth factors in the colon. The loss of the riflip probe, they said, meant that the gene was lost, or at least part of it was lost. Moreover, Bodmer found evidence that, when family members retained one copy of the gene from one of their parents, they merely grew polyps, but that when both copies were missing, the polyps turned cancerous. It was, in other words, a "two-hit" genetic cancer. Indeed, the researchers wrote, this finding "extend[ed] Knudson's ideas to FAP and, at least, to a major subset of colorectal carcinomas."[24] The evidence was beginning to accumulate that Knudson's farfetched ideas, hatched almost two decades earlier, were standing up in a widening range of cancers.

But in Utah, White was upset. By the time Bodmer's paper was published, White already had found the exact same riflip missing in a set of Utah families. (His report was submitted one month before Bodmer's was published. White's finding wasn't published until months later.[25]) But neither of the groups—in Utah or in Britain—had found the gene itself. That was crucial if scientists were to begin understanding how the gene, or the protein it produced, controlled normal cell growth, or what role it might play in common colon cancers. Vogelstein called just as White was gearing up for the final leg of the race for the gene. "[White] was looking for genes that were inherited, and I was looking for genes that probably were damaged after birth," says Vogelstein. "There was a likelihood they involved some of the same genes, so we decided to collaborate."

White shipped Vogelstein a set of riflips for chromosomes 17 and 18, and he instructed his lab to begin finding other probes for the two chromosomes. With a fuller complement of probes, Vogelstein went back into his lab and began picking off sections of chromosome 17, trying to get closer and closer to the exact spot where DNA was being deleted in the colon tumor tissue. That, he assumed, was where the gene must be hiding.

At the same time, Vogelstein and Fearon extended their search for deleted DNA to chromosome 5, where the FAP gene was thought to lie, and to chromosome 11, which contained the oncogene called *ras* that also had been associated with some cases of colon cancer. The researchers were certain that genes on all the four chromosomes were involved in the progression of healthy colon tissue into an end-stage fatal malignancy. "Our idea was that as the genes were altered—some lost, some mutated—the cells increasingly lost their ability to control their growth, turning, by steps, into an increasingly hostile tumor," Vogelstein says. Vogelstein wanted to find those genes. But he also began to suspect that deciphering the order in which the genes disappeared would provide cancer specialists with an invaluable tool.

In the September 1988 *New England Journal* issue that also contained the Skolnick paper, Vogelstein's lab reported the results of riflip studies in 172 tumor samples representing various stages of colon cancer development. The lab found *ras* oncogene mutations in about half of the precancerous adenomas and about half the carcinomas. They found chromosome 5 deletions, similar to those found in FAP families, in about one-third of the noncancerous adenomas and carcinomas. And they found losses in chromosomes 17 and 18 in only a few of the polyps and carcinomas, but in almost all the end-stage cancers. Vogelstein reported that it appeared that the *ras* oncogene and FAP gene were involved in activating the polyp growth, but that a faulty gene on chromosome 17 and another on 18 drove the polyp tissue into a deadly cancer.

"Our results support a model in which accumulated alterations affecting at least one dominant acting oncogene and several tumor-suppressor genes are responsible for the development of colorectal tumors," the Vogelstein team reported in the *New England Journal* paper.[26]

In Utah, White was becoming certain that the gene he was racing for on chromosome 5 wasn't just involved in a rare cancer disorder, "but probably was a gene that many people inherit in mutated form, which predisposes them to developing cancer," White said. He stepped up his efforts to identify the gene by collecting families with colon cancer. He even asked Skolnick for some of his families, too, hoping that the gene Skolnick was tracking was the same as the one on chromosome 5.

Meanwhile, Vogelstein was running full-throttle to complete the genetic picture of colon cancer. "He was a ball of energy and ideas," said his colleague Fearon, "and he was sort of whipping us into this frenzy of work."

Their lab was in a two-story brick building in what had been a strip shopping mall on the fringe of the Johns Hopkins campus. The university had outgrown its main cancer center, and, until a new one was built, it bought the mall, renovated the stores, and placed Vogelstein's team in one of them. Vogelstein had been in on the lab's design, but was unable to engineer windows into what had been a supermarket building. Even so, he liked the location. When he had gone to medical school, he lived in dormitories across the way and had regularly shopped in the supermarket where his lab was now located. He liked the symmetry of returning to his former haunt where he had first dreamed of being a medical researcher. And he didn't mind the lack of windows. He regularly arrived before dawn and left after dark. "This way I have no idea what I'm missing outside," he told one visitor to his lab. "I'm not even sure what's going on out there at all."

He began experimenting on tumor tissue samples removed from patients, trying to predict, based on chromosomal losses, at what stage of cancer the specimens were taken. If he could do this successfully, he would have a test to detect the inborn genetic severity of a tumor. "Cancer specialists could use this information to decide which patients should be treated with the strongest forms of chemotherapy, and which need not be treated so aggressively," he said.

Vogelstein also set in motion experiments to nail the exact identity of the genes on chromosome 17 and 18, leaving 5 to White. Indeed, by the time the 1988 *New England Journal* paper was published, Vogelstein already believed he'd found the gene on chromosome 17.

Earlier in 1988, Vogelstein had come across several reports by a research team under Arnold Levine, a Princeton University biologist, who had discovered a gene in mice that transformed normal mouse cells into cancers.[27,28] The gene, called p53, was thought to lie in a particular spot on chromosome 17. Vogelstein at first believed p53 might be a candidate for his chromosome 17 gene, but initial riflip tests didn't show a match between the deleted areas under study and p53.

The Baltimore researchers then began a taxing crawl down the length of the short arm of the chromosome. With a battery of probes, much closer to one another than those Vogelstein had previously used, the researchers bit by bit eliminated areas of the chromosome where the gene might lie. They did this by blocking out complete areas lost in a tumor and mapping them on the chromosome. Some tumors lost huge segments of chromosome 17, encompassing hundreds of possible genes, while others lost smaller bits. They found these areas by testing a series of twenty-seven sequentially located riflips that were lost one after another down the chromosome, until they found a riflip that wasn't missing. Once a riflip was found, the researchers knew they had gone beyond the border of a deleted region, and therefore they had traveled beyond the gene's location. For three months during the summer of 1988 they conducted excruciatingly detailed experiments until they found one small area that was consistently lost in every tumor tissue they tested. That was where the chromosome 17 suppressor gene must be located.

At first, Vogelstein refused to believe the finding. The spot they found was the exact one where p53 lay. By sequencing the nucleotides in the DNA they had located and comparing it to sequences of p53, however, Vogelstein was forced to accept his discovery. "The earlier experiment with p53 was wrong," Vogelstein said. "P53 was our gene, no doubt about it."

But to prove he had discovered a tumor-suppressor gene, Vogelstein needed to conduct one additional experiment. He examined the tumors from fifty-six patients. Cells extracted from healthy tissue of all the patients contained two copies of the p53 gene. But 75 percent of the patients' tumors had just one copy of p53, and that one copy was ever so slightly mutated. Vogelstein decided that just a single "point" mutation in the gene was enough to deactivate it. He was certain he'd found a Knudson "two-hit" gene. When present, the gene repressed cancer. When one copy was mutated and another was present in normal form, the cancer sometimes occurred, although not usually. But when a remaining copy of p53 was damaged or deleted, the tumor exploded into a full-blown malignancy.

The results of the experiments were published on April 14, 1989, in *Science*.[29] The report grabbed the attention of cancer researchers worldwide, for now scientists knew that there were at least two tumor suppressor genes, the one for retinoblastoma and one involved in colon cancer. In Cambridge, Massachusetts, Robert Weinberg of MIT's Whitehead Institute, a man who selects his words carefully, told the *Wall Street Journal* that Vogelstein had made a "dramatic finding."[30] Vogelstein, he said, had put "an obscure gene right in the cockpit of cancer formation." Moreover, Weinberg told the *Journal* that the discovery "confirms beyond a doubt, the concept of suppressor genes. I expect more of these genes will be found, and that some of them will be involved in more than one type of cancer."

Vogelstein told reporters that he hoped the development would lead soon to a test for routine screening of people with mutated p53 and the *ras* oncogene. Patients at risk for colon cancer could be

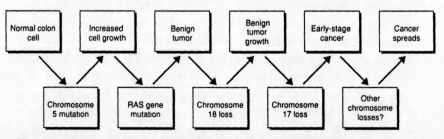

The progression to cancer as a result of the alteration of a series of genes.

screened for the earliest possible sign of mutations, while those who already have the disease could have their tumors screened for p53 alterations as an indicator of more advanced disease. Indeed, within weeks, researchers at Hoffmann-La Roche, a pharmaceutical and medical diagnostics company, were collaborating with Vogelstein to develop such a series of tests. Researchers at another drug company, Squibb Corp., wanted him to help them begin using the gene to develop new cancer therapies. The Squibb researchers wanted to find the protein produced normally by p53 and test whether it could serve as a prototype cancer treatment. They began experiments to insert the p53 protein into cancerous cells to see if the protein's presence could slow, perhaps even stop the cancerous growth.

But, even as those experiments were underway, Vogelstein knew that the p53 gene's importance was growing beyond colon cancer. In December 1988, researchers in Edinburgh, Scotland, had found deletions in chromosome 17 in twenty-three of thirty-eight tumors removed from the breasts of women.[31] When they checked the women's normal tissue, they found that the women had inherited a mutated or deleted bit of chromosome 17. Every normal cell contained an altered copy of chromosome 17 and one that was normal. The tumor tissue, however, contained two damaged copies of the chromosome. The scientists suggested that the women inherited a predisposition to breast cancer by inheriting an already-damaged chromosome 17. The cancer only occurred when the second copy was altered, too. The cancer was the result of Knudson's "two-hit" theory," the authors suggested in their report in the British journal *Lancet*.

Within weeks of their report, Vogelstein sent them copies of his p53 gene. Sure enough, it was the p53 gene that was altered in the breast cancer. Throughout the spring, researchers at other laboratories began sifting through DNA from cells of a variety of cancers they were studying. In Montreal, Webster Cavenee, who had become director of the Ludwig Institute for Cancer Research, had shifted his focus to brain cancer. Like Vogelstein and the other molecular biologists, he hoped the cancer studies would provide a doorway to understanding normal protein functions in the brain. By analyzing cancerous tissue removed from the brains of patients at nearby Royal Victoria Hospital, Cavenee and his research team had uncovered DNA deletions at several chromosomes including chromosome 17. When Vogelstein reported that the deletion in colon cancer on chromosome 17 involved the p53 gene, "we naturally looked to see if it was same bit of DNA we saw affected," he said. Indeed, it was. Cavenee found altered p53 genes in about 30 percent of the brain cancers he investigated. "We think [altered] p53 isn't needed to start the cancer, but when it is mutated, the cancer seems especially dangerous," Cavenee said in an interview.

Researchers at the federal government's National Cancer Institute then found p53 genes missing or mutated in many cases of lung cancer they studied. And by the end of the year, several teams of scientists, including Vogelstein's lab, said in a combined report that they had uncovered altered p53 genes in 60 percent of tumors of the colon, breast, lung, ovaries, cervix, adrenal cortex, bone, and bladder that they analyzed.[32] But p53 alone wasn't responsible for any of the cancers. As with colon cancer, the scientists found that each of the tumors contained deletions sprinkled about the genome at other chromosomal sites. To switch on a full-blown cancer, at least several different genes had to be mutated or lost. And, as Knudson had suggested years before, each of the two copies of the gene in an affected cell had to be lost for the cancer to arise.

Thus, by 1990 about a dozen molecular biology laboratories in the United States, Canada and Europe were racing to isolate the exact identity of suppressor genes specific to each type of cancer. Vogelstein's success was fostering an entirely new field of cancer research, the boundaries of which seemed almost limitless to the molecular biologists. "It's an exciting time," said Eric Stanbridge, a molecular biologist at the University of California, Irvine. "We're in the cataloguing stage right now, trying to compile the complete list of all the genes which when mutated or deleted can let loose malignant growth."

The scientists were paying Vogelstein the highest compliment they could: using his strategy and repeating his experiments in searching out genes in different cancers. But Vogelstein was being feted by the cancer establishment, too. In late 1989 he was awarded the Milken Family Medical Foundation Research Award and the Rhoads Memorial Award from the American Association of Cancer Research, and, in early 1990, he was given the highly coveted Bristol-Myers Squibb Award for Distinguished Achievement in Cancer Research—all for his pioneering gene deletion work.

At one of the federal government's chief cancer research labs, John Minna turned the focus of his work toward tracking down the suppressor genes involved in lung cancer. He speculated that lung cancer involved the alteration of at least a half-dozen and perhaps as many as ten or more different genes, including, in some instances, the gene that when mutated or lost triggers the rare eye cancer retinoblastoma. His labs were seeking to identify all the mutated genes, a venture he guessed might take the scientists into the next century. Minna and his group also suggested that some of the gene mutations might be caused by cigarette smoke, and the researchers began conducting experiments they hoped might someday finally identify the exact biological link between cigarettes and lung cancer.

Minna's group then undertook experiments to determine whether people can inherit some of the suppressor genes in already mutated form. Such people would have an inborn risk, or predisposition, to the cancer since their lung cells would require that one less gene be knocked out for a cancer to develop. "Although lung cancer is not generally considered to be an inherited disease, there are actually several pieces of evidence indicating a potential inherited predisposition," Minna and his associates wrote in an article published in late 1989.[33]

Minna also said he expected that the catalogue of tumor suppressors might hold as many as thirty or fifty genes that normally are involved in cell growth, which, when mutated, can unleash the wide range of cancers. "When these genes are healthy they act as tumor suppressors," Minna said, "that is, they naturally suppress uncontrolled cellular behavior by just doing their normal jobs. The cancer research is giving us a chance to isolate genes that play pivotal roles in normal healthy cell development. And with each discovery comes one more piece of the puzzle of human life. We hope what we're doing eventually will give us ways of preventing and treating cancer. But even if it doesn't help against cancer right away, without a doubt these discoveries will help fill in the genetic picture of life."

Indeed, Vogelstein next made a crucial finding proving Minna right. In a report in *Science* in early January 1990, Vogelstein and Fearon and their colleagues at Johns Hopkins said they had isolated the gene on chromosome 18 that also is lost in most colon cancers,[34] along with the p53 gene on chromosome 17. The researchers said the gene, which they dubbed DCC for Deleted from Colorectal Carcinoma, was missing or altered in 70 percent of cancers they inspected. But what was especially intriguing to Vogelstein and Fearon was that the DCC gene produced a protein that closely resembled proteins known to act as a sort of cellular glue, adhering one cell to another during cell replication and tissue growth. "It looks like DCC plays an important role in communicating how and where each new cell should grow," Vogelstein said at the time the *Science* paper was published. "When the protein is altered or lost, the glue is gone and normal growth is disrupted, giving rise to a malignancy."

In the space of one year, Vogelstein and his colleagues had reported finding two genes involved in the development of a human cancer. The p53 gene, of course, already had been discovered, although its importance had been unknown. But in finding the DCC gene, Vogelstein's team used the riflip strategy to place a previously unknown gene onto its exact location on the genome map. It now was clear that by hunting down suppressor genes, scientists could reveal the identity of unknown genes, thus contributing to the worldwide gene mapping enterprise.

All of which was very satisfying to Vogelstein. As he had hoped, the cancer research was providing new avenues for gene discovery. And, as important, it was elucidating the mechanics of cancer, too. Scientists at drug companies and in university and government labs were taking the p53 and DCC genes and attempting to use the protein the genes produced in test-tube studies to stop or slow cancer growth. Others, including Vogelstein, were collecting families in which cancers were common and checking to see if the mutated or altered forms of the p53 and DCC genes were being passed from parent to offspring. If that proved true, then it wouldn't be long before genetic tests were created to identify who has inherited an increased risk of cancer.

By early 1990 the prospect of screening for people with a predisposition for cancer still was so hard to believe that few scientists were prepared to say what might be done with the information. Vogelstein said it might be used to help prevent cancer or to help discover what in the environment triggers additional mutations. But by unveiling the genetic aspect of cancer, scientists had gained a research foothold where none existed before.

"When I began all this almost ten years ago, we knew nothing about the genetics of cancer," Vogelstein told a visitor to his lab in late 1989. "Now we have several genes that we know for certain are involved in cancer, and we are looking for the others, too. There's plenty to be done, much, much more to be done. But I'm astounded how far and fast we've come already."

CHAPTER 9

Genes
and the Heart

One day in late June 1982 Jan Breslow sat down at his desk at Children's Hospital in Boston to leaf through the latest copy of the weekly *New England Journal of Medicine*. He was feeling quite happy, even excited, about his own research. For four years Breslow had been trying to unravel the mysteries of the genes that control the processing of cholesterol in the body. He hoped ultimately to gain an understanding of why cholesterol seemed to predispose some people to coronary heart disease and heart attacks but not others. Four months earlier, in February, Breslow had succeeded in making a DNA probe that landed on one of these cholesterol-processing genes, known by a tongue-twisting name, the apolipoprotein A-I gene. In March, a young Greek molecular biologist newly arrived in the lab, Sotirios K. Karathanasis, had taken Breslow's probe and fished out a strand of human DNA that contained the apolipoprotein A-I gene. To the scientists' astonishment, the strand also contained a second gene involved in cholesterol processing, the apolipoprotein C-III gene.

"We were very excited," Breslow recalls. "No one would have predicted the A-I and the C-III genes were right next to each other." But now Breslow, Karathanasis, and Breslow's close friend and research partner, Vassilis I. Zannis, had a piece of DNA from a single chromosome that contained both genes, side by side.

Neither Breslow nor his colleagues had had time to lean back and speculate on the significance of the discovery. There hadn't even been time for Karathanasis to write up the discovery for publication. One task they faced was a search of the medical and genetic literature to find out if anyone had ever reported some mystifying disease or abnormality that might be explained by the fact that the two genes were linked.

As Breslow glanced over the table of contents printed on the cover of that week's issue of the *New England Journal,* he was suddenly brought up short. The lead article was titled "Familial Deficiency of Apolipoproteins A-I and C-III and Precocious Coronary Artery Disease."[1]

The article described two sisters who had been diagnosed as suffering advanced coronary heart disease while they were respectively in their early and middle twenties. The diagnosis was startling. Coronary heart disease is uncommon in women before passing menopause around age fifty; the sisters' cholesterol levels appeared normal and there was no evidence that the sisters' family suffered any of the rare inherited forms of coronary heart disease.

The team of specialists who reported the two sisters' plight in the *New England Journal* described an exhaustive battery of tests they made in an attempt to explain the young women's coronary disease. There was one unusual finding: A detailed analysis of the cholesterol in the blood indicated both sisters lacked a form of cholesterol known as HDL-cholesterol. Perhaps, the specialists speculated, this had something to do with their precocious coronary heart disease.

"It was absolutely amazing, I couldn't believe the coincidence," Breslow says.

Breslow put in a call to the lead author of the article, cardiologist and geneticist Robert A. Norum at the Henry Ford Hospital in Detroit. Breslow requested samples of the two sisters' blood cells for a genetic analysis. Both sisters, he found, had inherited identical defects in the two genes—apo A-I and apo C-III—that he and his colleagues had plucked out in early 1982.

Thus began a cascade of discoveries about genetic defects that predispose individuals to coronary heart disease, helping to solve one of the major mysteries of coronary heart disease—why some people succumb to heart attacks in their forties or fifties or sixties, while others live to their eighties without the slightest hint of heart disease. The new findings were about to make coronary heart disease, along with cancer, the first major scourges to yield their secrets to the new technology outlined by Botstein at that historic 1978 meeting in Alta. As with cancer, the heart disease findings meant that, for the first time, scientists would be able to distinguish those individuals genetically susceptible to disease from those genetically protected against it.

This ability to spot an individual likely to develop coronary heart disease would be far more than a scientific advance. Today, it is rare for the "coronary prone" person to be identified until his or her arteries are already narrowed by cholesterol-laden deposits. At this stage, little can be done except cease cigarette smoking, reduce the fats in the diet, engage in some exercise, and hope that the steady progression of coronary disease will slow down, postponing the inevitable heart attack. The patient's doctor is limited to sifting through a small armamentarium of drugs, prescribing one after another until, through trial and error, he finds one or two that seem to influence blood pressure, heart rhythm, and/or blood cholesterol levels favorably and which may—or may not—help stave off the heart attack. All in all, the treatment of coronary disease is a medical art with no guarantees.

This situation will change markedly as the multitude of genetic defects underlying coronary disease is uncovered. Pharmaceutical chemists will be able to design entirely new drugs, each targeted at overcoming specific genetic deficiencies that lead to coronary disease, thereby allowing doctors to tailor their therapies to individual patients. Genetically susceptible individuals can be identified at birth and reared to avoid the excesses of life that would prompt early coronary disease.

Coronary heart disease is the plague of the developed world, accounting for a quarter of all deaths in the United States and Western Europe. The disease is a gradual building up of deposits on the interior walls of the coronary arteries. The buildup of these clogging deposits—atherosclerosis—is threatening in any artery but it is especially dangerous in the coronary arteries, the lacy network of blood vessels that crown the heart (hence the name coronary). This network supplies the heart muscle itself with oxygen-rich blood.

As the deposits build up, perhaps over many years, they gradually reduce the flow of blood to the portion of the heart muscle "downstream" of the deposits. Often the flow is reduced so severely that, when a person exerts himself, the beating heart muscle is unable to obtain enough additional oxygen to meet the demand. The result is a chest pain known as angina pectoris. The clogging can become so severe that even the slightest movement produces the angina pain, rendering the victim crippled and bedridden.

More often than not, however, a more dramatic event occurs, the heart attack. The deposits may well have built up without producing any symptoms at all. Then at some unpredictable moment a blood clot (a thrombus) forms and lodges against the deposits, completely blocking the flow of blood. As a result of this coronary thrombosis, the downstream portion of the heart, suddenly deprived of oxygen, goes into spasms. The victim is stricken with a severe viselike chest pain. If a large portion of the heart muscle is damaged, the attack can be fatal in less than an hour.

The heart attack, and thus coronary heart disease, probably afflicted at least some humans even in prehistoric times. But the disease wasn't common in centuries past. It wasn't until 1912 that physicians recognized the distinct set of symptoms that characterize a heart attack, suggesting that doctors in the eighteenth and nineteenth century weren't so familiar with heart attacks that they could distinguish them from, say, strokes. In the 1940s, the vital statistics in the United States and Europe began showing a sharp increase in deaths from heart attacks. By the 1950s, public health experts realized that a pandemic of coronary heart disease was sweeping the Western industrialized nations. (At the peak of the pandemic in 1968, almost 675,000 Americans died of coronary heart disease, a third of all deaths. Death rates in Europe were equally high.)

The pandemic developed too slowly for any infectious agent, a virus or a bacterium, to be blamed. Instead, the outbreak appeared to stem from some change in the environment, probably related to the rising standard of living that accompanied industrialization. The pandemic didn't affect the underdeveloped nations nor, strangely, did it afflict the Orient. Even the industrialized Japanese seemed immune.

Baffled public health officials turned to epidemiology for clues to the cause of the pandemic. Epidemiology is one of the oldest medical sciences and unquestionably has saved millions of lives over the centuries, but it has its limitations. It is the comparative study of populations—not individuals—to see what distinguishes those groups who succumb to an epidemic disease from those who escape the disease. In one of the earliest uses of epidemiology, for example, the ancient Romans realized that people who lived near swamps were more likely than those living in drier locales to experience a debilitating disease characterized by severe fevers and chills. Thinking it was caused by the "bad air" emanating from the swamp they called it *mal aria*. Not every Roman who lived near a swamp developed malaria, and there was no way to predict who would and who wouldn't "catch" the disease. But the threat to the public health was important enough to prompt the Romans to drain the swamps around their city. Malaria wasn't eliminated but its toll on the Roman public was drastically reduced. Thus, without having the slightest inkling of how mosquitoes transmitted malaria protozoa, the Romans took the right action; swamp drainage is still one of the most effective weapons against malaria.

Using the same basic principles as the Romans, epidemiologists set out to see if they could find what had occurred to make Western societies more susceptible to the pandemic that Oriental societies had seemed to escape. In the West, the pandemic coincided with a sharp rise in cigarette smoking, an increase in obesity, and an apparent rise in the incidence of high blood pressure, all of which later proved to increase the risk of having a heart attack.

But one of the most suspicious life-style changes that accompanied the pandemic was a dramatic change in the diet occasioned by the rising affluence of industrialization. Not only did Westerners increase the amount of food they ate (accounting for the rise in obesity), but they also heaped in large amounts of animal fat in the form of steaks, roasts, butter, and cheese. Meat three times a day, usually pork or beef, became the norm for the general population rather than just for the rich. For the average American or European, fat began supplying as much as 40 percent of each day's total calories (and even more in many individuals), as much as double the proportion of calories coming from animal fats in previous centuries. In the Orient, despite industrialization, the Japanese continued to eat a traditional low-fat diet and continued to enjoy a low incidence of coronary heart disease. To epidemiologists, these changes in the West—an increasingly fat-rich diet and a sudden outbreak of heart attacks—could hardly be the result of coincidence.

There was one other key clue. When pathologists first started studying the deposits that clogged the arteries, long before anyone realized their role in heart attacks, they noticed immediately that the deposits were loaded with cholesterol, a fattylike substance that circulated in the bloodstream. It wasn't known whether the cholesterol actually formed the deposits or whether the deposits formed because of some injury to the artery and the cholesterol just happen to accumulate there.

But this hint that cholesterol might be involved in atherosclerosis prompted the epidemiologists, almost as an afterthought, to check the populations for cholesterol levels. To their surprise, average blood cholesterol levels in the Western populations were much higher than those of the Orient. And the more animal fat consumed by a population, the higher the average cholesterol level. Equally important, groups with the highest cholesterol levels had the highest incidence of heart attacks.

By the late 1970s, public health officials and heart specialists were warning that a high level of cholesterol in the blood appeared to promote the progressive clogging of the coronary arteries and increased the risk of what was termed a premature heart attack, that is one occurring in early adulthood or middle age rather than in the seventh or eighth decade of life. Dietary experiments had indicated that cholesterol levels could be reduced somewhat—about 15 percent—by reducing the amount of animal fat in the diet. (More drastic reductions in blood cholesterol levels required the use of pharmaceuticals.) And several experiments, involving hundreds of volunteers, showed that lowering cholesterol levels, by dietary changes or pharmaceuticals, reduced the incidence of heart attacks among those groups who had high levels of cholesterol.

These epidemiological discoveries stimulated the current government-backed campaigns urging every adult to have a blood cholesterol test. The programs are designed to reduce the national incidence of deaths and heart attacks from coronary heart disease. If they are totally successful and the average cholesterol level is reduced by 25 percent, the incidence of deaths from heart attacks in most Western countries would be cut in half, according to biostatisticians' calculations. For the United States, this would mean 250,000 fewer deaths from coronary disease each year.

But the epidemiological evidence, although leading to the right public health measures, leaves unanswered a question of vital importance to individuals: Which 250,000 Americans will be saved from death by reducing their cholesterol levels?

From the earliest days of the diet-cholesterol hypothesis, scientists were well aware that the pandemic wasn't sweeping through the population indiscriminately. Not every person who had a high cholesterol level developed and died of a heart attack—while many with seemingly normal cholesterol levels did. Moreover, the high-fat diet was found on almost every dinner table, and while some people who ate it developed high cholesterol levels and heart attacks, others developed neither. Moreover, among those who did develop coronary disease, clogging of the arteries didn't proceed at the same rate in everyone. Some men were being stricken before they reached age forty while other men weren't hit by the sudden chest pains of a heart attack until their early sixties. Women rarely developed the disease until after menopause, and most of the time it took ten years or more for the artery-clogging to progress to the point where a woman's risk of a heart attack was as high as that of a man.

Obviously, there was wide variation in individual susceptibility to coronary heart disease induced by dietary fat and cholesterol.

There is no question today that this varied susceptibility is genetically determined. "Diet determines the population's risk of heart disease, genes determine the individual's risk," explains University of Minnesota physiologist Henry Blackburn. Jan Breslow is fond of citing the paradox of Winston Churchill and tennis champion Arthur Ashe. Churchill smoked incessantly, drank heavily, overate, and rarely exercised, yet lived well into his nineties. Ashe, on the other hand, was a professional athlete, didn't smoke, remained trim, and lived what most would consider a healthy life-style. Yet he suffered two heart attacks before age forty.

"The Churchills of the world we presume have a genetic protection against heart disease, whereas the Arthur Ashes have a genetic susceptibility," Breslow says. The Churchills seem to constitute the 10 percent of the population that is genetically resistant to coronary disease while the Ashes comprise the 10 percent of the population at the other extreme, those who are almost certain to develop coronary disease regardless of what they do. The remaining 80 percent of the population constitute a spectrum of susceptibility that lies between these two extremes.

The clue to this spectrum of genetic susceptibility lies in the discovery more than a century ago of cholesterol piled up in the artery-clogging deposits. Cholesterol is emerging as the fulcrum upon which susceptibility to coronary disease rests. The body's handling of cholesterol is under genetic control. As with any genetically controlled system, outright defects in any of several specific genes can disrupt the system. And there are clear examples of families in which a defective gene prompts an early and rapid buildup of the cholesterol-laden deposits.

But there's another aspect to the genetic susceptibility for coronary heart disease or atherosclerosis. The entire cholesterol-handling system evolved in a dietary environment low in animal fat. Over the more than one million years of human evolution, a number of genetic variations occurred in the cholesterol-related genes. These variations were innocuous and of little import to survival when both food and life were short, and thus they persisted, passed from one generation to the next. But in the twentieth century the environment changed. Food became plentiful, particularly fat-laden meats and dairy products once considered luxuries. At the same time the virtual elimination of plagues like tuberculosis, diphtheria, and pneumonia allowed people to live long enough to accumulate cholesterol-rich deposits in the arteries.

The genetic variations, so benign in previous centuries, began to make their presence known. Some genetic combinations were unable to handle even a modest increase in dietary fat, particularly animal fat, rendering their inheritors susceptible to early atherosclerosis and early heart attacks—the Arthur Ashes. Other combinations of genetic variations were unfazed by the fat-rich diet, protecting their inheritors from the excesses of the twentieth century—the Churchills.

The molecular geneticists tracking down these inborn susceptibilities to coronary disease face a somewhat different problem from those hunting for the genes that underlie cancer or the genetic diseases. In cancer genetics or, say, Huntington's disease, the mechanism by which the disease occurs is unknown; the scientists know only that some hidden gene is producing a protein—or failing to produce a protein—that somehow leads to the disease. But the protein and its function are unknown. The geneticists are searching for the hidden gene in hopes it will lead them to the culprit protein and an understanding of how the disease develops, the "reverse genetics" approach that solved the mystery of muscular dystrophy.

In the case of coronary heart disease, the proteins involved in the processing of cholesterol are known. So far, eighteen more of these proteins have been ferreted out by decades of physiological research. What is not known is how these proteins vary from individual to individual and why these variations render some people susceptible to coronary disease and others apparently invulnerable to it. To unveil these variations in the proteins, the molecular geneticists need to pluck out the genes producing the proteins and essentially lay the genes of individuals side by side to see the differences. This strategy of going from the proteins to the genes might be called "forward genetics."

About half of the body's cholesterol is from the diet and about half is manufactured by the liver using, as raw materials, the fatty acids that have come into the body from the diet. Cholesterol is needed for a variety of jobs. It is the raw material for a number of hormones and it is a major constituent of the bile acids that break fats down into minuscule globules that can pass through the wall of the intestine into the bloodstream. But, perhaps most important, cholesterol is needed throughout the body to make the membranes that encapsulate each cell.

This last requirement means that cholesterol must be transported by the bloodstream to every nook and cranny of the body. Each cell, in turn, is constantly breaking down and rebuilding its membranes, taking in fresh cholesterol and disposing of "used" cholesterol. The used cholesterol has to be transported back to the liver for recycling and/or disposal.

This transport of cholesterol about the body is complicated by a simple chemical fact: Cholesterol does not dissolve in water. Technically, it is a fatty alcohol. In its pure state it looks and feels like a white wax. If dumped into the bloodstream in its pure state, it would simply float aimlessly about. To solve the problem of distributing this water-repellant cholesterol, nature over the eons evolved a complex system of protein carriers. These proteins do dissolve in water, that is, the molecules in the proteins literally latch onto water molecules. By wrapping a globule of cholesterol in a protein, the body manages to get the globule delivered to the cells by the watery part of the blood, the plasma.

The cholesterol transport system is a physiological labyrinth, made more so by the fact that the same system also transports the fats that come into the blood from the diet, fats also being insoluble in water. The distribution process essentially is a relay system with globules of cholesterol and fat being carried from one protein to another not unlike, to use an oversimplistic analogy, a basketball being passed from one player to another to move it down the court. It begins when one or more proteins—the exact details are still uncertain—grab a globule of fat and cholesterol at the wall of the intestine and haul it out to the fat storage cells where much of the dietary fat is dumped off. The now-diminished globule is taken to the liver, where protein receptors on the liver cells grab the protein-wrapped fat-and-cholesterol package and pull it into the cell for processing.

When one set of cells completes its particular bit of processing, the package is dropped again into the bloodstream, rewrapped in new proteins, to be picked up by another set of cells for further processing. At each step, the package, known as a lipoprotein ("lipo-" being the prefix denoting fat), becomes more and more dense. Hence, at any one time a variety of fat-cholesterol packages are circulating in the bloodstream—very-low-density lipoproteins (VLDL), intermediate-density lipoproteins (IDL), low-density lipoproteins (LDL), and high-density lipoproteins (HDL).

It is the penultimate package—the LDL package—that is the much-publicized "bad" cholesterol strongly linked to coronary heart disease. LDL is the form in which cholesterol is transported throughout the body to meet the cells' needs for making membranes. The final package—HDL—is the form in which the "used" cholesterol leaves the tissues and returns to the liver for disposal. (Because HDL signifies the extraction and disposal of cholesterol, it is becoming known as the "good" cholesterol, the more of it in the bloodstream the better for good health.)

The details of this cholesterol-transport system were just beginning to be understood in the early 1970s when two scientists in Texas—Joseph Goldstein, a fast-talking South Carolinian, and Michael Brown, a "laid-back" contemplative Pennsylvanian—made the first discovery of just how a genetic defect can disrupt the system and render its victims susceptible to coronary heart disease.

They had become close friends in 1966 as medical interns at Massachusetts General Hospital in Boston.[2] Both had become excited by the discoveries in molecular biology that were mushrooming out of the Watson-Crick unveiling of the double-helix structure of DNA. After finishing their internships, both young scientists plunged into the study of medical genetics at the National Institutes of Health.

In casting about for a genetic disease to study, Brown and Goldstein grew intrigued by an inherited disorder known as familial hypercholesterolemia (FH), characterized, as the name implies, by an abnormally high level of cholesterol in the blood. Little was known about the disorder except that it seemed limited to certain families; it was inherited in accordance with Mendelian laws. Those who inherited one copy of the mysterious and presumably defective FH gene had blood cholesterol levels two to three times higher than normal—and they usually suffered a heart attack by age fifty. Such heterozygotes—having two dissimilar copies of the same gene, one normal and one defective—are relatively common for an inherited defect: about one in every five hundred persons is an FH heterozygote.

The tragedy of FH is the children who are homozygous for the FH gene, having two defective copies of the gene. It happens occasionally that two FH heterozygotes will bear children together. The parents usually are unaware that they each possess a single copy of the defective FH gene, since it's uncommon for young adults to have their blood tested for cholesterol. There is a one-in-four chance that any child they bear will inherit each parent's defective FH gene. This happens about once in every one million children born.

Such an FH homozygous child has a blood cholesterol level six to eight times higher than normal. The tremendously excessive cholesterol is almost wholly the LDL form, the form that is churned into the bloodstream by the liver for delivery to cells throughout the body. The coronary arteries in these unfortunate children clog up with atherosclerotic deposits at an almost incredible rate. Many of the children suffer heart attacks before they reach puberty and, if untreated, will die of coronary disease before reaching their twentieth birthday.

After a brief interruption to pursue their genetics training at separate universities, Brown and Goldstein were reunited in 1972 at the University of Texas Health Science Center in Dallas.[3] They decided to resume their search for the genetic defect that produced such deadly levels of LDL cholesterol in FH. Being new to the Dallas center and junior in status, their regular working hours were filled with teaching and clinical duties. Research money was hard to come by. Their FH experiments had to be done largely on their own time with whatever money they could scrape up from other funded projects.

At the time there was no animal model for FH, so the two young geneticists were limited to experimenting with test-tube cultures of human cells. In the winter of 1972–73 they managed to obtain a sample of cells from a twelve-year-old girl being treated in Denver for FH. She was homozygous for the FH gene, having inherited two defective copies of the gene.

Their tactic was to bathe cultures of the FH cells and cultures of normal cells in a nutrient simulating the blood plasma with its packets of circulating cholesterol. Using radioactive tracers on the cholesterol, they watched for any differences in the way the two types of cells processed cholesterol. Both the FH cells and the normal cells, they quickly discovered, were able to manufacture their own cholesterol if the surrounding bath was low in cholesterol.

By summer, however, they began noticing a distinct difference between the normal and FH cells. When the surrounding bath was rich in cholesterol, the normal cells quit making their own cholesterol—but the FH cells didn't. Instead, the FH cells continued to pour out their own internally made cholesterol, regardless of how cholesterol-rich the surrounding bath was.

They soon spotted a second difference between the normal and FH cells. The normal cells seemed able to grab cholesterol from the bath and engorge it, essentially feeding their needs from the outside. The FH cells, however, seemed unable to do this; they simply didn't bind cholesterol and didn't engorge it. Being more or less "starved" for an external source of cholesterol, they continued to make their own supply internally.

From their experiments in 1973 and 1974, Brown and Goldstein began to piece together a hypothesis. The liver cells that process LDL appeared to have a feedback control mechanism built into them. Their surfaces are dotted with receptor proteins that "recognize" and grab LDL cholesterol, holding it so that it can be taken into the cell interior. When the cells' receptors indicate there is sufficient LDL in the passing blood, the receptors close. But if the receptors signal there is a dearth of LDL cholesterol, the cells begin making their own cholesterol internally. When the receptors begin acquiring the correct amount of cholesterol, the internal machinery shuts down.

In the FH cells, however, the gene for the LDL receptors is defective. Because of this defective gene—the FH gene—they lack the protein receptors and are unable to gauge the level of circulating LDL. As a result, the internal machinery is constantly turned on, pouring out a steady stream of cholesterol, despite the fact the passing blood is already loaded with cholesterol.

If a person is homozygous for the FH gene—both copies are defective—his cells are totally lacking in the LDL receptors and the cells' internal machinery pumps out a continuous supply of cholesterol. If the person is heterozygous for the FH gene—one defective and one normal copy of the receptor gene—his cells have half the normal number of receptors, and the cells' internal machinery produces half as much cholesterol as the homozygous cells.

In 1977, the definitive experiment was carried out, proving that what appeared to be true in the test tube was also true in the intact human. A young internal medicine specialist, David W. Bilheimer, who was caring for the FH homozygous children treated at the health center, enlisted three groups of volunteers: one group with normal cholesterol levels, a group who were heterozygous for the FH gene, and a group who were homozygous for the gene. Bilheimer injected the volunteers with LDL cholesterol tagged with a harmless radioactive tracer and then timed how long it took for the injected cholesterol to disappear from the bloodstream.

In normal volunteers, half the tagged LDL cholesterol disappeared in two days as the receptor-mediated feedback system regulated the LDL-cholesterol level. But in the FH heterozygous volunteers—those with one defective copy of the gene and half the number of LDL receptors—it took four days for half the tagged LDL-cholesterol to disappear, twice as long as normal.

And in the FH homozygous volunteers—those totally lacking in LDL receptors—it took six days for half the tagged LDL cholesterol to clear from the bloodstream. The fact that it was cleared at all was surprising, indicating there is some backup system available that can scavenge at least a modest amount of LDL cholesterol from the bloodstream.

The discovery of exactly how a single gene defect could render its inheritors susceptible to coronary heart disease, by fouling up the cholesterol processing system, brought Brown and Goldstein the 1985 Nobel prize in medicine. It also gave heart disease researchers solid laboratory evidence to clinch some epidemiological findings. The epidemiologists had begun to find that people with abnormally high levels of LDL cholesterol were susceptible to early clogging of the coronary arteries and "premature" heart attacks. The Brown and Goldstein discovery presented a clear-cut case of high LDL-cholesterol levels *causing* the rapid buildup of the deposits in the coronary arteries. Excessive LDL cholesterol was unquestionably atherogenic.

(It was the Texas scientists' ingenious use of homozygous FH cells and patients to bring out the extreme effects of a defective gene that prompted Nancy Wexler and her colleagues to launch their 1979 expedition to Venezuela in search of individuals who were homozygous for the Huntington's disease gene.)

The Brown and Goldstein discovery uncovered one genetic mechanism underlying coronary heart disease. But the Texas scientists did not uncover the defect in the FH gene. At the time, neither they nor anyone else had the tools to identify and study the FH gene itself.

The first identification at the DNA level of a genetic defect rendering someone susceptible to coronary heart disease was made by Jan Breslow and his colleagues, Vassilis Zannis and Sotirios Karathanasis.

Breslow is a tall, slim New Yorker whose slightly graying hair betrays his forty-seven years. Anyone chatting with him about his research will quickly realize he has the kind of mind invaluable to a scientist, an ability to organize information. His impromptu responses to even the most complicated questions have the clarity and simplicity of a carefully organized lecture, as though he were flipping through a mental file cabinet where each fact has been dropped into its proper folder.

Breslow had intended to become a pediatrician following his graduation from the Harvard Medical School in 1968 and spent two years of residency training at Children's Hospital in Boston. But like most young physicians whose training had been subsidized by the federal government, he faced two years of government service. He decided to try to gain a spot at the National Institutes of Health.

During interviews at the NIH he encountered Donald Fredrickson, a scientist who was studying a variety of inherited disorders characterized by high levels of fats and cholesterol circulating in the blood, known collectively as the hyperlipoproteinemias. There seemed to be a score of these disorders in fat and cholesterol metabolism, each presenting a slightly different clinical picture. Some involved extremely high levels of fat (triglycerides) in the blood but normal cholesterol levels, while in other disorders the reverse was true, and in still others both cholesterol and fats were abnormally high. Some disorders seemed to increase the risk of coronary heart disease, others didn't. Fredrickson had gained some fame among heart disease researchers for categorizing this confusing array of lipoprotein disorders into five major types, which he designated by Roman numerals.

"Fredrickson wanted a tissue culture laboratory and I had had some experience at Harvard in tissue culture," Breslow recalls. Fredrickson asked the young Harvard graduate to work for him after Breslow finished his pediatric residency.

After three years of working under Fredrickson, Breslow was offered a chance to return to Harvard to work in a clinic treating patients suffering from hyperlipoproteinemias. It was during his work at the Harvard clinic that Breslow became intrigued by one category of these disorders, which Fredrickson had designated Type III. Type III patients have abnormally high levels of both cholesterol and dietary fats circulating in their bloodstreams. Type III disorder usually is diagnosed in men when they are in early adulthood and in women ten to fifteen years later. By the time the disorder is spotted, the patients' arteries are already in the advanced stages of being clogged by fatty deposits.

The Type III disorder seemed ripe for research. In 1973, the year Breslow wound up his training stint at the NIH, one of the proteins that wrapped some of the packages of fat-cholesterol in the blood had been isolated. Being in the fifth group of such proteins isolated, it was dubbed apolipoprotein E, or simply apo E. There had been a growing number of hints from various laboratories that apo E was somehow involved in the rapid atherosclerosis that afflicted Type III patients. Perhaps, it was suspected, a defect in apo E was preventing the liver from clearing dietary fat and cholesterol from the blood, causing the two lipids to back up in the bloodstream, thereby speeding the development of their coronary disease.

To tackle apo E, Breslow in 1978 teamed up with another young

Harvard scientist, Vassilis I. Zannis, an expert in a technique for separating proteins according to their weight and electrical charge (electrophoresis). The two young scientists began isolating apo E from the blood of both Type III patients and normal volunteers and laying the proteins on Zannis's electrophoresis gels to see what differences existed that might account for the Type III disorder. Surprisingly, they found that the patients' apo E was slightly different from that of normal volunteers. More surprising was that fact that there seemed to be a confusing variety of apo E proteins scattered among volunteers and patients.

As they tried to make sense of this array of apo E proteins, Breslow and Zannis began to find the footprints of evolution. Sometime in millennia past there had been mutations of the original apo E gene. Since these mutations didn't cripple the gene and didn't interfere with reproduction, the mutant varieties persisted to this day. The mutations in the gene, of course, are reflected in the proteins produced by the gene. And Breslow and Zannis found that there seemed to be three relatively common forms of the apo E proteins, which they dubbed, respectively, E-2, E-3, and E-4. Each of these proteins is the product of a variant gene.

The E-3 version of the gene (and protein) seems to be the most common, at least among Caucasians, and the E-2 version the least common.

Since everyone has two copies of the apo E gene, one from each parent, there are six possible two-gene combinations of these apo E genes among humans. But since the E-3 version is so widespread, chances are high that most parents pass on two copies of the E-3 version to their children. Indeed, studies indicate that, in 56 percent of the population, both copies of the apo E gene are the E-3 version.

It is the rarest of the three versions, the E-2 version, that is under suspicion as an atherosclerosis susceptibility gene. Only 1 percent of the general population possesses two copies of the E-2 version. But when Breslow and Zannis checked the young adult patients with the Type III disorder in fat-cholesterol processing, they found 90 percent of them had two E-2 genes. The E-2 version of the gene is clearly implicated in these patients' rapidly progressing coronary heart disease. (Unfortunately, this isn't the total answer to the mystery of the Type III disorder. The vast majority of people possessing two E-2 genes don't develop early coronary disease. This suggests that a second mutant gene, still unknown, might also be involved in the Type III disorder.)

The E-2 gene offers an insight into how a simple genetic mutation might render a person susceptible to coronary heart disease. All three versions of the apo E protein consist of strings of 299 amino acids. The E-2 version is identical to the common E-3 version in all these amino acids except one; it has the amino acid cysteine in the spot where the E-3 version has the amino acid arginine. The substitution results from a change in a single base in the DNA of the apo E gene.

This single change in the DNA of the gene has a dramatic impact on the processing of fats and cholesterol. Ordinarily, in the early stages of fat-cholesterol processing, the fat-cholesterol globules coming in from the intestine are shuttled off to the fat-storage cells, where they dump a large part of the dietary fat. At this point the apo E protein wraps itself around the remnants of fat-cholesterol to transport them to the liver. In the liver, receptors "recognize" the apo E protein and grab the remnants for further processing.

At least, that is the way the system ordinarily works. But when people with the Type III disorder eat a high-fat meal, the remnants of fat-cholesterol pile up in the bloodstream to abnormally high levels. Moreover, it takes hours longer than normal for the remnants to be cleared from the bloodstream. It is now believed that the reason for the slow removal of the remnants from the bloodstream is the single amino acid difference in the apo E-2 protein. This tiny difference makes the apo E-2 protein unrecognizable to the receptors. The receptors either can't grab the passing remnants or they latch onto the remnants too loosely to process them at the normal rate.

The consequences of this slowed-down processing of the remnants can be disastrous. When anyone with the Type III disorder eats three meals a day, particularly if the meals are typical of the high-fat American diet, the bloodstream remains filled with the fat-cholesterol remnants throughout the day. With the coronary arteries being bathed in high concentrations of fat-cholesterol day after day, it's little wonder that the cholesterol-engorged deposits build up so rapidly in the arteries.

Breslow and Zannis were interested in far more than the apo E gene, however. They wanted to explore the entire spectrum of cholesterol-transporting proteins to uncover the genetic defects underlying the hyperlipoproteinemias. When they began their collaboration in 1978, the ability to pluck out a specific gene to examine it directly for defects was severely limited and difficult, particularly for the larger and more complex genes. Their strategy, by necessity, involved studying the proteins themselves for clues to the underlying and invisible genetic variations. When a protein is isolated, its amino acid sequence can be determined. Once this sequence is known, then it is possible to deduce the probable sequence of the DNA bases in the gene that produces the protein. Thus, examination of the protein can indirectly reveal defects in the gene.

By 1980, the year the Botstein-White-Skolnick-Davis paper was published on the riflip concept, the two scientists sensed an impending upheaval in medical genetics. "We realized we would have to learn the new DNA technology," Breslow says. They knew that to pluck out the "apo" genes they had to know how to make cDNA probes that would hybridize with a piece of an "apo" gene. At the time, making the cDNA probes was a tedious bit of biochemistry. It required about three hours to synthesize each base in a cDNA probe, and the probes were only a dozen or so bases long, just enough to grab the tip of a gene. (Today, the synthesis of cDNA probes is automated and machines can turn out probes hundreds of bases long.)

Breslow enlisted Harvard biochemist Alexander L. Nussbaum, a specialist in the nucleic acids, to help him synthesize cDNA probes that, he hoped, could hook on to the gene that produces an "apo" protein known as apo A-I. Apo A-I's function is to take the "used" cholesterol from cells throughout the body and cart it back to the liver for recycling. It is the main protein of the so-called "good" form of cholesterol, HDL (high-density lipoprotein).

Breslow returned to his own lab and began the search for the apo A-I gene. Because his team knew the sequence of amino acids in the apo A-I protein, they knew the sequence of bases in the gene. But the probes were short, only fourteen bases long, so they would hybridize with the tips of fragments of DNA that failed to contain the gene. Thus, the search was tedious, involving months of fashioning one probe after another, dropping each into a flask of human DNA, pulling a fragment, and sequencing it to see if it contained the apo A-I gene.

In February 1982, Karathanasis joined Breslow and Zannis. Within a few weeks, using one of Breslow's cDNA probes, the young Greek pulled out a fragment of DNA with a base sequence that showed it contained the apo A-I gene.

With this success in hand, the trio immediately launched a search for a second protein that is wrapped around HDL, a protein known as apo C-III. Once again, Breslow began fashioning cDNA probes, this time for the apo C-III gene. In May, Karathanasis took one of Breslow's probes and applied it to the fragment of DNA that contained the apo A-I gene. To the researchers' surprise and delight, the probe hybridized to the fragment; the fragment contained both the apo A-I and C-III genes almost side by side.

It was in the wake of this discovery that Breslow was startled to find Robert Norum's report in the *New England Journal of Medicine* describing the two sisters and their baffling but life-threatening coronary heart disease.[4]

Three years earlier the elder sister, then thirty-one, had been referred to Norum and his fellow specialists at Henry Ford Hospital in Detroit for shortness of breath and chest palpitations that had baffled her doctors. The symptoms seemed to have come on suddenly a month earlier and were steadily getting worse. Chest X-rays showed an enlarged heart, and an echocardiogram showed a poorly functioning lower left chamber of the heart.

Diagnosis of advanced atherosclerosis in a premenopausal woman would be extremely unusual. Young women seem to be protected against coronary heart disease by their sex hormones. Moreover, the woman's blood cholesterol level was in the low-normal range, well below that considered a risk for coronary disease.

Nevertheless, X-ray arteriography of her coronary arteries left no doubt; the young woman was in the advanced stages of coronary heart disease. She was immediately referred for coronary bypass surgery.

During his physical examination of the young woman Norum noticed several small yellowish bumps dotting her eyelids, neck, chest, arms, and back. The bumps, known as xanthomas, reminded Norum of similar yellowish bumps seen in people who suffered rare inherited disorders in which the blood is loaded with extremely high amounts of cholesterol. The woman said the xanthomas had been there since childhood. "She mentioned that her younger sister also had these yellowish plaques and had had heart trouble when her first child was born," Norum recalled.

The younger sister, then twenty-nine, had developed the xanthomas on her arms and eyelids when she was nine years old. But she had seemed healthy until, at age twenty-five, she delivered her second child. At the time, she developed symptoms of gradual heart failure. Her doctors diagnosed her problem as postpartum cardiomyopathy, that is, gradual heart failure induced by giving birth. Fortunately, drug treatment in the succeeding four years had improved her heart function.

The xanthomas plus the mention of the sister's heart problems immediately aroused Norum's suspicions that a genetic disorder might be involved. The younger sister volunteered to undergo X-ray arteriography. Her coronary arteries also were badly clogged.

"My partner suggested we check the blood lipids," Norum related later. The initial tests found that, even though the two sisters' total cholesterol levels were normal, the level of HDL cholesterol was barely detectable. The likelihood that a genetic disorder afflicted the sisters grew stronger. More definitive tests, looking at the various cholesterol-transport proteins, were ordered. Given what appeared to be an almost total lack of HDL, the level of apo A-I, the protein that enwraps HDL, would be of particular interest. Neither sister had detectable levels of apo A-I.

Norum and his colleagues enlisted the help of researchers in a half-dozen centers as far away as Halifax, Nova Scotia, and began a definitive study of not only the two sisters but also their immediate families. The focus of the study was apo A-I.

A picture began to emerge of a previously unknown inherited disorder. The parents, their son, and the two sisters' five children all had detectable levels of both HDL and apo A-I—but their levels were well below normal. It appeared that the parents each had one copy of a defective apo A-I gene and one copy of a normal apo A-I gene. This would account for the parents' half-normal levels of HDL and apo A-I. Their son had been lucky to the extent that he inherited only one copy of the defective gene. The sisters' children similarly had inherited only a single copy of the defective gene.

But the two sisters had lost the toss of the genetic coin and both had inherited two copies of the defective apo A-I gene.

They did have one other baffling laboratory finding. "Measurements of concentrations of several other apolipoproteins in our patients were remarkable in that apolipoprotein C-III could not be detected," Norum and his colleagues wrote in their *New England Journal* report.[5] "An absence of apolipoprotein C-III has not been reported previously." Perhaps the combination—the lack of both apo A-I and apo C-III—explained why the sisters developed early heart disease, Norum and his associates speculated.

Immediately after receiving Breslow's phone call in June 1982, Norum arranged for samples of white blood cells from the two sisters and their immediate families to be sent to Breslow's laboratory at Children's Hospital.

Karathanasis, Zannis, and Breslow took the sisters' DNA and cut it into fragments with a restriction enzyme. Then, with the aid of the new-found probes, they pulled out the fragments containing each sister's apo A-I and C-III genes. They repeated the procedure with DNA from normal subjects. The lengths of the fragments with the two genes from normal subjects and from the sisters and their families were markedly different in length. The fragments from normal subjects were roughly thirteen thousand bases long, but the fragments from the sisters were half as long, about sixty-five hundred bases. Clearly, the sisters had suffered some kind of mutation that had changed the base sequence in their Apo A-I and C-III genes, causing the restriction enzyme to cut the sisters' fragment in half.[6]

What kind of mutation could have made the sisters so vulnerable to rapid atherosclerosis? By cutting the sisters' DNA with a series of different restriction enzymes, the three molecular geneticists gradually narrowed the site of the mutation down to a particular point in the apo A-I gene. They plucked this fragment of the gene out and determined its base sequence. The fragment, they found, wasn't part of the apo A-I gene at all. Instead, it was a small piece of the apo C-III gene. In some previous generation, perhaps in the distant past, a piece of the apo A-I gene had traded places with a piece of apo C-III gene. This inversion had inactivated both genes, explaining the absence of these two proteins in the sisters' bloodstream and their almost total lack of HDL.

For the first time, there was solid evidence at the DNA level of a mutation that rendered its inheritors susceptible to premature coronary heart disease. As Breslow put it later, in more formal language, "Although this particular mutation is rare, these studies establish the principle that apo A-I gene lesions can cause low HDL levels and atherosclerosis susceptibility."[7]

The small team at Children's Hospital in Boston began to break up in 1984. Breslow accepted an appointment at Rockefeller University, where he organized the Laboratory of Biochemical Genetics and Metabolism. Zannis moved on to Boston University. Karathanasis remained at Children's Hospital as head of the Laboratory of Molecular and Cellular Cardiology.

Karathanasis, in early 1985, reported another surprising discovery.[8] Lying just below the apo C-III gene was a third "apo" gene, this one for apo A-IV. Apo A-IV's function is uncertain. The protein is made by the intestine. There are hints, Breslow says, that it may be there to grab free-floating cholesterol discharged by the tissues and cart it back to the liver for disposal, a function somewhat similar to that of apo A-I.

Breslow, Karathanasis, and others across the nation are now in the clinics, obtaining samples of DNA from heart patients and from normal volunteers. They are looking for other mutations and variations of these cholesterol-transport proteins that might predispose people to— or protect them from—early, rapid atherosclerosis.

The mutations discovered by Breslow and his colleagues cannot account, by any means, for all the millions of "premature" heart attacks that occur every year. There are more than a dozen genes involved in the transport of fat and cholesterol. Any of a dozen different mutations in any one of these genes could foul up the processing of fats and cholesterol enough to trigger the clogging of the coronary arteries. There may be hundreds of genetic variations that can predispose people to early coronary disease, some more common than others. Collectively, these genetic variations may well encompass almost everyone who suffers a heart attack or crippling angina pains.

Not long after the discovery of the defect inherited by the two sisters, Breslow made a remark that is likely to be echoed often in the future: "It's pretty awesome to think of a single gene abnormality that can accelerate the age for a heart attack by fifty years."

CHAPTER 10

The Map

few in biology or medicine were more delighted—and surprised—
by the discovery of the riflips than Victor A. McKusick at the Johns
Hopkins Medical Institutions in Baltimore. The riflip map proposed by
Botstein, White, Skolnick, and Davis was the key to a forty-year-old
dream of McKusick's, the mapping of the human genome.

McKusick is a tall, bony man who looks like what he is, the son
of a Maine dairy farmer. He arrived at Johns Hopkins on Washington's
Birthday in 1943 in the midst of World War II, with three years
of pre-med at Tufts University and plans in his head to become a
small-town physician. He had become enamored of medicine eleven
years earlier when, at age fifteen, he was forced to spend ten weeks
in Massachusetts General Hospital in Boston while doctors fought to
cure a stubborn bacterial infection that had created an abscess in his
armpit. "Sulfanilamide was just coming in at the time and that finally
saved the day," he recalls. (Sulfanilamide was the first of the sulfa
"wonder drugs" that preceded penicillin in use.) During the long,
boring days at one of the country's biggest teaching hospitals, the boy
got an inspiring inside look at the famous Harvard medical faculty at
work and decided to pursue medicine as a career. "There also was a
local physician [Guy Dore back in Guilford, Maine] whom I admired
greatly, so I went to medical school fully intending to return to Maine

to practice medicine," he says. But Johns Hopkins and its medical faculty, which was at least as famous as Harvard's, was too alluring. "Gradually, I just didn't leave. I became involved and enthralled with academic medicine," he says.

Genetics had fascinated McKusick since his undergraduate days at Tufts. At Johns Hopkins he seemed to encounter the consequences of damaged genes at every turn. During his medical internship, "I ran up against a new hereditary syndrome, characterized by intestinal polyps and melanin spots on the lips, inside the mouth and on the fingers," he recalls. His study of five patients who had inherited these strange marks of errant genes led to his being a coauthor of a key paper on the newfound syndrome—Peutz-Jeghers syndrome—published in the *New England Journal of Medicine*, a cap-feather for any fledgling young physician.

Genetics, however, was considered strictly a basic biological science. While it was important that physicians be able to recognize and diagnose a genetic disorder, the actual study of human genetic diseases wasn't considered a distinct field of medicine. Instead, each genetic disorder was left to the medical specialty best trained to treat its particular symptoms: muscular dystrophy was the purview of the neurologists, hemophilia the hematologists, and so on.

There were good reasons why the paths of medicine and genetics, while touching each other tangentially, hadn't yet converged. Unlike the biologists, whose science of genetics was based on experiments with fast-breeding fruit flies, mice, and corn plants, medical researchers could hardly resort to the inbreeding and crossbreeding of humans just to study the passage of human genetic traits. The study of human genetics consisted of little more than charting family pedigrees to see whether an inherited disorder was caused by a dominant or a recessive gene.

But the biggest obstacle separating medicine and human genetics was a seemingly mundane problem. Early on, the biologists had discovered how to stain the chromosomes of fruit flies and corn plants to produce distinctive bands of shading that could be seen under the microscope. The shading patterns enabled them to distinguish one chromosome from another. More importantly, by studying the changes in the shading patterns from one generation to the next, the biologists could see mutations where bits of chromosomes had been deleted, added, or rearranged—and they could see the consequences. Thus, the biologists would tell which of the fruit fly's four pairs of chromosomes contained, say, the gene or genes controlling wing shape or wing color or dozens of other inherited traits. Moreover, they knew from the changes in the shading patterns approximately where these particular genes lay on the chromosome. In this manner the biologists were able to compile rough maps of fly and corn chromosomes showing approximate locations of a variety of genes.

"In my college genetics course I was tremendously impressed with the chromosome maps you could draw for drosophilae [fruit flies] and for maize [corn] and which appeared in the genetic texts," McKusick recalls. "It always seemed to me that, even if you had no perception that it would be of applied value, it would be intellectually very satisfying if you could map human chromosomes."

But human chromosomes remained as difficult to study as the distant stars. A dye had been found that would turn the chromosomes a dark red so their silhouettes could be seen in the microscope. But the human chromosomes remained impervious to the stains that brought out the shaded patterns seen in the chromosomes of fruit flies and corn plants. When researchers looked at the chromosomes in the nuclei of human cells, they usually saw only a tangled mass of dark ribbons that literally resembled a can of worms. Human genetics, per se, offered little to the young intern who was interested in the treatment of disease.

To pursue a medical career, McKusick had to pick a recognized speciality. He chose cardiology and was soon a leading expert in the sounds of the heart. But inherited diseases continued to intrigue him. As a cardiologist, he soon found himself deep into the study of Marfan's syndrome, an inherited disorder of the connective tissues that often causes defects in the heart and blood vessels, particularly a weakening and eventual rupture of the aorta, the body's main artery. The Marfan's research, in turn, led him to other inherited disorders of the connective tissues, such as Ehlers-Danlos syndrome, the "India rubber man" disorder in which the skin is remarkably elastic. His first book, in 1956, was *Heritable Disorders of Connective Tissue*.

"People often ask me when I gave up cardiology for genetics," McKusick notes. "The fact is my activities in genetics and cardiology developed in parallel. Later it was more a matter of cardiology being phased down and genetics being phased up."

What prompted McKusick's "phase-up" of genetics was the development in the mid-1950s of cytogenetics. Scientists who specialized in the study of cells, the cytologists, began developing a number of tools that suddenly opened the way to study the chromosomes. Methods were found to separate the chromosomes from other parts of the cell's genetic apparatus and to lay the chromosomes out in an orderly way—karotyping—so they could be studied under the microscope. A technique for grabbing the chromosomes at different stages of cell division also was developed. The new techniques paid off in 1956 when the number of human chromosomes was confirmed to be forty-six (twenty-three pairs) instead of forty-eight, an error that had been made thirty years earlier because of the inability to distinguish the individual chromosomes.

The new cytogenetic tools meant that, for the first time, geneticists could correlate abnormalities in human chromosomes with genetic disease. A new science of medical genetics, embracing the study of human genetic disease at both the patient and the chromosome levels, formally emerged in 1957 when two departments of medical genetics were founded, one at Johns Hopkins under McKusick and the other at the University of Washington in Seattle under Arno Motulsky, a hematologist by training. "Both of us had established credentials in a [medical] specialty before we moved into this terra incognito of medical genetics and I have thought in retrospect that we undoubtedly felt more secure and had some credibility within the profession and the academic world because of those credentials," McKusick says.

This new science was quickly justified by a startling discovery in late 1958.[1] Six years earlier, a young French geneticist, Jerome Lejeune, had begun studying the chromosomes of patients afflicted with Down's syndrome, a form of mental retardation accompanied by physical abnormalities that include slanted eyes and a flattened nose. John Langdon Down, who had first described the syndrome, believed the victims were evolutionary "throwbacks" to an earlier Mongolian race and called the syndrome "mongolism."

Lejeune, however, was among those who suspected Down's syndrome was hereditary. The young geneticist launched a study trying to relate the Down's victims' physical characteristics to abnormalities in the chromosomes. In 1958, with the aid of Marthe Gauthier, an expert in the test-tube culture of cells who was able to wield some of the new cytogenetic tools, Lejeune looked at the chromosomes from three Down's patients, expecting to find one chromosome missing, as was often the case of deformities in fruit flies. Instead, he counted forty-seven chromosomes rather than the normal forty-six. The extra chromosome, he concluded in a paper published in early 1959, was the cause of Down's syndrome. (Within a year, it was confirmed that the extra chromosome was indeed the cause of Down's syndrome, that Down's patients had three copies of chromosome 21 instead of the usual two.)

In succeeding months chromosomal abnormalities underlying other disorders were discovered. Women with Turner's syndrome—short stature, thickened neck, undeveloped breasts and immature ovaries— were found to have only one X chromosome instead of the two possessed by normal females. At the same time, boys with Klinefelter's syndrome—tall, with enlarged breasts and small testes—were found to have an extra X chromosome and, thus, were XXY males instead of XY.

As McKusick would later remark, with the new tools of cytogenetics opening up the study of the chromosomes, "the clinical geneticist acquired 'his organ,' just as the cardiologist had the heart and the neurologist the nervous system."[2]

Any hopes that human genes could be mapped to specific chromosomes remained unfulfilled, however. The exceptions were the genes on the X chromosome, or more accurately, the defective genes on the X chromosome. Because certain obviously inherited disorders showed up exclusively in the males of a family, such as color blindness, hemophilia, or muscular dystrophy, the defective genes causing these disorders were automatically assigned to the X chromosome.

The cytogeneticists were unable to say exactly where on the X chromosome each of these defective genes lay. But they could gain an idea of the genes' positions relative to each other. This was done by linkage studies. One of McKusick's first projects as head of the new medical genetics division in the department of medicine, for example, was to find the linkage between the color-blindness gene and the gene for an enzyme called G6PD (for glucose-6-phosphate dehydrogenase). In 1958, a Johns Hopkins geneticist, Barton Childs, had tracked the G6PD gene to the X chromosome by virtue of the fact that in certain families a deficiency in this enzyme afflicted only the males. A mild form of this deficiency renders its possessor vulnerable to anemias when exposed to a variety of chemicals and drugs. The G6PD deficiency is found in the males of about 10 percent of American black families.

McKusick and two of his new breed of medical geneticists, Ian Porter and J. Schulze, decided to find out how close the G6PD gene was to the color-blindness gene, which also was known to be somewhere on the X chromosome. For this project "we needed families that had both color blindness and G6PD deficiency," McKusick recalls.

"We started out by going into the Baltimore schools and doing color-blindness tests on black boys," McKusick relates. Out of 3,648 black schoolboys who were tested, they found 134 who were color blind.[3] These 134 color-blind boys then led Porter and his colleagues to ten families in which some of the male children had both color blindness and G6PD deficiency. Other boys in these families had only one trait, either color blindness or G6PD deficiency, so it was clear that the two genes sometimes separated. The problem McKusick, Porter, and Schulze faced was to determine how often the two defective genes had traveled together and how often they separated during the formation of the mothers' ova. Thus, they needed to know whether the two defective genes were linked together in the mothers. Unfortunately, the mothers were of little help, since the women, having a normal X chromosome as well as the presumptively defective one, were neither color blind nor deficient in the G6PD enzyme. The geneticists tracked down the mothers' fathers and tested them for color blindness and G6PD deficiency.

By tracing the inheritance of the two defective genes through three generations—fathers to daughters to grandsons—they found that, among the families where the mother carried both genes, nineteen of every twenty sons who inherited the defective X chromosome were both color blind and deficient in G6PD. In 5 percent of the boys, the two genes had separated and they had inherited just one defect, either the color-blindness gene or the G6PD deficiency. Such "tight" linkage indicated the genes must be physically quite close to each other on the X chromosome, though exactly where the pair lay on the chromosome wasn't known.

Thus, gene mapping, such as it was, was limited almost solely to the X chromosome. "By 1968, there were approximately 68—the number is easy to remember—genes that were X linked, that is located on the X chromosome," McKusick recalls.

But 1968 was memorable for another reason: It was the year that one of McKusick's first doctoral students in medical genetics mapped for the first time a gene to an autosome, one of the twenty-two pairs of non-sex chromosomes.

The student was Roger P. Donahue. Donahue grew up in an academic family, spending his childhood around college campuses, particularly that of Texas A&M at College Station, where his father was an agronomist. From childhood he intended to become a scientist: he obtained his bachelor of science degree with a major in biology from the University of New Hampshire. Like many young college graduates, Donahue wasn't sure he had made the right choice of careers and in his uncertainty he enlisted in the air force, where he spent the next four years as a meteorologist. After his enlistment was up, "I applied to various graduate schools but I wanted Johns Hopkins because they could pay more on their fellowships," recalls Donahue, who now directs the chromosome laboratory for the pediatrics department at the University of Miami Medical School in Miami.

It was at Johns Hopkins while working on his doctorate in biology that he heard about McKusick's new doctorate in medical genetics. "I was the first one to go through the program," he says. "You had to spend your first two years in medical school taking the same basic science courses as future doctors. Then when the doctors went off to begin working with patients, we went to the laboratories.

"When I went to the chromosome lab, the first thing I did, like everyone else, was examine my own chromosomes," Donahue says. "In my case the number 1 chromosome looked different." Chromosome pair number 1 was easy to identify even in those days of being able to see only the dark red silhouettes: it was the longest of the chromosome pairs (which is why it is designated chromosome 1), and its upper and lower arms, as demarked by the pinch near the center, were of equal length, unlike the other twenty-two chromosome pairs.

One of Donahue's number 1 chromosomes looked as though it was coming unraveled. "This chromosome differs from the usual no. 1 by having a greater over-all length and a greater length of the long arm," Donahue and his colleagues would write later.[4] A section of the chromosome immediately below the pinched center was unusually long and thin "with, occasionally, an alternating pattern of light and dark staining, suggestive of coils." The lower half of the chromosome, in other words, looked like a string coming uncoiled.

"We didn't know in the sixties what was a normal variation [in the chromosomes] and what was abnormal," Donahue explains. A check of the literature, however, found reports of several other families in which a similar partially unraveled chromosome had been seen, suggesting that the "uncoiler element," as the strange configuration was dubbed, was a fairly normal or at least a benign chromosome irregularity. There seemed to be no obvious trait, disorder, or defect associated with the uncoiler element.

"Donahue had both the wit and the gumption to do a family study,

Roger Donahue's "uncoiled" chromosome 1 *(left)* beside a normal chromosome 1.

realizing this might be a heritable variation in chromosome 1," McKusick says.

Donahue's parents happened to be working in India at the time, but his mother was scheduled for a home visit. At his son's request, the father had a skin biopsy taken and put in tissue culture. Mrs. Donahue flew back to Baltimore, carefully carrying the package containing the iced-down tissue culture in her lap. In Baltimore she delivered the tissue culture to her son and donated a vial of her own blood.

"I looked at both parents' chromosomes and found that my mother was the carrier of the uncoiler element," Donahue says. "Mother arranged to draw blood from her brothers and sisters. I had two sons at the time and a third was born while the research was in progress." Thus, Donahue had three generations in which to track the inheritance of the strange uncoiler element. Of the eighteen blood relatives—Donahue, his mother, six aunts and uncles, five cousins, a brother, a sister, and three sons—ten had one copy of the uncoiled chromosome 1 and a cousin was a presumed carrier of it, since the cousin's son had the uncoiled chromosome 1.

All of this, of course, was interesting to Donahue and his family but hardly anyone else—until Donahue realized he could use the uncoiler element as a genetic marker for chromosome 1. Any gene that was found linked to the uncoiler element also would be on chromosome 1.

This realization brought two other scientists into the study of Donahue's strange maternal chromosome. One was immunogeneticist Wilma B. Bias, an expert in using the tools of immunology, such as antibodies, to study human genetics. The other was population geneticist James H. Renwick of the London School of Public Health, who spent half of his time in those days in Baltimore working with McKusick, doing the statistical analyses that are the backbone of genetic linkage studies, and the other half of his time in London.

"We said, 'Let's see if something else is linked to the uncoiler element,'" Donahue says. "It was kind of a shot in the dark." Without the vaguest idea of what genes might lie on chromosome 1, the group combed through all the genes known to be polymorphic, that is, genes that vary in one way or another from individual to individual. They settled on checking the blood groups. Human blood can be categorized or grouped by the proteins on the surface of blood cells. Each surface protein is the product of a gene, hence each blood group is a genetic marker. Moreover, there is wide variation, polymorphism, in the blood group proteins among humans. The best known of these polymorphic blood groups, of course, are A, B, AB, and O, the blood groups or "types" that hospitals and blood banks need to know to make sure a donor's blood is compatible with that of a recipient. But there also are a score or more of other human blood groups, some of them rare but many of them common, which can vary from individual to individual but are of little import in the transfusion of blood. Among these other blood groups is one known as the Duffy group, consisting of two common groups, Duffy-a and Duffy-b, and a few rare groups.

Donahue and Bias began testing Donahue's maternal relatives for the various blood groups, while Renwick worked out statistical analyses to see how closely each blood group was linked to the uncoiler element. It turned out that Donahue's own uncoiler element had come from his mother accompanied by a gene for the Duffy-a blood group. His brother also had inherited the same combination, uncoiler element plus Duffy-a, from the mother. But their sister had inherited a gene for the Duffy-b blood group from the mother, and she lacked the uncoiler element. The uncles, aunts, and cousins showed the same pattern: Everyone who possessed the uncoiler element also possessed at least one copy of the gene for the Duffy-a blood group. And every member who lacked the uncoiler element also lacked the Duffy-a blood group gene. The Duffy-a gene appeared to be linked to the uncoiler element, whatever it was.

The Johns Hopkins trio quickly turned up two other families in whom they could track the inheritance of the uncoiler element and the Duffy-a blood group in two generations, from a parent to three children in each case. These families, too, hinted that the two markers were linked. Renwick calculated that there was a 90 percent chance that the genes for the Duffy blood group lay on chromosome number 1, linked fairly tightly to the mysterious uncoiler element.

"In this paper we report what we believe is the first assignment of a specific gene locus to a specific autosome in man," began the paper that the Johns Hopkins team rushed off to the *Proceedings of the National Academy of Sciences* in September.[5]

Within months, a British team under Renwick's guidance mapped a gene to chromosome 16 and shortly afterwards an American team mapped a second gene to chromosome 16. In both cases, the genes were linked to an anatomical irregularity on chromosome 16—a visible translocation in one family and a threadlike structure called a "fragile site" in another family.

Even as Donahue was checking his family's chromosomes, another development that would play a role in the mapping of the human genome was taking place in New York. In the early 1960s, biologists had learned how to fuse human and mouse somatic cells to create mouse-human hybrid cells. In 1967, Mary Weiss and Howard Green at New York University found that, as these hybrid somatic cells divided through new generations, they began to lose their human chromosomes. Eventually, there was a collection of hybrid cells, each of which had only one human chromosome or, sometimes, just a piece of a human chromosome.

Weiss and Green used this phenomenon to isolate a chromosome with a specific gene.[6] They took a mouse cell that was unable to make an enzyme called thymidine kinase and fused it with a human cell that could make the enzyme. They then let the cells multiply in a nutrient that was deadly to any cell that lacked thymidine kinase. After several generations, all the cells died off except one clump of identical cells (a clone) that was producing thymidine kinase. The cells in this clone all contained the same single human chromosome. Weiss and Green reasoned quite logically that the human chromosome in the cells must contain the human gene for thymidine kinase. Unfortunately, it was impossible to identify the particular chromosome, so the human chromosome with the thymidine kinase gene remained anonymous—but only for a year.

The anonymity of human chromosomes was suddenly stripped away in 1968–70 by Torbjorn O. Caspersson at the Karolinska Institute in Sweden. Caspersson, a physician-turned-cytochemist, was already famous. In the early 1930s, he developed techniques for studying the nuclei of cells with the aid of ultraviolet light. This led him in 1940 to discover that DNA was concentrated in the chromosomes, thereby providing the first convincing evidence that DNA was the substance of the genes. Caspersson continued to experiment with techniques for visualizing DNA, and in the 1960s he began an effort to detect DNA differences that distinguished cancer cells from normal cells.[7] To find such differences, Caspersson first had to look for aberrations in individual chromosomes. Unfortunately, "the way appeared to be blocked by the deplorable state of chromosome identification in man at that time," he recalled years later. "Without an absolutely reliable method to recognize every individual chromosome, any effort toward a detailed chromosome analysis would be meaningless."

At the time, cytologists could identify human chromosomes only by differences in length. The three longest chromosomes were easy to spot individually and thus, were labeled chromosomes numbers 1, 2, and 3. The remaining twenty human chromosomes were categorized by length groups, group B containing two chromosomes of approximately equal length, group C eight equal-length chromosomes, and so on.

In 1968, Caspersson and his colleague, Lore Zech, hit upon a novel idea to solve this chromosome identification problem. They reasoned that, since genes had different functions, they must differ in their concentrations of each of the four base nucleotides, G, A, T, and C. They cast about and picked a chemical, acridine quinacrine mustard, which had an affinity for the G base. Equally important, quinacrine mustard glowed under ultraviolet light. If their assumption was correct, a chromosome stained with quinacrine mustard and viewed under ultraviolet light would glow in a pattern of bright and dim spots reflecting high and low concentrations of base G.

Indeed, that's exactly what happened. When the treated chromosomes from a single human cell were spread under an ultraviolet lamp, a spectacular array of bright and dark bands jumped out on each chromosome. In 1971, after studying the banding patterns on more than five thousand chromosomes, Caspersson was able to identify each human chromosome by its banding pattern. (Almost immediately, the anonymous chromosome containing the human thymidine kinase gene was identified as number 17 by Barbara Migeon and C.S. Miller at Johns Hopkins.)

McKusick's dream of mapping human chromosomes using banding patterns suddenly became reality. A host of other DNA stains were discovered in ensuing months, and by the early 1970s, cytogeneticists were able to easily identify each of the chromosomes by any of several banding patterns. The science of cytogenetics suddenly blossomed. Once the normal banding pattern was recognized, it was only a small step to pinpointing alterations in the normal arrangement of the bands. Deletions, additions, translocations, and duplications of bits of particular chromosomes suddenly became visible. By studying such alterations in the banding patterns of specific chromosomes and relating them to specific traits or genetic disorders, the cytogeneticists could get a rough idea of where a gene lay on the chromosome, that is, whether it lay in or near a particular band. If one gene could be located near a particular band and it was known, from family studies, to be tightly linked to another gene, then the second gene's approximate location also became known.

With the banding patterns, a genetic coordinate system began developing. In 1966, it had been decided to designate the long arm of each chromosome as "q" and the short arm as "p." An international conference in 1971 officially adopted the numbering system developed by Caspersson. In addition to numbering each chromosome, the bands on each chromosome were designated by numbers, beginning near the pinched center and working outwards along each arm. (Thus, when the genes for the ABO blood group were mapped to chromosome 9, the coordinates could be given as 9q34, that is, on chromosome 9, the long arm, region 3, in band number four, which is at the very tip of chromosome 9.)

The discovery of the banding patterns opened the gate to the mapping of genes to particular chromosomes. By 1977, McKusick and fellow geneticist Frank H. Ruddle of Yale could write: "In the last seven years new methods for staining the chromosomes ('banding techniques') have revealed distinctive regional landmarks permitting unique identification of each chromosome.

"Thus, in exploration of the genetic planet that is the cell nucleus of man, the broad outlines of the continents and some of the gross details of their topography have been known for some time. It is only since 1968—and mainly in the last five years—that cartographic details of the chromosomes and regions thereof, have been determined."[8]

McKusick had begun to keep track of the number of known genetic disorders and in 1966 had published the first edition of his catalogue, *Mendelian Inheritance in Man*. In that first edition slightly fewer than fifteen hundred genes were identified by their phenotype, that is, by their end product, the protein they produced or the disease they caused

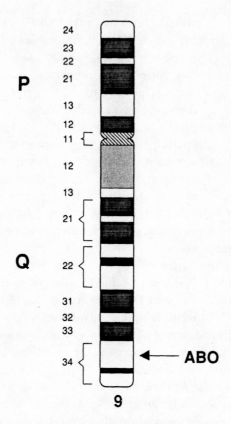

Chromosome 9 showing the location of the
ABO blood-type genes at 9q 34

(as, for instance, the Huntington's disease gene).[9] Only sixty-eight of these had been "assigned" to the X chromosome and, of course, none to any specific autosome. By 1973 the number of known genes had jumped to more than 2,300, of which 152 had been mapped to specific chromosomes (64 to the autosomes and 88 to the X chromosome), thanks to the discovery of the banding patterns.

Obviously, there was a growing need for a cartographic bureau, a central agency to keep track of the genes as they were mapped. McKusick's pioneering project, *Mendelian Inheritance in Man*, as extensive as it was becoming, was primarily a catalogue of genetic traits and disorders, described in detail. It included each gene's location on the chromosomes, if known, but it wasn't a gene map per se.

The task of compiling and continuously updating a gene map was taken on by Ruddle at Yale. Unlike McKusick, who entered genetics by way of medicine, Ruddle was a biologist, a native New Jerseyite, born in 1929, who did his undergraduate work in biology at Wayne State University in Detroit and obtained his doctorate in the science from the University of California at Berkeley.

Standing at six feet seven inches, Ruddle tends to physically overwhelm people at first meeting. Despite this, or perhaps because of it, he quickly and easily disarms new acquaintances with a gentleness and politeness of a past age, listening with seemingly rapt attention to their questions or statements and answering even the most naive questions with sincerity. This manner, laced with a kind of paternal concern, brings out the best in his students.

"I think he has a real genius for getting the best out of people who worked with him," says one of Ruddle's former postdoctoral students.[10] "He would be critical of the work but in a very helpful and constructive way. No one ever felt antagonized by his suggestions." Adds McKusick: "He is very clever in devising strategies for mapping the chromosomes of man. But those strategies wouldn't do him much good if he didn't have a large clutch of graduate students who want to work with him and provide essential manpower so that he can get something done."

In the late 1960s, Ruddle and his cadre of young biologists at Yale were deep into the use of somatic cell hybrids to isolate individual chromosomes. When Caspersson's banding technique burst upon the cytogenetic world in 1968–70, Ruddle's laboratory was one of the few that had the experience and the setup to instantly take advantage of the banding patterns to identify the chromosomes culled out by the mouse-human hybrid cells.

Within months after the 1971 Paris conference that officially numbered the human chromosomes and adopted the coordinate system, Ruddle began planning the first international workshop on gene mapping. Funded by the National Foundation-March of Dimes (now the March of Dimes Birth Defects Foundation), the workshop was held in New Haven in June 1973. "People already were pretty excited [about the new banding techniques], so we had a good number of people willing to work" on a gene mapping project, Ruddle says. The workshop attracted more than fifty scientists and resulted in reports of about fifty newly mapped genes, he recalls. The reports, kept in file folders, one for each of the twenty-four chromosomes (the twenty-two autosomes plus the X and Y chromosomes), became the New Haven Human Gene Mapping Library.

"When the workshops started in 1973, only 150 genes were mapped to specific chromosomes, and Thomas Shows of the Roswell Park

Memorial Institute in Buffalo, N.Y., chairman of the nomenclature committee, could remember the precise name and symbol of each," records science writer Maya Pines in a short history of the human gene mapping effort.[11] The entire human gene map, consisting of a schematic drawing of each chromosome showing the bands and the approximate location of each of the 150 genes, took up one printed magazine-sized page.

By 1980, five of the gene mapping workshops had been held and the number of mapped genes had tripled to more than 450, thanks largely to the new ability of the banding patterns to identify the chromosomes isolated by the somatic cell hybrids. The New Haven Human Gene Mapping Library, now financed by the federal National Institutes of Health, was computerized and accessible to anyone with a computer and a modem.

But it was the publication of the Botstein-White-Skolnick-Davis and Wyman-White papers in 1980, describing the riflips and their use as benchmarks to map genes, that sparked an explosion in the mapping of genes. Anyone wishing to locate a gene for a known trait could do so by finding out if a riflip of known location on a chromosome was inherited in tandem with the trait. If so, the gene for the trait must be tightly linked to the riflip. The faster that new riflips were located along the chromosomes, the faster new genes were mapped by this linkage. Within two years, the number of mapped genes rose to 600 and by the ninth international human gene mapping workshop (HGM-9) in 1986, almost 1,500 genes had been mapped to specific chromosomes. "Every night, during the HGM-9 meeting, Victor McKusick entered about 50 additions or changes into his own database of disease related information," Pines notes,[12] referring to McKusick's pioneering catalogue of disease-causing genes, *Mendelian Inheritance in Man*.

The advent of the riflips, however, added a new complexity to the gene map. The original Botstein et al. paper was essentially a proposal to build a genetic linkage map that would pinpoint genes tightly linked to riflips of known location. Obviously, a map showing the location of the riflips would have to be compiled first. Botstein, White, Skolnick and Davis had calculated that, for a riflip and a gene to be tightly linked, they would have to be within ten million DNA bases of each other. To be fully useful, a linkage map would have to identify a riflip every twenty million bases. They estimated that 150 riflips so spaced would cover the entire genome.

White had moved to the Hughes Institute at the University of Utah specifically to utilize the extensive Mormon families in compiling a riflip-based linkage map. He needed to study a proposed riflip's inheritance through at least three generations of a large family to be sure that it could be used as a benchmark for locating genes. "His strategy was to collect blood samples from 46 normal families of 'perfect' structure from his point of view (three generations, including four living grandparents and at least six—and sometimes 16—children), cut up their DNA with every available enzyme, and look for polymorphisms with every available probe," Pines explains.[13]

With his large collection of Mormon DNA, White became a major asset to an international cooperative effort to build the human genome linkage map. This international effort was centered around Jean Dausset at the Centre d'Etude du Polymorphisme Humain, or CEPH, at the College de France in Paris.[14] Dausset set up CEPH as a kind of central repository for DNA samples of multigenerational families from around the world. If any researcher anywhere found a probe that he suspected identified a riflip, he could borrow from CEPH a complete set of DNA samples from a three-generation family. He could then test his probe to see if the riflip it sought out was polymorphic and useful in tracking the inheritance of DNA through the generations. CEPH saved the researcher from having to seek out his own multigenerational family. The only requirement was that the researcher had to test his proposed riflip probe on all the DNA of all the families in the CEPH repository, not just that of one family, and he had to give both his probe and his results to CEPH. By mid-1985, CEPH had complete DNA samples from forty large families, twenty-seven of whom were Mormon families whose DNA samples were supplied by White.

With access to complete panels of DNA from both the Mormon and the CEPH families, White and a colleague at Utah, population geneticist Jean-Marc Lalouel, rapidly began drawing a riflip map of the human genome. In September 1987, White reported to the ninth human gene mapping workshop in Paris that he was nearing completion of a full riflip map. He had DNA probes for 475 riflips of known location. Although far more riflips than the 150 mentioned in the Botstein et al. paper, the riflips weren't evenly spaced along all the chromosomes but, instead, were scattered unevenly along only seventeen of the chromosomes. Nevertheless, it was the major part of a map that ultimately would be "an atlas, a Gray's anatomy, of our genetic makeup."[15]

A month later, a small biotechnology company, Collaborative Research Inc. in Bedford, Massachusetts, announced with considerable fanfare that it and researchers at MIT's Whitehead Institute had compiled "the first human genetic map" consisting of 404 riflips of known location, with at least one riflip on each chromosome except the Y chromosome.[16] In its annual report to stockholders Collaborative asserted, "We published the world's first primary map of the entire human genome, garnering a coveted cover-featured article in the prestigious journal *Cell* and received worldwide scientific acclaim for this accomplishment."[17] The potential use of the riflip probes "to diagnose the common diseases exposes us to a massive commercial market," the report added.

Collaborative's claim stirred White's anger. "We have more markers than Collaborative has but we don't think it's appropriate to call it a map of the whole genome," White told David Stipp of the *Wall Street Journal*.[18] "They still have holes in their map just like we do. That makes it like an atlas of the U.S. in which there's no coverage of Florida, Georgia and California." To top it off, Collaborative had used the CEPH bank of familial DNA, including some samples from White's Mormon families to make its map.

The semantics of the tiff, however, were more important than its substance. For the first time, there was talk of an organized, concerted attempt to map the entire human genome. McKusick, Ruddle, and other early gene cartographers had thought originally in terms of a gene map building up as the collective result of individual decisions by geneticists and molecular biologists to locate genes of particular interest. "Mapping the human genome, like any uncharted terrain, is a challenge to the human intellect," McKusick and Ruddle wrote in their 1977 paper in *Science*.[19] But at no point in this lengthy paper do the two gene mapping pioneers suggest any formally organized effort to map the genome.

Behind a change in this thinking was an audacious proposal that suddenly mushroomed out of the U.S. Department of Energy in 1986 and 1987. The proposal was to launch nothing less than a formally organized, federally financed effort to determine the sequence of all three billion G, A, T, and C base pairs that comprise human DNA, regardless of whether bases were parts of genes or not.

It's not quite clear where the sequencing project was conceived. One version of the story, by science writer Stephen S. Hall, traces its origin to an informal meeting of biologists in May 1985 at the University of California in Santa Cruz.[20] It was organized by the university's chancellor, Robert Sinsheimer, himself a molecular biologist, who had expressed some envy of the ability of physicists and astronomers to organize and gain funding for "mega-projects" such as multi-billion-dollar atom smashers and space telescopes.

Botstein was among those present at the Santa Cruz meeting. But two of the key participants were Walter Gilbert of Harvard and Leroy Hood of California Institute of Technology. Both of these biologists were deeply involved in "sequencing" DNA. In 1975, Gilbert and his Harvard colleague Allan Maxam had worked out a method for determining, one by one, the sequence of the bases in a strand of DNA. For this, he and Maxam shared the 1980 Nobel prize with Frederick Sanger of Cambridge, England, who had worked out a somewhat different method of sequencing DNA. Hood in 1985 had succeeded in developing and building a machine that would automatically sequence a stretch of DNA.

At the meeting, Sinsheimer posed the question of whether it would be possible to sequence the bases of the entire human genome. Botstein was skeptical of any such "mega-project" ever getting done. But Hood quickly picked up Sinsheimer's enthusiasm for a massive genome sequencing project. Gilbert was dubious at first. "In the course of the meeting I changed my mind and decided that in fact the technology was there and the project could be done," he later told Hall.[21] "And let it be said," Hall added, "that, for better or worse, once Gilbert believes something can be done, the devil take the hindmost. Gilbert began firing off memos to Sinsheimer about a 'human genome institute.' He had it all figured out: the number of staff (120), the annual operating budget ($10 to $12 million to start), even the square footage of the lab space (25,000)."

The idea of sequencing the genome also occurred about the same time to biophysicist Charles DeLisi at the U.S. Department of Energy.[22] ". . . DeLisi recalls looking up from a Government report on biotechnology one day and thinking that if you could only compare the genome of a child with those of its parents, DNA base by DNA base, you would have an unparalleled medical and research tool for studying mutations," Robert Kanigel reported later in the *New York Times Magazine*.

The Energy Department had a logical interest in sequencing DNA, DeLisi felt. It owned the four big National Laboratories that had evolved from the World War II atomic bomb project. The laboratories had experienced and talented cadres of biologists whose study of the biological effects of nuclear radiation had plunged them deep into genetic mutations, which, of course, are essentially alterations in the sequence of DNA bases. Thus, the DOE biologists were already deeply involved in sequencing stretches of DNA. The Los Alamos National Laboratory near Santa Fe, in fact, was the location of GenBank, an international data base of all the known sequences of DNA and RNA.

The sequencing project, however, had another, unmentioned appeal to the DOE. The department had been born in the energy crisis of the early 1970s, taking over and rapidly expanding the research function of the old Atomic Energy Commission. The federal purse-strings were opened wide to the newborn department in the Nixon Administration's much-touted push for energy independence. By the mid-1980s, the crisis had disappeared. Petroleum and natural gas were in oversupply and their dropping prices had scuttled the economics of the DOE's big energy projects such as the liquefaction and gasification of coal. As public and congressional interest in energy waned, so did the department's funds and future. The DOE was casting about for imaginative new projects that could appeal to the Congress and the public. An all-out effort to sequence the genome had the kind of scale and prestige the DOE needed.

DeLisi organized a meeting of scientists interested in DNA sequencing, held in March 1986 in Santa Fe, a few miles from the Los Alamos laboratory. Gilbert fired up the Santa Fe meeting with this declaration: "The total human sequence is the grail of human genetics."[23] The sequence, he declared at one point, was "the ultimate answer to the commandment, 'Know thyself.'"[24] The fifty or so scientists left the Santa Fe meeting with a new enthusiasm for the sequencing project.

The sequencing project might have languished in the DOE bureaucracy had it not been for Renato Dulbecco, a Nobel prize-winning cancer researcher at the Salk Institute in La Jolla, California. The week of the DOE's Santa Fe meeting, *Science* published an editorial by Dulbecco titled, "A Turning Point in Cancer Research: Sequencing the Human Genome." Claims McKusick, ". . . perhaps more than any other single factor, the editorial galvanized the scientific community and even the public, and also polarized the scientific community to some extent."[25]

Citing the recent discovery of the so-called oncogenes, genes that must be active before a cell can turn malignant, Dulbecco declared, "If we wish to learn more about cancer, we must now concentrate on the cellular genome."[26] In order to understand the genome, he wrote, "We have two options: either to try to discover the genes important in malignancy by a piecemeal approach, or to sequence the whole genome of a selected animal species." The "piecemeal approach" of finding "the genes important in malignancy" was essentially the McKusick-Ruddle-White gene mapping approach adapted to finding oncogenes and tumor suppressor genes. The drawback to this approach, Dulbecco argued, was the likelihood that the oncogenes would vary widely, being different in different tumors, in different organs, and in different species. To find such wide variety of oncogenes would "require a vast investment of research," he said.

"I think it would be far more useful to begin by sequencing the cellular genome," Dulbecco asserted. Once the sequence of the bases is known, then probes for any and all genes could be prepared and researchers could then determine which genes were essential to cancerous growth. "If we wish to understand human cancer, it [the sequencing effort] should be made in humans because the genetic control of cancer seems to be different in different species," he declared.

Finally, Dulbecco wrote, "An effort of this kind could not be undertaken by any single group; it would have to be a national effort. Its significance would be comparable to that of the effort that led to the conquest of space, and it should be carried out with the same spirit."

Dulbecco laid out the two gene hunting strategies at the center of what was to become a most contentious debate: gene mapping versus DNA sequencing. The mappers were the Lewises and Clarks who wanted to reconnoiter the genetic continent and locate the valleys, lakes, rivers, and fields (i.e., genes) that could be exploited. The sequencers wanted to walk across the genetic continent with transit and chain, recording the topography of every measured mile on the assumption that eventually the continent's detailed topography would reveal its secrets.

The first clash between the mappers and the sequencers occurred three months later at a meeting at Cold Spring Harbor. During the meeting, Gilbert wrote on the blackboard, "$3 billion." It was the cost of sequencing the human genome at one dollar per base pair, which was what sequencing DNA cost in 1986. Such a figure wouldn't have caused an eye to blink at the Pentagon or at NASA. But to the biologists, accustomed to thinking of a one-million-dollar grant as large, the figure was astonishing. The scale of project was beginning to sink in. With existing technology, it would take thirty thousand man-years of laboratory work to sequence the human genome, that is, one thousand scientists working for thirty years. Once finished, the sequence would fill the equivalent of thirteen sets of the *Encyclopedia Britannica*, even if each base was designated only by a single letter.

To the DNA sequencers, it inspired visions of unprecedented amounts of money being poured into developing new machines, new computer software, and new techniques for sequencing DNA, a virtual golden age of sequencing. But to the gene mappers, the prospect of a DOE-directed sequencing effort was a nightmare. They already were well funded by the National Institutes of Health, an agency where advisory committees of academic scientists, rather than government bureaucrats, dominated the dispensing of funds. The mappers feared that, if Congress bought the sequencing project, it could well divert gene mapping and other funds from the NIH to the DOE's sequencing effort. Such diversion would force the mappers into the sequencing effort, where their research would be subjected to the direction of a genome czar and a messy DOE bureaucracy instead of the NIH's "do your own thing" approach.

Before the gene mappers could recover their wits, the DOE was racing ahead with a full-scale sequencing project. DeLisi managed to get a recommendation for a twelve-million-dollar appropriation inserted into the fiscal 1988 federal budget. To head a genome center at the department's Lawrence Berkeley Laboratory in Berkeley California, the DOE recruited geneticist Charles Cantor of Columbia University, coinventor of an improved method of gel electrophoresis that would be extremely useful for sequencing. By early 1987 the DOE was launching Phase I of the project, in which the biologists would simply use restriction enzymes to slice each chromosome into equal length segments, carefully keeping track of where each segment lay on the chromosome.[27] In later phases, the DNA sequences of the segments would be determined. To start off, Columbia University would take on chromosome 21, Los Alamos chromosome 16, and the Lawrence Livermore Laboratory in Livermore, California, chromosome 19. At the same time, Los Alamos and Lawrence Livermore would work to speed up the rate of automated sequencing, while the National Science Foundation would fund development of Hood's sequencing machine at Caltech.

This initial chopping up of each chromosome was known as "physically" mapping the chromosomes and was a procedure fundamentally different from gene mapping. The physical map shows the position on the chromosome of each of the segments without regard to whether the segments contain genes or not; it is a necessary prelude to sequencing the DNA of a chromosome. The gene map, of course, shows the location of genes on the chromosome and is an end in itself. Failure to realize the difference between the two kinds of maps misled many outsiders, including most of the lay press, into thinking there was a single project being discussed, an effort to "map the human genome." To most of the outside world it appeared that the only controversy was whether "the human genome" project should be taken on by the DOE.

There was, to be sure, a turf battle underway by mid-1987. Although the NIH bureaucracy publicly remained uncommitted on the DOE effort, there was little question that it sided with and encouraged the gene mappers' attack on the project. The attack was launched from MIT. In early 1987, Botstein called the DOE effort political.[28] "I do not believe that there is any strong scientific justification for knowing the sequence of the entire human genome," he declared. "The motivation for doing it is frankly political, or 'science' political, more than it is that science is being held up by our lack of knowledge of every nucleotide [base pair] in the genome." Microbiologist David Baltimore, the Nobel laureate who then headed MIT's Whitehead Institute, told an NIH meeting that "a mega-project is a way of handing scientific programs over to politicians." A year later, Baltimore called the DOE project "a ploy to raise money, a project justified by its public relations value, not its scientific value."[29]

The most vehement opponent of the DOE's sequencing project was the much-respected MIT biologist, Robert Weinberg, whose labs were located at the Whitehead Institute. Ironically, Weinberg was a major figure in the discovery and elucidation of the oncogenes, the discovery that had spurred Dulbecco to propose the sequencing project in the first place. But Weinberg was clearly annoyed that his work was being exploited to legitimize gene sequencing. "I'm surprised consenting adults have been caught in public talking about it. . . . it makes no sense," Weinberg said of the sequencing project in early 1987.[30]

"Many think it foolhardy and retrogressive to argue against a project that promises to yield a mountain of new data," Weinberg told readers of *The Scientist* later in the year.[31] "With the U.S. Department of Energy and the National Institutes of Health now fighting turf battles for various pieces of the action, the decision to go ahead seems already a *fait accompli,* as inevitable as taxes on April 15, death and all the rest of life's certainties."

The project made no sense because most of the genome didn't make sense, Weinberg argued. "The limited data that we currently possess suggests that 90% to 95% of the base sequences in our DNA are biologically meaningless, having no ability to encode proteins or regulate gene expression," he asserted. "Much of this material seems to represent evolutionary detritus, dead ends, side alleys and parasitic sequences that are extrinsic to the functioning of our genes." Moreover, the MIT biologist argued, just knowing the sequence of bases in a lengthy segment of DNA would be of little help in identifying genes within that segment. Weinberg is much renowned for the poetry of his scientific descriptions, and he brought his skill to bear in his arguments. A gene, he explained, was composed of small bits of DNA separated from each other by meaningless stretches of DNA, "a small archipelago of information islands scattered amidst a vast sea of drivel." The biologists, despite their newfound powers to explore for genes, still couldn't look at a DNA sequence and distinguish one of these "information islands" from the "drivel" around it; they simply couldn't tell where a gene began and where it stopped, he argued.

The DOE initiative, nevertheless, picked up support. The sequencers began calling congressmen's attention to reports from Japan that engineers and biologists there were launching a government-backed effort to automate the sequencing of DNA and had cut the cost of sequencing the genome by 90 percent–ten cents a base. If the Americans didn't get moving on such a prestigious and important project, the Japanese would certainly beat them to it, legislators were told. And even some of the gene mappers, although opposed to a full-scale sequencing of the genome, conceded that physical maps to be developed in Phase I of the DOE project would be useful, providing benchmarks of known location to which nearby genes could be linked in family studies.

But the anti-sequencers carried their case to the National Academy of Sciences, whose operative arm, the National Research Council, commissioned a study of the issue. The NRC committee, which included McKusick, reported back in February 1988.[32] Diplomatically, the committee recommended that both mapping and sequencing should be the goal of a fifteen-year-long effort with federal funding to reach two hundred million dollars a year. But it was quite evident that the gene mappers had carried the committee. "The committee recommended that researchers map first and sequence later, for two reasons," McKusick reported. One reason was that sequencing required a much greater advance in technology than mapping, he explained, and thus would take longer to get underway. The other was that the physical maps of the chromosomes that necessarily preceded the actual sequencing could be used immediately for gene mapping.

Thus, "map first, sequence later" would be the strategy with the quickest payoff in such practical results as locating the genes underlying cancer and the hereditary disorders.

The NIH moved almost immediately after the NRC committee issued its report. NIH director James B. Wyngaarden set up a special Office of Human Genome Research to oversee and advise on a special NIH human genome mapping effort. Wyngaarden cast about for a genome czar who would be acceptable to the bureaucracy-fearing biologists and at the same time have the prestige to eclipse the DOE program. He found such a czar in James Watson, the codiscoverer of the DNA double helix, who agreed to take on the job in addition to his duties as director of the Cold Spring Harbor laboratory.

In the fall of 1988, top officials from NIH sat down with their counterparts at DOE to sign a "joint memorandum of understanding."[33] DOE would continue its genome sequencing effort and NIH would concentrate on gene mapping.

The United States now had two genome projects underway. Even experienced Washington science reporters found themselves confused. Many reported that Watson had been named to head "the genome project" as though the entire federal effort, rather than just the NIH portion, fell under Watson's new office. As late as January 1989, in a lengthy story on "the biological community's $3 billion human genome project," the *Washington Post* devoted two full columns to Watson and the NIH program; the DOE received a scant one sentence: "The Energy Department, which also is interested in genetics research, has a separate, smaller genome budget."[34]

As early as 1987 gene mappers like McKusick and Ruddle in the United States and Sydney Brenner in England had begun to worry that government bureaucrats were about to break up the informal international gene mapping effort that had developed around the human gene mapping workshops that Ruddle had initiated in 1973. The British and the Europeans were just as deep into gene mapping as the Americans, and they had no intention of being left out of any effort to map or sequence the genome. European governments, hearing the noise from Washington and Tokyo about a genome sequencing project, also had started to talk of stepping up their gene mapping and gene sequencing efforts. Neither the European scientists nor their American colleagues wanted to become participants in some genetic Olympics to see which country would be the first to map and sequence the genome or even a single chromosome. Such competitions would result in large-scale duplication of effort, in that several laboratories might tackle the same chromosome or stretch of DNA simultaneously while ignoring other parts of the genome.

The biologists and geneticists began talking of setting up their own coordinating council, divorced from the bureaucracies and politics of their respective nations. This led to the privately financed Human Genome Organization, or HUGO.[35] "After being discussed privately for at least a year, this organization was conceived on April 30, 1988, during a symposium on genome mapping and sequencing at Cold Spring Harbor," McKusick reported. "It was the idea mainly of Sydney Brenner of Cambridge University, who suggested the name and its felicitous acronym." HUGO was formally organized five months later when a founding council of forty-two scientists gathered in Montreux, Switzerland. Eschewing government monies, the scientists sought and received funding for the Montreux meeting from the Howard Hughes Medical Institute.

HUGO plans to set up offices in North America, Europe, and Asia to serve as clearinghouses for information on gene and riflip locations, DNA base sequences, and other data. It also will take over from Ruddle the job of funding and organizing the Human Gene Mapping workshops. "Coordination by HUGO is needed, not only among nations, but among disciplines—for example, between scientists who map genes and those who perform sequencing—and also among those who study the genomes of other species, such as the mouse," McKusick explained.

Weinberg, among others, remained skeptical.[36] "I fear," he told readers of the scientific journal *BioEssays*, "that the important decisions have already been made and that the great sequencing juggernaut will soon begin its inexorable forward motion, flooding our desks with oceans of data whose scope defies conception and our ability to interpret meaningfully. . . . We can only hope that the costs of these sequencing efforts will not impoverish efforts to solve the real problems of biology!"

Whatever the debate over the value of sequencing the genome, the mapping of the genes continues. On June 17, 1989, at the closing session of the 10th International Human Gene Mapping Workshop, held at its birthplace in New Haven, seven hundred more genes were assigned positions on specific chromosomes.

CHAPTER 11

"Siss im Blut"

What David Housman and James Gusella remember about the fall of 1983 are the calls that came flooding into their labs at MIT and Harvard.

Gusella's research team had just announced they had found a marker for the gene that causes Huntington's disease. The lay and scientific press widely celebrated it. Researchers around the world studying a host of other untreatable diseases besieged the Cambridge, Massachusetts, biologists. Among the first to call were two Americans, Janice Egeland (studying manic-depression among the Amish in Pennsylvania) and Linda Nee (studying a severe and premature form of senility in eastern Canada).

They hoped to bring the gene-hunting technique described by Botstein and his colleagues into the realm of diseases of the mind.

For much of the twentieth century, especially the 1930s, as Freudian theories held sway, behavioral afflictions such as depression and schizophrenia were largely attributed to damage to the pysche, usually as a result of emotional wounds inflicted at a young age. But beginning after World War II, many psychiatrists were so frustrated by their inability to treat many of their patients, they began to suspect biochemical glitches—perhaps the result of faulty genes—in the workings of their patients' brains. This notion gained favor through the 1950s and 1960s, as doctors found that some mental health problems could be relieved—although never cured—with new drugs. The medicines' temporary benefits threw psychiatry into turmoil as mental health practitioners argued the relative importance of environment and biology. Nowhere in medicine was the debate over nature versus nurture more contentious.

By 1959, when Janice Egeland first visited the Amish communities in Pennsylvania, there was a vocal, but still small school of psychiatry convinced that most mental illnesses were caused by bad chemistry—an inherited flaw that perturbed the normal transmission of nerve signals about the brain. Recent studies had found that first-degree relatives—parents, siblings, and offsprings—of patients with manic-depression were many times more likely to develop similar problems than were relatives of otherwise healthy individuals. Some studies found that identical twins, who share the exact same genes, were four times more likely to share manic-depressive problems than were fraternal twins. But critics of the genetic hypotheses argued that close kin and identical twins also shared the same family and cultural environments, and that they might also have been similarly exposed to other external agents, perhaps even a virus.

The genetic proponents countered with adoption studies. When these researchers looked at children separated at birth from their biological parents, they found striking similarities in the mental health of the offspring and their original parents.

"For some of us, the nature-nurture argument was just a lot of baloney," says Theodore Reich, a psychiatrist at Washington University School of Medicine in St. Louis, who, in the early 1960s, was one of the first in the field to search seriously for genetic factors in mental disorders. "The clustering of problems within families was too strong and occurred too often to be ignored. Certainly, these people were sharing similar biologies, which made them especially susceptible. And, yes, they shared similar environments, which heightened, or perhaps triggered their problems. It wasn't just this or that, it was both—genetics and environment."

Reich and others began collecting large, extended families, looking to see if some of the mental health problems segregated in a Mendelian fashion. Manic-depression was the first disorder tackled since it was the easiest to identify, and also because it is relatively common; about one in one hundred people develop the illness, usually by the time they are thirty-five years old. People diagnosed with manic-depression exhibit sharp swings in mood that involve periods of devastating depression alternating with episodes of ecstasy, restlessness, and delusions of grandeur. Since the 1930s, researchers had found hints that this syndrome, also called bipolar disorder because of its pronounced fluctuations in behavior, was a result of a gene that was housed in the X chromosome. This was because researchers found that the disorder almost never passed from father to son. (Males, who have an X and a Y chromosome, only pass their Y to their sons; if they pass along the X, their offspring will always be female.) Some of these studies also suggested that a gene for manic-depression might be located near a gene that causes color blindness, since that gene had been traced to the X chromosome, and the problems sometimes seemed to be inherited together. But by the late 1950s, no bipolar gene had been uncovered, nor had researchers found a family where the disease's passage through generations suggested a distinct pattern of inheritance—either dominant or recessive.

These ideas were bobbing about without any hard evidence to satisfy science when Egeland, a graduate student in medical anthropology and sociology at Johns Hopkins University, began what she thought would be a brief bit of research among the Amish. At the time, Victor McKusick had just set up his genetics program at the university. He told Egeland, who was one of his students, that the Amish might provide material for a variety of genetic studies. But he warned her that no one had been able to enter their guarded, closed society. Egeland was quiet and serious by nature, and McKusick believed she might gain entry where others had failed.

McKusick saw the Amish families as a "geneticist's dream" because they are the product of the kind of inbreeding scientists produce in the laboratory when trying to pick apart traits in animals or plants. The Amish are an ultraconservative branch of the Mennonite Church that split off in the late 1600s and emigrated to the United States in the early 1700s. The fifty or so pioneers who settled south and west of Philadelphia were originally from parts of Switzerland and the southeastern Rhine valley. They spoke a dialect of high German, which in the United States became known as Pennsylvania Dutch.

McKusick believed the inbreeding among the following generations provided a classroom example of what geneticists call "the founder's effect." The gene pool from a few dozen founders had been isolated for centuries, passed on intact to the twelve thousand or so living descendents. A particular trait could be traced backward without concern that it might have come from the blood of an outsider. In fact, because the genes from these founders were undiluted, they would have especially profound influences on the living, allowing researchers to pick out inheritance patterns of traits much more easily than in the general population.

The Amish villages and farms in Pennsylvania's Lancaster County were only a few hours by car or train from Baltimore, and Egeland began visiting the area, slowly gaining the confidence of a few families, living in their homes for days and then weeks at a time. The trips to the countryside were peaceful interludes from the city and her graduate studies, and she was drawn to the farmers' simple life-style, their pacifist philosophy, plain dress and plain talk. Moreover, the trips were exotic—she later said that she felt as if she'd been transported centuries back in time. The Amish spoke an archaic mix of English and German and eschewed modern technology, traveling in distinctive horse-and-buggy carriages; their homes were without phones or electricity. She was exhilarated by the research possibilities, but this inner peacefulness she derived from the visits was equally seductive.

"It was a very special time in my life," Egeland said in 1987. "There is a very great level of caring for one another and truthfulness."[1]

Between 1959 and 1967, Egeland spent long periods among the Amish. She produced a catalogue of their genetic disorders, plus a genealogy of a few dozen families with 7,353 entries. Her work was helped by the fact that the Amish gathered together for worship in "districts" of about thirty to fifty families, each of which had a "scribe" or historian whose job was to keep a record of family doings through the generations.

By 1972, she had become immersed in the isolated society. She was to say later that she felt "married" to it. The Amish also felt close to her. When her father was dying of leukemia, they donated 955 pints of blood. As she got closer to the people, the Amish shared their gossip, passing along tales of individuals whose un-Amish behavior set them apart. One spring when it was time for planting, Egeland heard about a young man who suddenly packed his suitcase and headed off on vacation. Another time she heard about a young fellow who, on an impulse, ordered a new Mercury, with whitewall tires. Several others, she was told, went on uncharacteristic buying sprees, were unusually boastful or talkative, or even engaged in violence. A few were known to drink. Some committed suicide. Because their conduct was so out of character, Egeland guessed the cause might be due to some mental illness.

By examining her genealogies, she saw the faint outlines of familial clustering among some of these individuals. Her interest piqued, Egeland investigated medical records at a psychiatric hospital in nearby Hershey, collecting the names of the Amish admitted for any mental disorder. When she contacted several of the patients' families to see if any close relatives or ancestors were similarly affected, Egeland was startled by their responses. "The Amish already believed something was being passed along in certain families and they weren't at all surprised by my inquiries," says Egeland. "They have a saying about it—*Siss im Blut*—'It's in the blood.'"

In 1976 Egeland, who by then was on the faculty of the department of psychiatry at the University of Miami School of Medicine, received a federal grant to study the prevalence of manic-depression among the Amish. The goal was to see if, in certain families, the disease segregated according to Mendelian law, as if caused by a gene.

Egeland set up shop full-time in a small office in Hershey, a town mostly known as the headquarters of the chocolate company. For weeks at a time, she drove out in the countryside, dropping in at the farms or in the small villages, gossiping and picking up news about anyone whose demeanor might be a tip-off to the disease. Among the Amish, alcoholism, marital tension, sexual indiscretions, even loud talk were perceived as deviant, and the few individuals with such problems stood out starkly from their peers.

Over the next four years she found only 112 people, in thirty-two separate families, whose "strange behaviors" could be classified as specific mental health problems. Of those cases, thirty-two individuals were identified as manic-depressive. Each proved to have family histories of the disease going back several generations. Egeland and some associates then interviewed all the thirty-two individuals' first-degree relatives, about one thousand people in all, since each of the thirty-two had, on average, about thirteen people in his or her nuclear family. Egeland focused her research on a handful of especially large extended families, and then, on one in particular, in which she found four generations of what appeared to be bipolar behavior. In any case "the family was loaded with problems," Egeland says. "Something was definitely wrong."

The family's great-grandfather, who died in 1898, had been diagnosed as having an obsessive-compulsive personality, a condition in which an individual is fixated by certain ideas or actions. Of his eight children, six were diagnosed with psychiatric illness during their lives. Eight of his thirty-five grandchildren and twenty-five of 191 great-grandchildren also were diagnosed as mentally ill. A search backward into the family's records uncovered a male ancestor born in 1763 who was described as often suffering from depression; his brother committed suicide.[2]

Indeed, suicide was uncommon among the Amish, but Egeland discovered it peppered among the family trees of the mentally ill. Egeland found that, between 1880 and 1980, there were twenty-six suicides. So unusual was it for the Amish to take their own lives that each instance was recalled by friends and family with great detail. For instance, Egeland documented that of the suicides, twenty were hangings, four were gunshots, and two were drownings. In one instance, she found that a man left a note, tied a stone to his body, and, with his plain dark Amish hat in his hand, jumped into the deep water of a local quarry. Months later he was found and identified by the initials in his hat.[3]

The detailed descriptions helped Egeland ascertain that all but two of the people who committed suicide suffered a major mental health problem and that twelve were bipolar. All twenty-six individuals were from just four families, and all four families were "loaded" with mental health problems. In one family, which Egeland labeled Pedigree 214, there were seven suicides, all among the direct descendents of a man who had died in 1893. Because the cause of his death was disputed among his descendents, Egeland spent several years interviewing all his hundred or so grandchildren until she was convinced that, indeed, he had taken his own life. In another family, called Pedigree 265, which included five of the suicides, Egeland was able to find that an ancestor committed suicide in 1831 and that another close relative's death about then also was suspected to be suicide.

By 1983, Egeland's file cabinets in Hershey were stuffed with details documenting her claim that, within a few families, a gene or genes were leaving an effect. The next step was to find the genes, if, indeed, there were any at all. She needed to find a marker, such as a gene whose location already was known, that was consistently inherited with bipolar illness. Egeland first checked out the X chromosome hypothesis by comparing the bipolar cases to color blindness, but no correlation was found. Next, she was advised to take blood from one family to see if the affected members shared HLA antigens, which were known to reside in chromosome 6.

"I was getting all kinds of advice," Egeland recalls. "Researchers were very excited by the possibility of finding a gene, and the Amish looked to be an especially fertile place to find it. People were saying, 'We don't know where this [gene] is; it's a needle-in-a-haystack. Why don't you check serum proteins, why don't you check everything.' . . . The end result of those early genetic linkage studies conducted from 1979 through 1982 was that we could not demonstrate any positive linkage for any of the conventional markers. . . . [We] found our backs against the wall . . ."[4] She was discouraged and depressed and uncertain where to turn next.

About then Jim Gusella at MIT reported finding the Huntington's disease gene using a riflip marker. "The cornfields can be so isolating. There were times I said, 'Oh, I'm going to leave, go back to the ivory tower.' Then when the probe technology came along I realized we were just at the beginning; there was no turning back," she says. "It sounded like a miracle technology."

At a medical meeting in 1983, Egeland heard Nancy Wexler speak of her trips to Venezuela and the euphoria she and her colleagues felt when the scientists transformed her family study into a breakthrough of landmark proportion. "I was totally inspired," Egeland says. Wexler suggested Egeland get in touch with Housman, who by then was no longer pursuing the Huntington's gene but, instead, was cloning riflips in an attempt to produce a complete gene marker map of chromosome 11. Housman agreed to help Egeland and assigned several postdoctoral researchers in his laboratories to the task. Mostly, Housman expected the families' DNA would help him find more probes. But the search for a specific gene would also fire the enthusiasm of his research team.

Thus, the "postdocs" at first viewed Egeland's family pedigree as lab material in which to gain experience using the new gene-hunting technology. Daniela Gerhard, a young molecular geneticist who had emigrated as a girl from Czechoslovakia in 1967, took the lead in the laboratory. She instructed Egeland to get blood samples from as many people as possible within a pedigree, noting which of them definitely had bipolar disorder and which didn't. Egeland focused on three of the thirty-two families with bipolar illness, calling them, collectively, Pedigree 110. As soon as a batch of the blood samples reached the labs at MIT, Gerhard began mixing DNA from the families with riflips from chromosome 11. The goal was to find a probe that consistently stuck to the DNA of the affected people, but never matched those who were healthy. Such a probe would indicate that a gene on chromosome 11 was involved in the disease.

Several years later, Gerhard was to say that she had little expectation anything would come of the studies she was performing. "It was a fishing expedition," she says. "The chance that we'd have linkage was one in a million. I expected to do the study, and then write a short report saying we'd looked for the manic-depression gene on 11 and it wasn't there. Then, within a few months, we found linkage with one of the early probes. I think it was in 1984. I was certain it was an error. I did it over and got the same result."

A statistical analysis showed that the likelihood against the linkage being due to mere chance was one hundred to one. Housman huddled with Gerhard, and two other postdocs, David Pauls and Kenneth Kidd. They decided they had a fairly strong finding but not as strong as would be needed to hold up against a barrage of scientific skeptics.

And they expected a lot of skepticism. The discovery, if indeed it was one, would be the second time in less than a year that researchers associated with Housman had, from a scientific point of view, pulled a rabbit out of the hat. It was, in David Pauls's words, "just a lucky shot."[5] Housman and Gusella were still smarting from the jibes by some that finding the Huntington's disease gene from work they had begun together had been serendipitous.

Housman had to admit that finding the gene to Huntington's disease so quickly was "unbelievable good luck." Even so, the discovery had brought him and Gusella the laurels due a legitimate scientific advance of extraordinary merit. Finding a second gene so soon after another brief and fortuitous search, he felt, might actually hurt his credibility rather than add to his stature. At least, in the case of Huntington's, there was no argument that the culprit was a single, defective gene. But psychiatry had yet to be convinced that a single gene could induce the complex and shifting mental behaviors characteristic of bipolar disorder. Not only would the finding have to stand up to close scrutiny from the molecular genetics field, but it was certain to be attacked within the halls of psychiatry, Housman says.

In Pennsylvania, Janice Egeland was incredulous. Housman told her what they had found was "suggestive" of linkage, but they needed to tighten it. "Dr. Housman made clear to me that they could hope to find some probes that would move closer to where a gene might be, but that the immediate quick fix would be for me to proceed into the field and obtain more blood samples from this particular pedigree," Egeland said.[6]

Fieldwork for Egeland was, quite literally, traipsing over manure-covered backroads to the farm fields where members of the family lived. She and her research associate, Cleona Allen, would pack clothes for several days, and ride into the countryside outside Hershey in her tan Subaru station wagon. Often they had to get to a farmer at dawn before he began milking. Sometimes they brought with them a box of pretzels as a goodwill gesture.

After about a year, Egeland had "bled" thirty more family members, bringing Pedigree 110 to eighty-one and the number of affected people to nineteen. Back in Cambridge, Gerhard expected the tests on the additional blood samples might loosen linkage. "I was away when the final analysis was done," Gerhard says. "David [Housman] came back to me in the lab when I came in and said the likelihood of linkage was even stronger." The odds against the results being due simply to chance jumped to one thousand to one.

The MIT researchers couldn't refute their own discovery. The link between the riflip probe on chromosome 11 and the gene became stronger the more the scientists analyzed it. At one point, a member of the Amish family, a woman about forty who had been considered healthy, suddenly began exhibiting symptoms of bipolar disorder. Back in Cambridge, the scientists had been troubled by the woman's healthy diagnosis up to that point, because analysis of her DNA suggested that she had the gene. Then Egeland called the MIT team to report the woman's changed status. "Imagine how we felt," says Housman. "Here we were in Cambridge and we could predict that this woman was going to get sick."

There was, in fact, additional evidence to support the notion that the gene was on chromosome 11. A scan of the library of other genes already known to reside on the chromosome found one gene encoded for an enzyme involved in the production of dopamine, a chemical used by the brain to help carry impulses between nerve cells. Perhaps, scientists guessed, the Amish family's gene somehow interfered with the manufacture of dopamine.

In late 1986, after months of analyzing and reanalyzing the data, the research team submitted the results to *Nature,* the British scientific journal. Within weeks the editors of the journal accepted it. In late February 1987, *Nature* reported that "a dominant gene conferring a strong predisposition to manic depressive illness" was located on chromosome 11.[7] But even before they wrote up their results, the researchers knew that what they had found couldn't be the only gene, or the only cause, for manic-depression. They had talked to other researchers who had been collecting smaller families and suggested that they try to find linkage with chromosome 11 probes, too. Although two more reports were published in the same issue of *Nature,* the two other research teams were unable to reproduce the Amish findings.[8,9]

Instead of undermining the claims of the Amish gene, the scientists said the contradicting reports indicated that manic-depression was het-erogenous, in other words, that it could be caused by more than one gene. Also, since only 63 percent of those in the Amish family who showed signs of having the gene were actually sick, the researchers said their finding suggested that, as many believed, only a suscepti-bility to manic-depression could be inherited. Something else in the environment or the inheritance of other genes had to be present to trigger the illness.

The confounding reports didn't dampen enthusiasm among some top people in the field. "This is the first demonstration of a possible genetic basis for one of the major mental disorders," Dr. Darrel Regier, director of the division of clinical research at the National Institute of Mental Health, told *Time* magazine when the *Nature* report came out. "The [Amish] study ushers in a new era of psychiatric research."[10]

Even so, hardly anyone in the field was prepared for what happened next. Unbeknownst to the MIT scientists or Egeland, two other sci-entific teams, one headed by a New York researcher, and another headed by a Belgian, were closing in on another gene for manic-depression. Only three weeks after Egeland's discovery garnered worldwide attention, *Nature* carried still another report that scientists led by the New York researcher, Miron Baron, had found another gene for the disease tightly linked to a riflip probe located on the X chro-mosome.[11] And two months later, the Belgian group, led by Julien Mendlewicz of the Free University of Brussels, reported evidence of a manic-depressive gene in the same area of the X chromosome in a separate group of families.[12]

"The reports blew away most of the doubts anyone still held that the disease is rooted in genetics," says Elliot Gershon, a leading re-searcher at the National Institute of Mental Health. "Of course, every-one now waited to see if the results could be reproduced."

The report by the team led by New York psychiatrist Baron was especially revealing, for it was the first to confirm the earlier suspicion about the X chromosome. Since 1980, Baron, an intensely confident young man, had been scouring the X chromosome for a gene. "I was convinced it was there, and that finding it was a matter of doing the right kind of looking," he says. "I thought I could succeed where others didn't."

Baron was well aware of earlier reports linking the disease in some families with color blindness and with studies showing that the disease rarely was transmitted from father to son. He had come across the notion in the late 1970s, while working on his residency in psychiatry. Like others before him he was surprised by how often his patients had family members with similar mental health problems. Reading the history of frustrated gene studies, he kept his own investigations low-key, merely dabbling in the idea. "When the era of the molecular genetic probes dawned in 1980, suddenly the prospects seemed tremendous," Baron says. "It was a real turning point."

Baron was convinced that, if there was a gene, it was hiding on the X chromosome. Several studies from researchers in Israel suggested families of Middle Eastern descent might be harboring it. The researchers, two doctors in Jerusalem, Rahel Hamburger amd Batsheva Mandel, had noted a "crowding" of manic-depressive problems in several non-European Israeli families. After contacting the doctors, Baron believed he would uncover several "perfect" pedigrees in Israel. The Israeli families were large and lived within a small geographic area, making them accessible for study. Moreover, they had unusually low rates of alcoholism or drug abuse, problems that can confound diagnoses of bipolar disorder.

Between 1982 and 1985, Baron and the Israeli doctors cast a wide net to capture the pedigrees they wanted. Patients who were admitted to the Jerusalem Mental Health Center were queried about the health of their first-degree relatives. Surveys were sent to the Israeli army to identify recruits with color blindness, or those who suffered from an enzyme deficiency that caused a type of anemia—the defective G6PD gene known to reside along the X chromosome. Families in which only fathers and sons were sick were excluded. "I believed that manic-depression, or illnesses that looked like it, were caused by other genes [besides one on the X chromosome]," says Baron. "But I was looking for one on the X."

By 1986, the researchers had located five large families encompassing 161 adults, 47 of whom were ill with bipolar illness or some related disorder. Blood from the families was sent for DNA analysis to laboratories at Yale University. The test results came back showing the strongest linkage ever uncovered between a riflip probe and a disease. The possibility that the results were due only to chance ranged from a low of ten million to one as high as one billion to one.

"I was surprised," says Baron. "When I read Egeland's report, my first reaction was that one of us had to be wrong. But then I realized maybe we both were right. The disease has to be the result of several genes."

For both Egeland and Baron, the next job was to find the gene itself, a task that required the collecting of more families and the use of more riflip probes. Press attention in New York helped both groups. "There was a lot of fanfare that I thought was very distracting at first," says Baron. "But it turned out to be instrumental. The media blitz caught the attention of a lot of people, and we were hit with dozens of phone calls from families and their doctors. Everyone wanted to know if what they had was due to a gene, and, in return for us helping them, they agreed to participate in the studies."

Within a few months, Baron was tracking down extended pedigrees from thirty families that had contacted him, and he sent out a call for riflip probes from the top DNA scientists in the United States and abroad. "Ray White was very, very helpful," Baron said in late 1989. "He sent probes and he gave us advice and a lot of support. It looks like, in these families, there may be still another gene causing manic-depression. We're going to have to test dozens, maybe hundreds of probes from all over the genome, because we have no idea where this other gene might be. It could take years. Of course, we could be lucky and find it tomorrow."

While Egeland and Baron were assembling their pedigrees with the specific goal of finding a gene, Linda Nee, a social worker at one of the federal research centers in Bethesda, Maryland, was trying to figure out what to make of an extraordinary family she first encountered in 1978. Indeed, one day the family, almost quite literally, was dropped into her lap.

Nee was working at the National Institute for Mental Health, when a letter arrived from a doctor in Canada who had worked at the institute years earlier. The physician had a colleague who was upset by some recent developments in his wife's family. Within a span of just a few years, two sisters and a brother—all in their fifties—had been diagnosed with various forms of premature dementia. They were losing their memory and their ability to function independently. It appeared as if all four siblings had Alzheimer's disease, an illness that in 1978 was only beginning to gain widespread interest, and for which there was no treatment.

Nee, in fact, had little knowledge of the disease, but a quick review in the medical archives revealed a smattering of reports suggesting that the disease sometimes could be inherited. The lack of family data wasn't surprising; most people developed the problem in their latter years, by which time their parents had already passed away. Nee was interested enough, however, to ask the doctor to contact his colleague.

Would the family be willing to come to Bethesda to be examined?

The response was quick. Within a few weeks the four siblings from the family were in Bethesda. The oldest sister was fifty-six years old and healthy. She brought with her a fifty-one-year-old brother, who was unable to speak; a forty-nine-year-old sister who was mostly withdrawn; and another sister, forty-seven years old, who was just beginning to show signs of the ailment. Nee had hoped to ascertain from the family whether they knew of any other clusters of dementia among their relatives, and whether there were any records of ancestors who, based on stories or medical documents, were believed to be similarly afflicted.

In an article in the *Washington Post Magazine*, Nee described her surprise when the family members unraveled before her a two-foot-long sheet of paper documenting four generations of a pedigree, loaded with members who had been stricken by the same disabling disease. "Within four days," Nee told reporter David Van Biema, "we had the date [the malady] came to the continent. We knew it affected both males and females equally, we had a pretty good idea about the average age of onset, and we knew we had a family that was extraordinarily motivated and responsible. If there was ever a moment, early on, when we felt like saying 'Eureka,' that was it."[13]

Details of a past shadowed by the disease rose up out of a parade of parents and their children dating back to 1837. The four siblings sitting before her in Bethesda were descendents of three brothers who sailed to the Canadian east coast province of New Brunswick from Britain's Northumberland countryside. Over the years, the brothers' family grew; some stayed east to farm, while others moved westward and north. They mainly prospered, some becoming professionals, and, in each generation, some portion of the clan's senior members prematurely died, usually by the age of about sixty or so, often after a decade in which their minds, bit by bit, lost their hold on reality.

A neurological examination of the three ailing siblings who had come to Bethesda confirmed that they suffered Alzheimer's disease. Although the disease was first characterized in 1907 by a German doctor, Alois Alzheimer, only in the 1970s did doctors begin to assign the diagnosis to a rising wave of middle-aged and elderly patients whose mental capacity, in the course of a few years, was overwhelmed by a deepening senility. Patients lost the use of their memory and their intellectual talents slowly slipped away, as did their ability to perform simple daily tasks: making meals, brushing their teeth, or even tying their shoes were skills lost in a mist of confusion. Family and friends were transformed into strangers. The condition was irreversible and untreatable.

Over the years scientists studying brain tissue from deceased patients uncovered some of the biochemical wreckage left behind. Alzheimer's disease victims lack a sufficient supply of the chemical acetylcholine, which helps transport nerve signals within the brain. They have abnormally high amounts of a calcium-like protein called amyloid deposited in bits of plaque that clog the brain's nerve channels. Some brain tissue has excessive concentrations of aluminum. But scientists have no way of determining how these abnormalities came about. Were the patients exposed to some detrimental agent in the environment? Was it a virus? Or was it the debris of normal aging in people with a certain genetic makeup? Unlike in heart disease, where doctors could thread thin optic wires into the arteries and see the accumulation of plaques that led to a heart attack, there was no way to look into the nerve transmission system of a living person's brain and study the destructive process taking place. Moreover, scientists were unable to mimic the disease in animals, and, thus, couldn't study it in the laboratory.

From time to time, over the years, doctors noted small clusters of the disease within families. Some researchers estimated that the disease was inherited in about 10 percent of all cases. But the families were small, and, often, older relatives who had the disease already had died—denying scientists the opportunity to study biological similarities from generation to generation. Thus, the Canadian family presented great possibilities. "We went up north and started beating the bushes for people in the family, for people who remembered who among their close relatives had the disease, and, of course, we looked for the disease in as many first-degree relatives as possible," Nee says.

By 1982, Nee had located a branch of the family with 531 relatives, 52 of whom had the disease.[14] Nee and others had collected information going back six generations. At first, the researchers were uncertain about whether certain family members in the second and third generation, people who had been dead for decades, had Alzheimer's disease. But the fourth generation was chock full of clearly identifiable cases of the disease—based on analysis of medical records, personal letters, and other anecdotal documents. An examination of the pedigree from the fourth to sixth generations clearly "looked" as if the disease progressed through the family as if due to a dominant-acting gene; within each nuclear family at least 50 percent and sometimes more of the offspring of one affected parent had the disease.

"Just looking at the pedigree spread out before us made us believe we had evidence of a segregating gene," Nee says. The finding sent a shiver through her, she recalls. The family was the first to provide evidence that one single gene was associated with the disease.

Nee and her colleagues decided to strengthen the genetic link by searching for evidence that people in the earlier generations also had Alzheimer's disease. Bit by bit, by talking to family and friends, and sifting through local records and diaries, they found that the nuclear branches containing numerous affected individuals were direct descendents of people who had died of illnesses that strongly resembled Alzheimer's.

For instance, Nee was able to get a collaborator in Great Britain to unearth documents suggesting that a grandson of the family's original couple had died in Newcastle in Great Britain after suffering years of Alzheimer's disease–like problems. The man, in the family's third generation, had ten children, three of whom, it was finally determined, also died of the disease. The three siblings, in turn, had twenty-four children, five of whom were known to have the disease for sure, and several others in whom it was suspected. (Others died early in life, before symptoms could arise.)

The affected children's offspring in the family's sixth generation were the ones in whom the 50 percent penetration of the gene was the strongest, mainly because there was more reliable data available on their health status—many were still alive or had passed away only a few years before. "We felt from the very beginning, from the day the first four members of the family came to us with a crude pedigree, that we probably had evidence of a dominant gene," says Nee. "But after we began to fill in who in the earlier generations was sick and who wasn't, well, it seemed we had proof about as strong as you'd want it."

Nee and colleagues at the Bethesda research facilities began to look for the gene. They sifted through blood taken from the family members, looking for the traditional markers Egeland had tested in her Amish family. An early study at first suggested some linkage to the HLA blood antigens on chromosome 6, but that finding failed to hold up under closer scrutiny. The researchers collected spinal fluid from some of the families, but they were unable to find any distinctively flawed chemicals that could be used as a marker.

It was in late 1983 that Nee contacted Gusella. Like Egeland, Nee says, "We had just about exhausted all the conventional markers." Gusella was impressed by the sheer size of the pedigree, and by Nee's close relationship to its members. From experience, he knew how difficult it would be to get the blood samples needed to perform a riflip experiment. The task of doing the DNA analysis was given primarily to Peter St George-Hyslop and Rudolph Tanzi. St George-Hyslop had been studying Alzheimer's disease at the University of Toronto and was recruited by Gusella to bring his expertise into the Harvard labs, which hadn't dealt with the disease before. One of the first things St George-Hyslop did was contact researchers he knew outside the United States who also had been tracking the disease in large families. "Gusella figured we'd strengthen our odds of finding a gene if we had the largest number of people from as wide a geographic range as possible," St George-Hyslop says.

Over the next few years, Gusella and his team struck collaboration agreements with the other researchers. One in France had pulled together an Italian family based in the southern region of Calabria that had forty-eight reported cases of Alzheimer's disease over eight generations. Another was a group that had assembled a family of Russian origin with twenty-eight cases of the disease. And, finally, there was a German family collection of six generations with twenty members affected.

"The sheer volume of people made us believe we'd find a gene, but, of course, we had to narrow down where we'd begin or we'd find ourselves just taking shots in the dark, trying [riflip] probes from this chromosome or that chromosome without any rhyme or reason," St George-Hyslop says.

Then, the serendipity that had propelled the search for genes throughout the 1980s happened again. Scientists studying brain tissue from elderly Down's syndrome patients found evidence of the amyloid plaque deposits previously seen only in Alzheimer's disease patients. As previously explained, Down's syndrome is caused by a genetic mistake, probably at conception, in which an offspring's cells contain three copies of chromosome 21, instead of the normal two. Few Down's syndrome patients live into old age, but those who do often develop a senility that looks much like Alzheimer's disease. St George-Hyslop and his team decided to test the DNA of the four huge pedigrees with probes from chromosome 21.

While the DNA probers undertook their hunt, another team working for Gusella began trying to piece together the structure of the amyloid protein found in the brains of the Alzheimer's victims. In 1986, this team found that a segment of the protein was made by a gene that, as luck would have it, also resided in chromosome 21. The finding convinced the gene hunters that chromosome 21 was where they should look for a shared gene among the various Alzheimer's pedigrees. And in the autumn of 1986, the statistical analyses began pouring in—a series of riflip probes on 21, in a region near where the amyloid gene was located, was linked to the DNA from the blood of family members with the disease.

The Harvard teams wrote a series of research papers describing the linkage and the localization of the amyloid protein gene and sent them in the winter of 1986–87 to *Science* magazine. In a paper titled "The Genetic Defect Causing Familial Alzheimer's Disease Maps to Chromosome 21,"[15] the researchers claimed that they'd found the approximate place where a gene defect was located. They said they didn't know how many of the three hundred thousand cases of the disease that occur each year in the United States were due to the genetic defect. But they declared optimistically that the uncovering of a probe linked to the disease would eventually lead to the discovery of the gene itself. And, that, they said, "might yield new insights into the nature of the defect causing familial Alzheimer's and, possibly, into the [cause] of all forms of Alzheimer's disease." *Science* magazine published the study's results on February 20, 1987, just six days before Egeland, Housman, and Gerhard reported their Amish manic-depression gene discovery in *Nature,* and one month before Miron Baron's finding was made public.

Science is filled with examples in which seemingly unrelated events converge to produce insights that, when added together, lead to a quantum leap in knowledge. To the outsider the three gene discoveries reported within one month of each other looked for all the world to be the result of one of the grandest coincidences in years, and certainly the most important to happen in psychiatry in decades. But, scientists say they often see researchers in different labs located at different parts of the globe make similar discoveries within days or weeks of one another without even knowing of each other's work. "What you're really seeing sometimes is the final, not so surprising, result of a maturing of concepts and technologies," says Gusella. "When new tools become available, people who have been looking for a way at getting an answer all say, 'Let's try that.' It's not surprising, but it draws the attention of those who were unaware that the new tools are available."

For the growing legion of mental health practitioners and research-ers convinced something biochemical was amiss in many of their pa-tients, the finding was the breakthrough they'd been waiting for. A gene at the root of the problems of the Amish, the Israelis, and the Alzheimer's disease families proved that biological defects were at work. The race was on to find in other cases what Egeland, Baron, Nee, and their colleagues had uncovered.

Finally, there was hard scientific evidence backing the hypotheses that, at least in some instances, "the problem is not something the parents did to their children but rather something they gave to their children—their genes," wrote David E. Comings, a geneticist at City of Hope Medical Center in Duarte, California.[16] At least a dozen psychiatric research teams were formed throughout the world to track down large families with higher-than-normal instances of depression, schizophrenia, and even suicide and panic and eating disorders.

Among the many in psychiatry startled by the turn of events was Ann Bassett, a young doctor doing a residency in psychiatry at the University of British Columbia in Vancouver. In 1987, Bassett reported an observation that knocked the field of psychiatry back on its collective heels, and sent research into schizophrenia down a completely new track. The report was based on a patient she happened to treat while on call at the university hospital's emergency room. The patient was suffering from schizophrenia, widely considered by psychiatrists to be one of the worst and most vexing forms of human madness. The disease, which affects about one in one thousand people worldwide, is incurable and often resists treatment. Drugs work only sometimes, and psychotherapy rarely works at all. Patients are often first diagnosed with a sudden personality change, as if hit by a bolt of lightning, in their late teens or early twenties. Next to nothing is known about what causes patients to be plunged suddenly and for the rest of their lives into a condition characterized by periods of deep depression, hallucinations, and paranoia. For decades some psychiatrists suspected the problem might be inherited. But many psychiatrists felt the behaviors also might be due to some deep emotional scar. Even as late as the 1960s, some psychiatrists talked of the "schizophrenogenic mother," described as an overbearing type whose powerful maternal behavior somehow damaged a developing youngster's psyche.

But by the 1980s, psychiatrists were convinced the disease was the result of a chemical imbalance that put certain people at high risk for developing the condition. Researchers worldwide were collecting families, trying to ascertain whether a pattern of inheritance existed. With the discovery of the riflip probes, some scientists began assembling their pedigrees, hoping to uncover a gene.

The patient admitted to the emergency room in Vancouver was

brought in by his parents, who were at their wit's end. The young man, of Asian decent, was twenty years old, and for weeks he had locked himself in his room, talking to himself and laughing. He had been going to college, but recently had quit, complaining that people were laughing at him. "He had all the symptoms and they had been coming on for a while," Bassett says. Over the next few weeks, she began treating the young man in her office. In the course of a conversation Bassett commented to the young man's mother that the boy looked quite different from his parents. His forehead jutted out slightly, subtly affecting the placement of his nose and eyes. "She told me that, in fact, the boy looked a lot like her brother, who also had schizophrenia," says Bassett. "Well, it was the kind of remark that just makes you stop and say, 'Wait a minute.'"

Bassett then examined both men and uncovered several other similarities. The uncle also had developed symptoms when he was twenty. Both men had a constellation of similar subtle deformities—on their ears, fingers, and toes. The boy's left kidney appeared abnormal; the uncle's left kidney was missing. Bassett then had the men's DNA analyzed under a microscope. The finding wasn't surprising: They shared a distinctive alteration in the DNA. They both had an extra piece of chromosome 5 that somehow had lodged itself into chromosome 1. "We thought it was a remarkable finding, something no one had ever seen, or at least reported, before," Bassett says. "The only explanation is that the abnormalities, including the schizophrenic illness, were the result of their shared chromosomal abnormality."

In the summer of 1987, the annual meeting of the American Psychiatric Association in Chicago, as could be expected, was abuzz from the reports about manic-depression and Alzheimer's disease. The potential for genetic research dominated the discussions. To the psychiatrists, almost all of whom were unschooled in the complexities of gene hunting, the possibilities presented by the DNA probes seemed limitless. The reports excited the attendees as nothing else had in years.

In this atmosphere Bassett presented her findings about the two Vancouver men, saying that she had uncovered a compass for a schizophrenia gene. "While it is now possible to test almost the entire genome for linkage by means of molecular DNA markers, such a study would require much greater resources than a test of one particular candidate region," Bassett said at the meeting. "The subject family provides the first clue to link schizophrenia to a specific region (chromosome 5) of the human genome. . . . If such studies find linkage it may become possible to isolate a major gene that predisposes to schizophrenia."[17]

Within weeks, research teams in London and at Yale began testing genetic probes from chromosome 5 in a collection of families with a concentration of schizophrenia. The British team, led by Robin Sherrington and Hugh Gurling, had been studying the congregation of schizophrenia within three generations of five Icelandic families and two in Great Britain. Of 104 people alive within the seven pedigrees, almost unbelievably, 39 of them had diagnosable schizophrenia. The families, heavily laden with schizophrenia, were sought out by Sherrington and Gurling beginning in 1986, when it became apparent the DNA probes could be used in psychiatric illnesses. "When we heard Ann's report, we knew where to start looking, and we started looking almost immediately," says Gurling.

By the following spring the team hit paydirt. A probe on chromosome 5 linked so tightly with the DNA extracted from the diseased members of the family that Sherrington and Gurling were convinced they had the gene. But they also knew they were racing against a team at Yale that was testing the exact same riflip probes in another pedigree found in Sweden, in which researchers had seen 31 cases of schizophrenia among 157 living members. The Yale researchers were led by Ken Kidd, who had first worked in David Housman's lab on the Amish gene project back in 1983 and 1984. By the summer of 1988, a year after they began testing probes from chromosome 5, the Yale team found strong proof that there was no linkage between chromosome 5 probes and schizophrenia.

While the findings seemed contradictory, the two research teams reached the same conclusions about what the two reports meant. "Schizophrenia, just like manic-depression and maybe Alzheimer's disease, too, is the result of more than one gene," says James Kennedy, a member of Kidd's Yale team. Says Gurling, "There's a gene for schizophrenia, we're convinced of that. There may be a gene somewhere else, too."

In November 1988, the two teams' reports were published in *Nature*,[18,19] although the results already had been widely disseminated throughout the field. In a commentary article accompanying the two contrasting findings, Eric Lander, a molecular biologist who holds positions at Harvard and the Whitehead Institute in Cambridge, Massachusetts, said that, while the "riddle of schizophrenia" has yet to be solved, "the papers pave the way for a genetic approach to splitting schizophrenia into a collection of distinct diseases." While noting that the disease almost always looks to be the same from patient to patient, the separate findings suggested that schizophrenia, like manic-depression and Alzheimer's disease, was actually the result of several biological defects that produced similar-looking symptoms.

Lander, while not widely known to researchers in psychiatry, was becoming a guru of sorts in the new DNA technology. He was renowned for his work in developing statistical analyses of linkage studies using the new probes. And he was becoming one of the best explainers of the technology to scientists untutored in molecular biology. The argument he made in the *Nature* essay was widely repeated, for it cogently stitched together an intellectual quilt under which conflicting notions in psychiatry could lie side by side.[20] Moreover, Lander's essay, coming as it did from an objective scientist uninvolved in specific gene hunts, provided psychiatry with an influential bit of support. "As with many fields of medicine, a major turning point in psychiatric genetics has occurred with the notion, first suggested by Botstein and colleagues" that riflips can be used to uncover the genetic factors at work in those with an inherited disease, Lander wrote in the essay.

By this time, however, psychiatric researchers needed no more scientific reports to prod them into action. Herbert Pardes, then director of the New York State Psychiatric Institute, a prestigious research facility associated with Columbia University, wrote in early 1989 that "psychiatry is being transformed by genetic research." Pardes also commented that "molecular genetics is as exciting as any science in medicine today, notably in its impact on psychiatry."[21]

By 1989, psychiatric researchers under Pardes's direction at the institute were scrambling for families to study in order to pin down the genetic basis of manic-depression, obsessive-compulsive behaviors, panic disorder, and anorexia and bulimia (the eating disorders) —everything from suicide to bed-wetting. Officials at the National Institute of Mental Health for the first time provided grants to researchers to spur the collection of families. Michael Conneally, the Indiana University geneticist who helped find the Huntington's disease gene, won a grant to track down as many as fifty families, in the United States and abroad, in which at least two living siblings were diagnosed with Alzheimer's disease. Conneally says, "It's reached the point, at least in the mental health field, where the DNA technology is the simple part of the job. What's limiting our efforts is a rich supply of families. That's really where the hard work is right now."

By decade's end longtime Alzheimer's disease researcher Marshall Folstein had turned almost all his efforts to uncovering new families.

"We're at the grind-it-out stage in research," says Folstein, a psychiatrist at Johns Hopkins University School of Medicine. "If we're going to find the gene on chromosome 21, or somewhere else, we need more, more, more families. It's going to take a while."

Indeed, even a dismaying finding reported in August 1989 didn't slow the search. Janice Egeland had continued to collect more people in her Amish family and follow the family members' health. But several individuals thought to be healthy then became sick. When the statistical analysis was recomputed, the once-tight linkage began to unravel, suggesting that the gene wasn't on chromosome 11 after all. "What it means is we've got a lot more work ahead of us than we all first thought in 1987," says Elliot Gershon of the National Institute of Mental Health.

Researchers hope the increased number of families will help them close in on the genes, and also locate still other genes. Meanwhile, some of the scientists began speculating about developing predictive tests based on the markers. If indeed the scientists could agree that two or three riflip probes for each disease could be used to identify the presence of a defective gene involved in a specific disease, then they could perform blood tests for the presence of the genes. The tests could be used in families with a history of mental illness to pinpoint who among the relatives carries the gene, and who is at risk of developing the illness. Researchers hope to use such information in long-term studies to pick out why some gene carriers get the disease and others don't. "It will provide a powerful way to figure out finally what in the environment triggers a mental health problem in an individual," says Pardes, who has since become head of Columbia University. "Once we have the biological defect—the gene—we'll be able to separate it and those things in the environment that make someone who is born with a predisposition develop the disease."

But the prospect of such tests makes some of the researchers anxious. "You know, only the scientists know that these tests are coming down the road because they want them for their research," says Ann Bassett, who by 1989 was working for Pardes in New York. "Families will learn about it and then doctors will find out about them, too. And then what happens when that kind of information gets out? I mean, what if I tell you that you carry a gene for manic-depression? Does that mean you'll get the disease? And what if other people find out, at your job or your health insurer? What are they going to do with that information. All this is going to raise some very tough questions, none of which, by the way, is being discussed very much. But it will be."

In any event, the new thinking in psychiatry—that genes play a pivotal role in causing disease—is already helping some of those where a disease seems to run in the family. Nineteen-year-old Susan Isaacs of New York has found a sort of peace in the knowledge that the anguish that has tortured her for most of her life "is something I inherited." Isaacs, who asked that her real name not be used, and her family were recruited in 1989 by researchers at New York Psychiatric Institute. At four years of age, Isaacs had been diagnosed as clinically depressed. She was deeply "phobic," afraid of noise and of water. "I couldn't go anywhere," she recalls.

By the time she was twelve, she was in therapy and being treated with little success with a battery of antidepressant drugs, tranquilizers, and antipsychotic medicines. Over a course of five years, she saw four different doctors who diagnosed her as manic-depressive. But especially troubling to her were "obsessions and compulsions," Isaacs says. She would get "violent ideas" about her friends and family that wouldn't subside. "I had these thoughts that overtook me completely and that I couldn't get out of my mind, and, of course I felt I was going crazy. When I was younger I would have these ideas that I was having psychic premonitions about what was going to happen and I had myself convinced it was going to happen. But now I realize it was just an obsessive thought that I couldn't let go of."

When she was eighteen, Isaacs went for treatment at Columbia University and was directed to researchers seeking families for the genetic studies. A researcher asked Isaacs about the behavior of other members of her family. "Of course, we'd never really paid attention," says Isaacs. "But my father has obsessions too, although not nearly as severe as mine. It's more like he's very, very rigid about things, very inflexible. He's often depressed." Researchers also found out that Isaacs's older brother is "often very nervous, very anxious about things," says Isaacs. Interviews with Isaacs's parents uncovered that Isaacs's paternal grandmother was similarly affected. "She had to do certain things, like kiss you good-bye, a certain number of times," Isaacs says. "We just thought she was weird." Finally, says Isaacs, as a result of the researchers' interviews, she discovered that several of the grandmother's brothers and sisters had problems, too. A brother was known to ask people the same question over and over again, while a sister was treated for depression.

Based on the findings, the family was brought into the institute

for treatment. Isaacs's brother and father were treated with antidepressant medicines, and "seem a bit more relaxed, a lot less anxious,"
Isaacs says. As for Isaacs, "I still have quite a bit of anxiety, but not
as much as before. I feel, well, it's not all my fault. It's one less thing
to be obsessed about."

Genes
in a Bottle

In early 1985 Arthur Schaffer, a thirty-three-year-old research assistant at Washington University in St. Louis, drove out to the county workhouse on a rather unusual mission. He was out for blood.

"When I went in they patted me down just like you see in the movies," he recalls. Assured that Schaffer carried no hidden files or weapons, the guard escorted him to a small, plain room to meet the object of his visit, a thirty-eight-year-old inmate serving a few months for petty theft. As the two talked, a nurse from the infirmary came in and proceeded to draw a test tube full of blood from the prisoner, which she then handed to Schaffer. His mission completed, Schaffer left.

It was the second time researchers from Washington University had shown an unusual interest in the man. Five years earlier, he had been interviewed by researchers working under psychiatrist C. Robert Cloninger. Cloninger had come up with a new insight into alcoholism that was beginning to gain wide attention among mental health researchers. He had proposed that there may be two types of alcoholism, one determined largely by the environment one grows up in and the other determined by an inherited susceptibility to impulsive drinking.[1] The theory had sprung from an unusual study of alcoholism in Sweden. In the late 1970s Cloninger wanted to confirm that Americans, like the Swedes, suffered these two forms of alcoholism. To that end, he and his associates had gone into the alcoholic wards and clinics in the St. Louis area between 1978 and 1983 and interviewed more than five hundred confirmed alcoholics. The alcoholics then led them to more than one thousand relatives, 70 percent of whom were interviewed about their drinking habits and life-style.

By 1985, however, Cloninger had good reason to go back and contact the St. Louis alcoholics and their families. The psychiatrist was one of the few researchers in the mental health field who realized that the riflip map postulated by Botstein et al. in 1980 would be a powerful tool for finding the genes underlying mental and personality disorders. A bare two years after publication of the Botstein et al. paper, when few physicians had heard of riflips and genome mapping, Cloninger and two colleagues, Theodore Reich and Shozo Yokoyama, had published a detailed description of the new technology in a psychiatric journal. "Recent advances in our understanding of the structure and organization of the human genome have strong implications for investigations of the diagnosis and etiology of psychiatric disorders," they told readers of *Psychiatric Developments* in 1983.[2] The new technology, they said, provides "methods to detect major gene effects, resolve cultural inheritance from biological inheritance, and identify the influence of multiple risk factors on the development and expression of psychiatric diseases."

By 1985 Cloninger felt it was time to begin applying the new technology to a psychiatric and behavioral disorder, alcoholism. To use the riflip technique, Cloninger, of course, needed samples of white blood cells from families of alcoholics where the siblings, parents, and possibly grandparents also had alcohol problems. To find such informative families, Cloninger turned to the original five hundred alcoholics and their first-degree relatives who had been interviewed between 1978 and 1983. They found that families of one hundred of the alcoholics contained at least two other family members who had alcohol problems. Cloninger dispatched a team under his research assistant, Art Schaffer, to locate the alcoholics and their families and draw blood samples from them.

Schaffer had been unable to locate one of the alcoholics who, at the time of the first study, was in his early thirties and already a severe alcoholic. So he drove out to a middle-class neighborhood that was beginning to show the shabbiness of decline to see the man's father, a widower. "The father was in his seventies and very friendly," Schaffer recalls. "The house was full of grandkids running all over the place. I guess he cared for them in the day."

"When he told me his son was in jail, he said it like it was no big deal, like it happened all the time," Schaffer says. The researcher told the father that he needed to conduct interviews with the prisoner's immediate family and to obtain blood samples from them. "He was very cooperative and offered to call all his kids for me."

The father gave blood, although "he didn't admit to having trouble with alcohol. But his family told me later that he did have alcohol problems," Schaffer says. As Schaffer contacted other members of the family, it became clear why Cloninger wanted to use the family in his riflip-based study. In addition to the father, the man in jail had a twenty-one-year-old brother who was a severe alcoholic and an eighteen-year-old brother who already was drinking heavily. Of his five sisters, one was an alcoholic and two others had been diagnosed as depressed. The man's nephews and nieces weren't yet old enough to have any alcohol problems, although two nephews were in their mid-teens, and "I think they might have begun to experiment with alcohol," Schaffer says.

Over the long months of interviews and blood collections, Schaffer and his team would see hard evidence of the democratic nature of alcoholism. "Some of the people didn't have enough money to pay their utility bills," he recalls, citing instances of families living in the middle of St. Louis depending on wood-burning stoves for heat and cooking. At the other end of the financial spectrum, one of the largest families in the study, consisting of more than seventy living members, was that of an alcoholic corporate executive whose parents and children also had problems with alcoholism.

The blood samples are now in Cloninger's laboratory, where the forty-six-year-old scientist is in the midst of sifting through the DNA of the alcoholics and their families, looking for the riflip markers that will distinguish the alcohol abusers from those unhampered by alcohol. Barring the luck that accompanied the finding of a riflip marker for the Huntington's disease gene, Cloninger may well be in for a long haul. In 1989, he and his colleague, Eric J. Devor, counted twenty-three genes that various scientists had proposed as candidates for alcoholism susceptibility genes and which had been located by the new riflip technology.[3] And it is quite possible, even likely, that none of these genes will be the one Cloninger is seeking. Indeed, it isn't even clear yet whether a susceptibility to alcoholism involves one gene or several genes.

"The genetics of alcoholism is at a pivotal point in its history," Devor and Cloninger told readers of the *Annual Review of Genetics* in 1989.[4] "... the years of effort that have gone into establishing the hypothesis that genetic influences in alcoholism do indeed exist have set the stage for new efforts to detect and map alcoholism-susceptibility genes," they declared.

The search for the alcoholism-susceptibility genes has broader implications than uncovering the causes of one of society's most costly diseases. It marks the movement of the new gene-mapping technology into disorders of behavior and personality. The effort is the boldest and most difficult effort yet to tease apart the influences of the genes from the influences of the environment.

Cloninger had come a long way from Beaumont, Texas, the Gulf Coast town noted for its oil refineries and chemical plants, where he was born, as he likes to note, on 4/4/44, the son of an English teacher. "I was always interested in biology and science, and by the time I was ten or twelve years old I was telling people I wanted to be a doctor," he says. In his undergraduate studies at the University of Texas in Austin, he undertook a wide-ranging liberal arts curriculum designed for premed students, which encompassed philosophy, psychology, and anthropology. He entered medical school at Washington University in St. Louis in 1966 and decided on psychiatry as his specialty. "I've always been intrigued with human behavior," he says.

The former Texan didn't wait to gain his board certification in psychiatry to launch his career. During his third year in medical school, one of the school's leading psychiatric researchers, Samuel B. Guze, took him on as a research fellow and thrust him into a study of antisocial personalities of female criminals and their family and social backgrounds. By the time he was ready to begin his residency training in psychiatry, he had already published articles with Guze in leading psychiatric journals.

The study of the criminals brought two facts home to the young psychiatrist. One was that alcohol abuse played a key role in the problems of the crime-tainted families. Alcohol was particularly linked to the more violent crimes committed by family members.

It also struck Cloninger that, if anyone wanted to separate the genetic influences on alcoholism and antisocial personality from the environmental influences, he would have to know far more about genetics than just the Mendelian laws of inheritance taught in medical school. It would require a thorough understanding of the complicated mathematics used by population geneticists to track the passage of a hidden gene through large families. The same year he received his board certification in psychiatry Cloninger received a ten-year-long research scientist development award from the National Institute of Mental Health to learn the basic tools of population or "quantitative" genetics.

In the summer of 1978, in the midst of his genetics training, Cloninger received a letter from Umea, Sweden, which would set the course of his scientific life for the next several years. "I had taken a sabbatical and was studying genetics at the University of Hawaii when I got this letter from Michael Bohman," he recalls.

Bohman, a Swede, was head of the child and youth psychiatry department at the University of Umea medical school. He and his fellow psychiatrists had tapped into what, for a geneticist, was a treasure trove of data. The highly socialized Swedish government keeps extensive records of its citizens' lives. It has centralized court records of all crimes committed by its citizens from petty to capital. In an attempt to combat alcoholism, local temperance boards keep track of incidents involving drunkenness and of drink-related misbehavior, fining those who repeatedly are intemperate and forcing alcoholics into hospitals for treatment. The socialized medical system has detailed records of every Swede who has sought medical care or who has gotten a medical certification required for taking sick leave from work.

Most valuable of all, as far as the Umea psychiatrists were concerned, were records that stemmed from a social policy begun in the early 1930s when the government took on the role of placing children born out-of-wedlock into adoptive homes.[5] The law required that every child born to an unmarried mother be assigned to a child welfare officer, who then oversaw the child's adoption. The government went to great lengths to confirm the paternity of each child, even to the point of having blood tests done, so the fathers of most of the children were known and recorded. Thus, the government had extensive records of both of the adoptees' biological parents and their adoptive parents.

The Umea psychiatrists had been monitoring the psychological and psychiatric fates of the adopted children ever since the mid-1960s and they had gained access to the extensive court, temperance, and medical records of the adoptees. They also had access to the same records of the adoptees' biological parents and adoptive parents (with numbers substituted for names to maintain privacy).

By the early 1970s there were more than twenty-three hundred adopted children who had grown into adulthood—in 1972 they ranged in age from twenty-three to forty-three years—when such behaviors as alcoholism and criminality are manifest.

The records stored in Stockholm provided an unprecedented opportunity to try to separate genetic influences from environmental influences on the lives of the adoptees. The researchers could attempt to make this distinction by studying the criminal, medical, and drinking records of the adoptees to see whether they more closely resembled the lives of the biological parents or the lives of the adoptive parents. If the adoptees tended to repeat the patterns of their biological parents, it would be a strong indication of a genetic influence, since the adoptees hadn't been exposed to the environment of their biological families. Any tendency to follow the fates of the adoptive parents would indicate an environmental influence, since the children didn't carry any of their adopted parents' genes.

Sweden had another asset that made it an ideal natural laboratory for studying genetic influences. Unlike the United States, with its huge influxes of immigrants over the years, Sweden was about as ethnically and racially homogenous as any nation can be. Moreover, a single, middle-class standard of living prevailed, there being few rich and few poor. Thus, the environmental effects of race, ethnicity, poverty, and extreme wealth were minimized, and genetic differences among the Swedes would stand out more clearly than in a more heterogenous population.

Bohman and his colleagues, however, lacked one vital tool to exploit this vast data resource. They had neither the experience nor the training to use the mathematics of population genetics that could separate the subtle effects of genes and environment on behaviors like criminality and alcoholism. Bohman had spotted one of Cloninger's papers on methods of analyzing the transmission of psychiatric illness in families and had written the American for advice on how to tackle the study of the Swedish adoptees.

"I wrote back a twelve-page letter suggesting steps he could take to do the analyses," Cloninger says. "I thought that was the end of it. But Bohman wrote back saying he couldn't find anyone to do the analyses and asked if I would come to Sweden and do it." Quite by coincidence, a Swedish biologist, also on sabbatical, was working in the same Hawaiian genetics laboratory as Cloninger that year. He was Lars Beckman, who happened to be president of the University of Umea, Bohman's institution. Beckman arranged funds for Cloninger to spend several weeks in Sweden working with Bohman. The visit quickly blossomed into a long-running collaboration between Cloninger's laboratory in St. Louis and Bohman's group in Umea, during which Cloninger trained some of the Swedes in population genetics.

By the early 1980s, the Stockholm Adoption Study, as the Cloninger-Bohman study is formally known, began to make some sense out of a century of confusion on the heritability of alcoholism.

Alcoholism, on the face of it, is an environmental disorder in the sense that no one becomes an alcoholic unless he or she picks up a bottle and drinks. But for centuries there's been an almost intuitive conviction that some people, once they start drinking, are inherently prone to chronic alcoholism. At the turn of the century, the American eugenics movement included "drunkards"—along with prostitutes, criminals and the feeble-minded—among the social misfits who should be isolated and sterilized to prevent their genes from being passed on to new generations of misfits.[6] The eugenicists' claims that chronic drunkenness was inherited were discredited in the 1930s, replaced by a widely held conviction that alcoholism was a character "weakness." That view, in turn, was replaced in the 1950s by the conviction that alcoholism was a disease.

Despite this changing view of alcoholism, there was no denying one outstanding fact: "Drunkard parents did indeed have drunkard children. They had them about four or five times more often than did parents who were not alcoholic," Donald W. Goodwin, a psychiatrist at the University of Kansas Medical Center in Kansas City noted in a 1985 review of the history of genetics and alcoholism.[7] But, Goodwin wrote, by the middle of the twentieth century, "running in the family had stopped meaning any inheritance whatsoever. It meant that children saw what their parents did and did the same, precisely like learning French or voting Republican. Blue eyes were inherited; alcoholism was not."

By the 1970s, however, the pendulum was beginning to swing back. Perhaps there was a genetic underlay to alcoholism. Fraternal twins seemed more likely to both become alcoholics than different-age siblings, and identical twins were more likely than fraternal twins to become alcoholics, Goodwin noted. In short, the more genes two siblings shared, the more likely they both would become alcoholics.

But there was also one facet of alcoholism that didn't fit either type of inheritance, genetic or learned. Alcoholism seemed to be a predominantly male disorder. There are four to five times as many alcoholic men as alcoholic women. It was the sons of alcoholic fathers that seemed to be most prone to alcoholism, not the daughters. The daughters of an alcoholic father had just as much chance of inheriting a hypothetical alcoholism gene as the sons, yet few of the daughters became alcoholics.

This was the state of confusion over the heritability of alcoholism when Cloninger and Bohman began their collaboration in 1980 to analyze the familial patterns of alcoholism in the Swedish adoptees. They first tackled the male adoptees.[8] As a group, these men seemed representative of the general male population; about 17 percent had court, temperance board, and medical records indicating problems with alcohol. But the alcohol abuse seemed to be skewed toward the adoptees whose biological fathers had records of alcohol abuse; 22 percent of these men had alcohol problems. In contrast, only 14 percent of the men whose biological fathers were free of alcohol problems were themselves alcohol abusers. This was the first hint at some kind of genetic influence in male alcoholism.

It was when Cloninger and Bohman decided to look at the severity of the alcohol abuse that they came up with even stronger hints of a genetic influence. They singled out those adoptees who had three or more drinking offenses registered at the temperance boards, a criminal record, and/or hospitalization for alcoholism. Here they made an unexpected finding: 18 percent of these moderate-to-severe alcoholics were born of biological fathers who also had records of severe alcohol abuse.

In these moderate-to-severe alcohol abusers, the researchers could see a pattern. The sons like their biological fathers tended to begin drinking heavily in their teenage or early adult years. Their drinking was impulsive and uninhibited and they frequently got into fights and often ended up with criminal violations. Moreover, these adopted men seemed to fall into early alcoholism regardless of whether their adoptive parents drank or were teetotalers; their home environment seemed to have little influence on their proclivity for alcohol.

To be sure, being born of a biological father who was a severe alcoholic didn't preordain a son to also become a severe alcoholic. Four-fifths of the adopted men born of such fathers *did not* become alcohol abusers. Nevertheless, it appeared that if an adopted boy's biological father had been an early-onset, severe alcoholic, the boy had a nine-fold higher risk of becoming an early alcoholic than did the son of a biological father who didn't abuse alcohol.

The researchers labeled these early-onset, severe alcoholics as Type II alcoholics. They represented about a quarter of the adoptees who were alcohol abusers.

The remaining three-quarters of the alcohol-abusing adoptees seemed to suffer a milder form of alcoholism, dubbed Type I. These adopted men might have begun drinking in their late teens or early twenties but they only gradually slid into alcoholism. A few slipped into severe alcoholism but most seemed to suffer drinking problems without any criminal violations, or at least not violations resulting in jail sentences.

Whether these late-onset Type I alcoholics were born of alcohol-abusing biological parents didn't seem to make much difference in whether they, themselves, became alcohol abusers—unless the boys had been adopted into a home of an unskilled worker, a stratum of Swedish society where drinking is commonly a part of home and social life. Having a biological father and/or mother who was a mild alcohol abuser plus growing up in a home where drinking was socially acceptable increased the risk of these boys slipping into mild alcoholism in their late twenties or early thirties. If, on top of this, the adopted men had been hospitalized as infants while awaiting adoption—an environmental stress—there was a greater chance they would slip into severe but late alcoholism.

In sum, environmental influences seemed at work among the late-onset, mild Type I alcoholics—and genetic influences seemed to be at work among the early-onset, severe Type II alcoholics.

This discovery of the two types of alcoholism—one environmentally based and the other genetically based—would dictate Cloninger's research for years to come and would change alcoholism research on a broad front. It was a key clue to the reasons for the confusion that had surrounded alcoholism for decades.

But Cloninger and Bohman still had to tackle the other major mystery of alcoholism, that of the daughters of alcoholics. Not surprisingly, alcoholism was far less common among the adopted women than among the adopted men: 3.4 percent versus 17 percent. Being born of a biological father who was a mild, Type I alcoholic seemed to slightly increase the risk of a daughter becoming a mild alcohol abuser. Otherwise, there seemed to be little correlation between the women's susceptibility to alcoholism and the drinking habits of either their biological or adoptive parents.

Obviously, a major question surrounded the adopted women: Why weren't the biological daughters of the severe, early-onset Type II alcoholics also Type II alcoholics? After all, they had the same chance of inheriting the putative alcoholism-susceptibility genes as the sons.

Cloninger had picked up a clue to this mystery in his earlier studies of criminals and their families in the St. Louis area.[9] In some families where there was a tendency for the men to become criminals, there was a tendency for the women to suffer Briquet's syndrome or what was once called hysteria. This was a psychiatric illness, seen most often in women, which manifested itself in multiple vague complaints, such as frequent headaches, nausea, stomach and bowel problems, fatigue and menstrual difficulties. Briquet's syndrome, along with hypochondria and so-called psychosomatic illnesses, are broadly classed as somatoform disorders (from the Greek *soma*, for body).

In his study of the criminals, Cloninger had concluded that somatoform disorders, such as Briquet's syndrome, were the female counterpart of a tendency towards criminality in men. Was there a similar link in alcoholism?

To find out, Cloninger, Bohman, and Bohman's colleagues, Soren Sigvardsson and Anne-Liis von Knorring, delved into the medical records of the adopted women. The correlations suddenly emerged.[10] Almost 17 percent of all the adopted women had two or more sick leaves per year, a considerably higher proportion than women in the general population (11.5 percent). They also had more psychiatric problems and took longer sick leaves for those problems than non-adopted women. Moreover, the psychiatric problems seemed to be concentrated in a few of the adopted women: 144 or 17 percent of the women accounted for about half of all the sick leaves taken by the adopted women.

As they sifted through the women's medical records, the researchers began turning up a remarkable parallel to alcoholism seen in men. There appeared to be two different types of women with somatoform disorders, just as there were two different types of alcoholism. Moreover, one type of "somatizer," as the suffering women were called, was highly likely to have had a biological father who was a severe, early-onset alcoholic, the Type II alcoholic in which there was a strong genetic influence.

The biological daughter of a severe Type II alcoholic was likely to be what the researchers named a "diversiform somatizer," a woman whose bodily complaints were highly diverse. A typical case was a thirty-three-year-old secretary with one child who had taken forty-seven sick leaves since she entered the work force at age eighteen.[11] "She was reared in harmonious circumstances in a middle-class family in a large city," the research team reported. A year after she had gone to work, "she began to complain of fatigue, nausea, abdominal pain, and viral infections. . . ." At twenty-one she married a man twice her age who turned out to be an alcoholic. During the one-year marriage she began to take sick leaves for depression, fatigue, and abdominal complaints. She took on a new job at age twenty-seven, but "she continued to have complaints of fatigue and tension, associated with abdominal pain, joint pains, dizziness, and frequent minor infections." At age thirty-three, she went to a hospital emergency room with a complaint of rapid heart beats, but an ECG was normal.

Such a diversiform somatizer differed from the other type of somatizer, called a "high-frequency somatizer." The latter had far more sick leaves but her physical complaints were concentrated in the back and abdomen, and she had frequent psychiatric complaints. The biological father of this type of somatizer was most likely to have had a record of violent crimes but no history of being treated for alcoholism.

For Cloninger, the Stockholm Adoption Study began laying the foundation for a new view of the genetics of alcoholism. There were indeed strong hints that a susceptibility to a moderate-to-severe form of early-onset alcoholism—Type II—could be inherited by males. The same genes that rendered the males susceptible to this type of alcoholism seemed to render females susceptible to a life of never-ending vague bodily complaints.

At the same time that he and Bohman were studying the Swedish adoptees, Cloninger and Reich were launching a major study of alcoholics in the St. Louis area. By 1983, they had contacted more than five hundred alcoholics and were interviewing their families for evidence of alcoholism. With the findings of the Stockholm study in hand, they decided to look into the families of 195 of the St. Louis alcoholics who had been hospitalized for alcoholism.[12]

In the St. Louis families, Cloninger, Reich, and Sheila B. Gilligan also saw two distinct types of alcoholism among the males. The fathers, brothers, and uncles of the hospitalized male alcoholics who also abused alcohol "were characterized by the inability to abstain completely from ethanol or by trouble stopping drinking. They became involved in more fighting and reckless driving while intoxicated, and they were treated more frequently for alcohol abuse" than the male relatives of hospitalized women alcoholics.[13]

On the other hand, the fathers, brothers, and uncles of the hospitalized women alcoholics who also abused alcohol tended to be like their female relative in the hospital. Their problems with alcohol began at a later age, "and abuse was characterized by a loss of control over

drinking, feeling guilty about level of drinking, and alcoholic bend-
ers," defined as going on a forty-eight-hour drinking spree "with de-
fault of usual obligations."

The distinction between the two types of alcoholism was even
clearer when the researchers divided the families into two groups,
regardless of the sex of the hospitalized alcoholic relative. In one type
of family, which they labeled "female-like," both sexes tended to
suffer the milder Type I form of alcoholism. In the other "male-like"
families, the alcohol abuse tended to be the moderate-to-severe Type
II alcoholism, "characterized by earlier age at onset of alcoholism and
higher incidence of antisocial personality."

Cloninger and his colleagues also applied the mathematics of population genetics to the St. Louis families to see if they could pick up
any hints of unseen alcoholism-susceptibility genes being transmitted
within the families. The analyses showed that in the "female-like"
families there wasn't any major effect of a gene or genes. But the
analysis of the "male-like" families failed to rule out a major genetic
effect at work, particularly among the male alcoholics. It was only a
hint but an encouraging one.

The evidence of two types of alcoholism, one of which had a strong
genetic component, was becoming convincing. But how did the genes
exercise their power to turn some people to early severe alcoholism?
Cloninger began to suspect the answer lay in the nervous system.
Humans come into this world with a brain and nervous system that
was assembled by the genes they inherited from their parents. How
this nervous system reacts to external stimuli, to lights, sounds, tex-
tures, tastes, and to situations, is genetically determined. Each stim-
ulus that the nervous system reacts to—whether it be a mother's caress,
a startling noise, a parent's scolding, or a teacher's attempt to train
the system to read and write—is, of course, environmental. The product of this interaction is the personality. To the extent that anyone
can measure such interactions, the personality is about half inherited
and half environmental in origin.

If the genetically based nervous system is involved in alcoholism,
then there should be personality differences between the two types
of alcoholics. Cloninger began exploring the personalities of the two
types of alcoholics and found differences.[14] The milder Type I alcoholic, with his loss of control and guilt feelings, seems to be guided
by external circumstances. His heavy drinking likely started with, and
was encouraged by, an environmental influence such as drinking with
friends over lunch or after work. He is "emotionally dependent, rigid,

perfectionist and introverted," Cloninger found. He also is anxious, eager to help others, sympathetic, and always on the lookout for social cues from others. He tries to stay out of trouble and is low on "novelty seeking," that is, he doesn't tend to be carefree.

The Type II alcoholic—the type that Cloninger believes may be subject to a genetic influence—is dramatically different. He's an impulsive, antisocial risk-taker who begins seeking alcohol in his teen years. He is excitable and disorderly with a high "novelty-seeking" trait. He is low on "harm avoidance," that is, he is uninhibited, carefree, and energetic. As a result, when he's drinking, he frequently ends up in fights and is often arrested.

In 1987, Cloninger stuck his scientific neck out and proposed a broad theory of the genetics of alcoholism based on the personality differences between the Type I and Type II alcoholics. There are three personality traits that distinguish the two types of alcoholism, he explained. And each trait has a biological—i.e., heritable—basis. One trait is "novelty seeking," which is "a heritable tendency toward frequent exploratory activity and intense exhilaration in response to novel or appetitive stimuli."[15] The biological basis for this trait is the activation of a chemical in the nervous system called dopamine. It is one of the chemicals that transmits electrical impulses from one nerve cell to the next. When the activity of the dopamine steps up, the individual gets a feeling of reward. Alcohol, Cloninger noted, has been found to excite nerve cells in the dopamine system. Experiments have shown that sons of alcoholics may experience a much stronger response to dopamine—and hence a much greater feeling of reward—than sons of nonalcoholics. In other words, males born with a nervous system that reacts strongly to dopamine may be susceptible to alcoholism because they get a bigger charge (reward) from alcohol than others with a less reactive dopamine system.

A similar situation exists with the other two traits, Cloninger argued. Another part of the nervous system, built around a nerve-transmitter called serotonin, inhibits activity; it is a kind of brake. When this system is excited, it gives one a feeling of aversion; it is what helps a person learn to avoid punishment or novelty. People who inherit a highly active serotonin system tend to be fearful, inhibited, and shy. In these individuals, alcohol tends to block the inhibition and they are less anxious upon drinking, thereby they tend to develop a psychological dependence on alcohol—a trait seen in the Type I alcohol abusers. Those who are born with a rather sluggish serotonin system are uninhibited and carefree—a trait of the Type II alcoholic.

The third trait—reward dependence—is a kind of maintenance system, which keeps the nervous system from forgetting that it just received a reward or avoided a punishment. The neurotransmitter for this system is known as norepinephrine. People born with a high sensitivity to norepinephrine tend to respond strongly to signals of reward, such as social approval, and persist in reward-seeking behavior. Individuals high in reward-seeking behavior are sensitive to social cues (a trait of the Type I alcoholic). Those born with a low sensitivity to norepinephrine would be socially cool, independent, and self-willed (a trait of the Type II alcoholic).

In short, Cloninger proposed in his 1987 theory that the alcoholism-susceptibility genes actually were the genes that determined how the nervous system reacts to the outside world. The Type I, environmentally influenced alcoholic is high in reward dependence (norepinephrine sensitive), high in harm avoidance (serotonin sensitive), and low in novelty seeking (dopamine insensitive). The Type II alcoholic appears to inherit a nervous system that produces a high novelty-seeking tendency, while being low in harm avoidance and low in reward dependence.

Now Cloninger had to test his theory.

He and his Swedish colleagues decided to see if personality could predict whether a child might be susceptible to alcoholism later in life. By the mid-1980s the Umea psychiatrists had been following the adopted children for several years. In the late 1960s they had made behavioral assessments of almost five hundred of the children when they were ten or eleven years old.[16] They interviewed the children and their teachers and examined their report cards and health cards. They checked such facets of personality as orderliness, ambition, diligence, trustworthiness, conflict with teachers and other children, initiative, and popularity.

In 1987, Sigvardsson went back into the files to see if any of the adoptees had developed alcohol problems at age twenty-seven. He found that only two of the 198 girls had records of alcohol abuse—but 30 of the 233 boys had such records. Deliberately remaining ignorant of Sigvardsson's findings, Cloninger and Bohman took the behavioral assessments made seventeen and eighteen years earlier and rated the children on their harm avoidance, novelty seeking, and reward dependence traits, the traits that distinguished the two types of alcoholism. They then "predicted" which of the adopted children ran a higher-than-average risk of developing problems with alcoholism when they became adults. Their predictions were checked against what Sigvardsson had found to be the actual fates of the children.

Considering that they were trying to predict what would happen to children seventeen years in the future, after they had experienced all kinds of unpredictable social buffeting, the risk calculations were remarkably reliable. For example, a low rating in harm avoidance suggests a higher risk for alcohol abuse. Cloninger and Bohman spotted sixty-four boys who, at age ten or eleven, were low on harm avoidance; twenty-two of these adoptees had records of alcohol abuse at age twenty-seven. Similarly, eighty-nine of the young boys had a high rating on novelty seeking, another high-risk trait, and nineteen of these boys had records of alcohol abuse seventeen years later. Generally, the predictions based on personality traits at age ten missed many of the boys who would later have trouble with alcohol (the sensitivity of the prediction was low)—but when Cloninger and Bohman did predict a boy had a moderate to high risk of becoming an alcohol abuser in adult life, they were right more than 90 percent of the time (that is, in the jargon of the statisticians, the prediction had a high specificity).

After a decade of research, Cloninger had succeeded in identifying, in the midst of what had been a confusing morass of conflicting data and theories, a form of alcoholism which appeared to arise from a genetic susceptibility.

The research already has had an impact in the discovery of a link between one behavioral disorder and alcoholism. The disorder is known as Tourette's syndrome after Gilles de la Tourette, the French scientist who first recognized it in the nineteenth century. It is known to be an inherited disorder with a rather unusual and startling symptom. It starts in childhood with involuntary motor tics or spasms of the facial muscles. But as the child grows, the tics are accompanied by grunting or barking noises. Many of those afflicted then develop an uncontrollable compulsion to suddenly shout scatological obscenities, a symptom known as coprolalia.

About 1980, medical geneticist David E. Comings at the City of Hope Medical Center in Duarte, California, and his wife, Brenda Comings, a psychologist, took a special interest in Tourette's syndrome among schoolchildren.[17] "When Brenda and I first became involved in studying this disorder, we thought it was an extraordinary rare disease of the type that is of interest mostly to medical geneticists," Comings recalled in a recent article. Initially, they encountered only one or two cases of the syndrome a month. "After our first public discussion of the syndrome on television, a rising tide quickly became a flood, one a week, two a week, then four a week, and now, for the past two years, eight to twelve a week." By carefully monitoring the boys in three different elementary schools, they found that one of every one hundred boys displayed some symptoms of Tourette's syndrome, including stuttering, compulsive behaviors, and learning and reading disabilities, as well as involuntary tics.

It was when they looked into the family pedigrees of boys with Tourette's syndrome that the Comingses began to uncover evidence of a putative gene that appeared to underlie a variety of obsessive-compulsive behavior disorders and, to their surprise, a higher-than-expected incidence of alcoholism. In one study, they found that almost 19 percent of the relatives of twenty Tourette's boys suffered alcohol problems, compared with 4 percent in control-group families.

As they studied the families, they began to narrow the alcoholism problem to a select few, where as many as half the relatives had alcohol problems. David Comings cites, as an example, a sixteen-year-old boy who started abusing alcohol at age fourteen. His mother was an alcoholic, and her father had a chronic motor-tic and was an alcoholic. His mother had one sister who was an alcoholic and, most interestingly, she had another sister who had an obsessive-compulsive disorder. When one looked at the pedigree of this family, it seemed as though the mother had inherited one copy of a gene from her father that underlay both alcoholism and Tourette's syndrome, and she had passed this one copy on to her son.

The Comingses teamed up briefly with Cloninger and Devor to look at the pedigrees of the Tourette's boys' families, and in 1984 the four reported finding firm evidence of a major gene at work in these select families. They labeled it the "Ts" gene, for Tourette's syndrome.

Noting Cloninger's discovery of the two types of alcoholism, Comings wrote in 1989, "I have been repeatedly struck by the similarities between the psychological profile of Type II alcoholics and that of many individuals with TS [Tourette's syndrome]. . . ."[18] Using the broad definition of Tourette's syndrome to include compulsive-obsessive disorders, both the syndrome and Type II alcoholism begin before age twenty-five, and victims share high sensation-seeking and risk-taking personality traits and a low reward-dependence trait.

Type II alcoholism, however, is predominantly male, while the TS families included many females with obsessive-compulsive disorders—and here the Comingses found a surprise. "When we ask female relatives of our TS patients whether they have any compulsive behaviors, a frequent answer is, 'Yes, eating,'" David Comings wrote.[19] Eating disorders included both over-eating to obesity and under-eating in the form of anorexia nervosa. "I suspect that in some cases the gene for TS, for male Type II alcoholism, and for the female type of familial obesity are one and the same and that the primary problem is an appetitive compulsion that takes the form of alcoholism in men and of overeating in women," David Comings asserted. "These are not absolutes—the men can also overeat and the women can over-drink. I am also not saying that all Type II alcoholics or all obese women carry a Ts gene—just that some may."

The pedigrees of some of the families began to make sense if one assumed that those who were alcoholic, obese, or suffering some kind of obsessive-compulsive disorder had inherited one copy of the putative Ts gene, that is, were heterozygous for the gene—and if those with full-blown Tourette's syndrome had inherited two copies of the gene, that is, were homozygous for the Ts gene. Comings cited the case of a family of one TS boy. His father was an alcoholic and had an attention-deficit disorder. His mother had panic attacks and weighed 360 pounds. The boy's maternal grandfather was a manic-depressive and his maternal great-uncle was an alcoholic. One copy of the Ts gene had apparently descended on each side of the family to the boy.

Comings noted that people with such genetically based behavior disorders are rarely helped by psychotherapy but often respond dramatically to chemical (i.e., drug) therapy. If these individuals could be identified with the new genetic probes, he suggested, billions of dollars in social programs might be saved.

Back in St. Louis, Cloninger is now trying to find the alcoholism-susceptibility gene or genes that underlie Type II alcoholism. Thanks to Arthur Schaffer and his team, Cloninger's group now has blood samples and pedigrees from the families of one hundred alcoholics in the St. Louis area who were selected for the study because at least two siblings in each family were alcohol abusers. Most are Type II alcoholics, although some Type I alcoholics are included. The St. Louis researchers are now using riflip markers to track the inheritance of various genes through each of the families, hoping to find one or more markers that distinguish the alcoholics (and, in the case of females, the somatizers) from the nonalcoholics. It is a long and tedious task, which may take several years—or which may hit paydirt at any moment, as happened with Huntington's disease in 1983.

In April 1990 one team of scientists said they found that a riflip linked to gene that controls dopamine indicated that the gene was associated with severe alocholism.

In a report in the *Journal of the American Medical Association*, scientists from the University of California, Los Angeles, and the University of Texas Health Science Center in San Antonio said they found the dopamine gene riflip in 77 percent of the brain tissue samples they studied from thirty-five deceased alcoholics. The same riflip was present in only 28 percent of the brains of people who weren't alcoholics.

The riflip previously had been found by other reseachers to reside on chromosome 11. The gene itself is believed to produce a dopamine receptor on nerve cells. The scientists said they weren't certain if the dopamine receptor gene they were tracking was a defective version of a healthy gene, but they declared that something about the gene altered the way nerve cells handled dopamine and might therefore explain at least some cases of alcoholism.

While acknowledging that their work was preliminary and needed to be reproduced, Ernest Noble of UCLA said the riflip might be used to detect whether someone is susceptible to alcoholism. Noble said he expected the gene to be only the first of at least a dozen or so genes that might predispose people to alcoholism. "This is just the beginning," Noble said.

Devor and Cloninger wrote of their methods and their hopes in 1989, "It is to molecular biology, gene mapping, and linkage analysis that research on the genetics of alcoholism is turning. The task is enormous, the pitfalls many, and the cost high, yet the potential payoff is staggering—a fundamental understanding of the genetics of an addiction and a realistic program of intervention based on that understanding."[20]

CHAPTER 13

Predictive Medicine

n ancy Wexler danced with surprise and joy when, that day in the spring of 1983, she was told the gene for Huntington's disease had been located. Certainly, the gene's exact identity would soon be found, she thought, and a treatment, perhaps even a cure, would be possible, although that was probably still years away. But she knew that immediately—within a few months at the most—scientists could take the DNA probe that pinpointed the gene's location and use it to find out whether or not people at risk, like herself, were destined to develop the devastating and ultimately lethal disease.

Within days it was clear that the gene test would certainly stand as one of the century's greatest medical achievements, for it would be the first diagnostic test ever to predict the onset of disease decades before the first symptoms arose. That test would be the first of what could be many tests for a variety of other genes associated with several diseases. Finally, there was a way for some people to know what their body had in store for them. For that alone Wexler felt great satisfaction and pride.

But perhaps sooner than anyone else, Wexler also understood that the gene test would carry with it unprecedented complications. Yes, the people who found out that they hadn't inherited the gene would finally be relieved of a "terribly burdensome anxiety, fear, and worry," she said. For them the nightmarish game of "genetic roulette" would finally be over. They could carry on their lives, free of the cloud that shadowed them.

But those who learned they had the gene "might be hurt in ways impossible to gauge," she warned friends and colleagues. Within days of the discovery of the gene's location, Wexler was cautioning people that the test's result might drive otherwise healthy and symptom-free people to suicide. Almost before she could catch her breath, Wexler began making plans to protect those who would be the first people ever to learn the consequences of their genetic destiny. In conversations with researchers and doctors, she argued that special counseling programs must first be put into place before anyone was given the test.

Even so, Wexler was optimistic about the test's impact. Asked if she would take the test herself, Wexler was coy. In an interview in *Columbia*, a magazine of Columbia University, Wexler said, "When it first became available, my sister, father, and I had no question about taking it. My sister and I thought, 'Isn't this fantastic! We could have children! My father could stop saving money for nursing homes and retire.' When we thought more about it, however, my father was first to say, 'Wait a minute! I don't want to know if either of you has a bad outcome. One bad outcome and we're all three dead.' Both my sister and I would have to have a good outcome for all of us to be free."[1]

Indeed, by decade's end, Wexler was deeply sobered by the experience of testing dozens of people at risk for Huntington's disease. Six years had passed and researchers still hadn't isolated the gene itself; the task was taking longer than she expected. Many people at risk for Huntington's disease could discover whether they had the gene, but scientists had made no advance in providing treatment or in finding a way of preventing the gene from carrying out its grisly mission. At least finding the gene might provide these people some hope of treatment; but as yet there was none. "It's not a good test if you can't offer people treatment," she had decided.

Now associated with Columbia University and the New York Psychiatric Institute, Wexler is sitting in a small and cluttered corner office with a view of the Hudson River as it flows southward to New York's harbor. Usually cheerful, and quick to crack a joke, she becomes quiet and cold when asked about the test.

"There are problems we never could have predicted," she says. "We hear that insurance may be denied to people who know they have the gene. Confidentiality is a big issue, even within families, sometimes between brothers and sisters, because some people want to know and others don't." Initially she thought that knowledge itself, even if it meant discovering one had the gene, would provide great relief to people at risk. Those who found out they had the gene, she originally felt, could better plan their lives—taking advantage of their time remaining. But, she now realizes that the uncertainty doesn't end, it merely changes. "Instead of worrying if they have the disease, people who are told they have the gene now worry when they will get it," she says. "The test still provides some people great freedom, but it also can be very destructive. People have spent their lives shielding themselves against the disease. Bad news will break that shield apart and what will they replace it with? Knowledge alone doesn't provide the support you need to live your life, you need to know there's hope."

At Boston University, Richard Myers, a geneticist and psychologist who runs a Huntington's disease test program, echoes Wexler. "It's a dangerous test," he says.

Twelve years after David Botstein and Ron Davis first proposed the notion of the riflip probe in the meeting with Mark Skolnick in Alta, Utah, predictive tests like this one for Huntington's disease are proliferating. None are as extreme as the Huntington's disease test, which can look into the crystal ball of DNA and see almost a half century into the future. But all the tests are freighted with a heavy cargo of benefit and pain, much of which was unanticipated by the gene hunters or those who first administered the tests.

Using the probes, doctors can conduct tests on an unborn child's DNA to detect the presence, for the first time, of about a dozen hereditary illnesses, including muscular dystrophy, cystic fibrosis, and hemophilia, as well as lesser-known problems such as retinoblastoma, polycystic kidney disease, and retinitis pigmentosa, an inherited eye disease. For many families, these prenatal tests provide answers to desperate questions. Many couples at risk of passing along a defective gene had decided to forgo having their own children rather than risk giving birth to a child afflicted with a hereditary illness. But with the advent of the tests, many of the couples finally dare to take the risk. Those who learn that the fetus's DNA harbor the deleterious gene, and feel they can't carry to term an affected offspring, abort early on. Others who decide to deliver an affected fetus use the information to steel themselves, financially and emotionally, for having a disabled child. And, for those who learn their unborn child is free of disease, "there is immeasurable joy," says Patricia Ward, a genetics counselor at Baylor College of Medicine in Houston, Texas, who daily deals with couples at risk of bearing an affected child.

"For some, after so many years without any hope, it takes a while to believe they suddenly can have mastery over the health of their future children," Ward says. "[The tests] give people the ability to decide, and remove a powerlessness that pervaded their lives."

Some researchers quickly realized that the power provided by these predictive tests would dwarf any of medicine's other diagnostic tools. Mark Skolnick told the *Wall Street Journal* that he was convinced the spate of gene marker discoveries "would revolutionize medicine."[2] And in an article in late 1984, Ray White proclaimed the potential power of the new gene discoveries as "far-reaching, extending not merely to genetic disease; predispositions to several illnesses and responses to infectious agents and to drugs should become more predictable." He boldly prophesied that "we will soon be in a position to reduce the toll of human genetic disease."[3]

But, even so, he saw the potential for problems. "Intriguing questions are beginning to emerge as we contemplate other applications of knowledge of individual genotypes," he wrote in an article in *The Lancet*. "Is it ethical or even useful to reveal the future when no therapy exists? Insurance companies will want to know of people at increased risk genetically of heart disease. Should they be allowed access to such information? Will there be a statutory requirement for a man and a woman to share genetic information before they marry and could failure to disclose genotype become grounds for divorce? Will 'my genotype made me do it' become an acceptable defense in criminal cases?"

And others were less sanguine about even the medical benefits. "Put yourself ahead 10 years or even five, and ask yourself what it will be like to live in a society where medicine can diagnose everything and treat very little," John C. Fletcher, a bioethicist at the National Institutes of Health, also told the *Journal*. "Without treatments, prenatal diagnosis and abortion may become the only or the most-used [outcome]. Society has yet to decide if that's how it wants to deal with disease. But it's going to have to decide, and do it soon."[4]

Almost from the outset, the Huntington's disease test surprised geneticists. Nancy Wexler used her newfound influence, gained from being a driving force behind the gene's discovery team, to help resist calls for immediate testing of people at risk of having the disease. At a genetics meeting in Salt Lake City in 1985, Arthur Beaudet, a genetics researcher and doctor at Baylor College of Medicine, raised a ruckus when he took the floor of a crowded session and charged that Wexler and her colleagues were wrongly blocking the release of the test by refusing to share the gene marker. Beaudet said the researchers were being paternalistic. In an interview at the time he said, "They have to have more faith that people will use the test responsibly. When science makes an advance you can't necessarily dictate how it's going to be used."

Nonetheless, Wexler was able, initially at least, to convince geneticists to hold off until counseling programs were in place and a few medical centers had experience in using the test. Three centers were designated to begin testing; one associated with Jim Gusella at Massachusetts General Hospital in Boston, one at Johns Hopkins Hospital in Baltimore, and another at Columbia University in New York City. The centers had surveyed the nearby population of Huntington's disease families, and between 60 and 80 percent of those people at risk for having the gene said they would come forward for testing. But, instead, many people held back. At Hopkins, only 73 of 450 people who said they would be interested in the test showed up for initial counseling programs. Of the 1,500 people at risk in the New England area, only 32 signed up after the first year. In 1989, Boston University School of Medicine was one of several new testing centers. But while it received 200 inquiries about the test, less than a half-dozen people took it.

"I get calls about every other day from people asking about the test," says Boston University's Richard Myers. "But hardly any of them actually take it."

The principal reason is that people are "frightened" they will lose their health, life, and disability insurance, says Myers. In 1989 the test at Boston University cost about three thousand dollars. Myers says, when some people asked their health insurance plans if they would reimburse the price of the test, the insurers said they would cover the costs only if they were told the results. "The fear, and it's legitimate, is that the insurer will limit coverage or drop them completely if it turns out they have the gene," says Myers. "It's a terrible situation because the person with the gene will need the protection of insurance even more, not less."

Thus, Myers warns callers about "the insurance problem," he says. As a result most people who contact him begin putting money aside to cover the cost of the test themselves. Myers guesses this may be one reason for the low number of test requests. Myers also warns people that once they know they have the gene, they may be denied health and life insurance in the future even if they don't seek insurance reimbursement for the test. "People ask me what they should write on an employment or insurance application when it asks them if they have a health problem; should they write they have this gene that will cause a problem in ten or twenty years?" Myers says, "I'm still not sure what advice to give. But I'm sure my explanation, that they may jeopardize their insurance, is why a lot of people aren't taking the test."

Myers says that while he is upset about the insurance issue, he understands the quandary the companies are in. In a conversation with one insurer Myers says he was asked, "Is it fair for someone who knows they are going to die prematurely to load up on a million dollars in life insurance?" Says Myers, "I didn't like what the people at the company said, but they had a point. You know, we felt people would have all sorts of reasons for not taking the test. They might not want to know, maybe people in their family who had to provide blood for analysis wouldn't be helpful or were unreachable. But the insurance thing isn't something we expected or were prepared for. Yet, it's become the greatest barrier to testing we know of. People feel very scared about letting this information loose. They're worried who will get hold of it."

For some people, the need to know, however, overcomes any such concern. Several families have told their stories publicly, some, perhaps, to unburden themselves. Some hope that, by sharing their experiences, they will help others understand the unfathomable—what it's like to peer inside yourself and see what the future holds.

In a 1988 article in *Johns Hopkins Magazine,* Paul and Karen Sweeney, a couple in their late twenties with four children, say they no longer could live with the uncertainty of whether Karen had the Huntington's disease gene. Not knowing had become so unbearable, they moved from Alaska to Virginia to be among the first to get the test. Hopkins was offering the test free of charge as part of a government study, but the study's rules, put together by Wexler and others, required that those being tested must live nearby to be available for frequent psychological counseling.

Karen's mother had developed Huntington's disease at thirty-three while pregnant with Karen, and the disease had stalked Karen's life ever since. "It's the waiting and wondering that destroys," Karen told the magazine. "It kills from within."[5]

Karen's mother died when Karen was eleven years old, and during her teenage years Karen became convinced that she had inherited the gene from her mother. She says she lost interest in the future, became alienated, took drugs, and got into trouble with the law. Fortunately, she landed in a foster home with a couple who helped her work through the emotional trauma.

As with all children of a Huntington's disease victim, the chance that Karen had been passed the bad gene was fifty-fifty. Her mother had one defective gene on one copy of her chromosome 4—the one that caused the disease. But she also had a second copy of chromosome 4 that contained a healthy version of the same gene. Since children receive just one of each parent's two chromosomes, they can inherit either the chromosome containing the problem gene or the chromosome with the normal gene. Before Nancy Wexler and her colleagues found a marker for the aberrant gene, there was no way to know which chromosome a child inherited, until the disease's symptoms first surface sometime after the age of forty. Even so, for scientists to use the marker in a predictive test, they must be able to study DNA from several family members—at least one of whom must have the disease.

When Hopkins began offering the test, Karen and Paul had been together for ten years. For a long while, the Sweeneys say, concern over Karen's fifty-fifty genetic risk took a backseat in their lives. But by the time they arrived at Hopkins, the disease had become a constant source of anguish. They were keeping a close watch for the first sign of problems—an absentminded slip, a glass dropped to the floor. Before they received results of the test, Paul said he felt "bitter and betrayed at the thought of losing her," and their marriage was loaded with stress. "I'm crazy about her," Paul told the magazine, "but sometimes I'm afraid to love her. . . . It eats at me."[6]

The Hopkins geneticists warned the Sweeneys that they might not be able to tell the difference between the family's healthy and aberrant chromosomes, a problem that sometimes happens if there aren't enough family members from which to study the co-inherited riflip as it travels through the generations. Sweeney's brother, eleven years older than her, was beginning to show signs of the disease and could provide DNA that contained the faulty gene. Her father and great-uncle, who didn't have the gene, would also provide blood for a DNA test. After several months of analyzing the DNA with the gene marker, geneticists were finally able to pick apart which chromosome carried the disease gene, and they told Karen they would be able to tell whether she had received it from her mother.

The Hopkins magazine recounted the Sweeneys' visit to the hospital for the test results, and their reaction:

Karen Sweeney paces the room, unable to sit still. Paul sits braced, poker-faced. They exchange greetings with their counselor, Kimberly Quaid, and Quaid asks the question one last time: "Are you sure you want to know?"

Yes, they want to know.

And yes, the news is good. There's a 95% probability: No Huntington. Karen and Paul cry. They jump up and down. Karen screams. "Are you sure?" she keeps asking. "Are you really sure?"

She screams again. "After 28 years of not knowing," says Karen later, "it's like being released from prison. To have hope for the future . . . to be able to see my grandchildren."[7]

A year later, in an interview in the *Chicago Tribune*, Karen says she is, of course, pleased she had the "inner strength" to take the test. "I can't tell anyone to take the test," she said. "But I think we faced death—looked it square in the eye—and are better people for it. . . . I couldn't have continued to live my life the way it was before."[8]

Jason Brandt, a psychologist who runs the Hopkins test program, says that by the end of 1989 he had provided similar good news to forty-three of the fifty-nine people who received results. (For eleven people, the geneticists were unable to provide an answer because the markers were unable to distinguish one chromosome from another.) Brandt says he is struck by the "extraordinary" fact that most people tested don't have the gene. He speculates that "people who have the gene, in fact, already may be having symptoms that they ignore or deny. Often people will call up for testing and we can even tell on the phone they have the disease—there is sometimes a mild slurring of speech that is distinctive. They may know they have the disease, but are holding out hope that they don't. For them, this test could be disastrous. It would take away the hope, which, of course, is what keeps them going."

Like Myers at Boston University, Brandt says he was thrown off balance by the difficult dilemmas that have arisen. "We thought we were real smart cookies," he says. "We thought we had anticipated every possible snafu when we started this. Then the first five patients in the door had us hollering for help."

Right away the Hopkins testers had to play King Solomon, of sorts. An identical twin came in for testing, but said her twin sister didn't want to know the results. Identical twins share the same genes, so one's result would be the same for both. When the second twin finally was convinced to come to the testing center, it was obvious to the Hopkins people that she already was showing signs of the disease and they told that to the sisters. "It was a shock for both of them," says Brandt, "and, of course, no test was needed."

Sometimes people who want the test ask the Hopkins team to help them recruit reluctant family members to donate needed blood. In one instance, a woman wanted the test, but her unaffected father refused to provide the blood necessary for the analysis. Brandt says the father "cared deeply about the daughter and didn't want her to be hurt by any bad news." He finally agreed to provide a blood sample on the condition that his blood not be used to test his other children. "He wanted to have some control over how his DNA was used," says Brandt. "It wasn't an unreasonable request—we all want to maintain control over our bodies. But, of course, nobody has had to deal before with how to regulate information gained from our blood. It's a tricky issue."

Brandt says that, while some people have no doubts about their desire to be tested, even good news can resonate in unexpected ways through a family. Karen Sweeney says that she feels terribly guilty whenever she thinks of her brother with the disease, wondering why she escaped and he didn't.

Huntington's disease testers are especially concerned by requests for test results from third parties, people other than those who are at risk. Early on, the Hopkins team was angered by a request from a doctor to test one of his patients before he performed a life-saving bone-marrow transplant. The patient was dying from leukemia, a cancer of the blood-forming cells in the bone marrow, and only a marrow transplant could help. But the transplant would be expensive and difficult, and the doctor wondered whether it was worth the trouble if the patient had the disease gene. The Hopkins team denied the doctor's request, but, says Brandt, "We were shaken by the request."

In another case, a set of parents about to adopt a child at risk for carrying the gene wanted the child tested before the adoption went through. Michael Conneally, the Indiana University geneticist, was called in to counsel the couple. "I explained to them that as a minor, the child could not be tested," he says. "It would be taking away the child's right to decide whether or not to know. We feel the child's right to know always should override the parents' desire. We know of parents who are being driven crazy by the need to know if their children have the gene and we [counsel] them all that it just wouldn't be right to test the kids."

In probably one of the most worrisome instances of third-party involvement, an airline wanted to know if one of its pilots had the gene. Michael Hayden, who runs a Huntington's disease test center at the University of British Columbia in Vancouver, says the air transportation company was worried about its liability and the safety of its other employees. But the pilot had refused to be tested because he felt the information might untrack his career. So the employer asked Hayden if he could perform the predictive test and not tell the pilot the results, were the company able to obtain the pilot's blood under false pretenses.

"The employer's proposal, of course, was denied," says Hayden, noting that he didn't believe the disease would cause the pilot to suddenly lose control before its symptoms were apparent.

Perhaps one of the most sensitive issues, however, involves use of the Huntington's disease gene marker in a prenatal test. Many U.S. test sites won't discuss the issue publicly, perhaps because for many people abortion is such an inflammatory subject. The prenatal test issue is especially difficult when, as often happens, the couple having the child don't want to know if one of them has the gene.

Consider the dilemma raised by such a scenario where one of the prospective parents has a 50 percent chance of having the gene:

A husband and wife have just learned that they are expecting a baby, their first. They are, of course, elated, but their joy is tinged by more than the usual anxieties of expectant parents. The tiny embryo, probably no bigger than a spot of dust, may harbor a defective gene that will lie quiescent until, in about forty years or fifty years, it explodes into one of the most tragic and devastating diseases. That's because the wife's mother has Huntington's disease, and the wife may have inherited the gene from her. The wife does not want to be tested herself. But a fairly simple prenatal test can tell the couple before the fetus is ten weeks old whether or not it has the same 50 percent chance of having the gene as the wife. Should they take the test? And, if they do, do they abort the fetus if it is found to have the 50 percent risk?

The scenario involves what geneticists call a nondisclosing exclusion test. Any child of a person at 50 percent risk would normally carry a 25 percent risk of having the gene, or half that of the parent. But geneticists can use the gene marker to determine if the child carries chromosomes from unaffected grandparents. If so, the child would not have the gene. In other words, this test can tell if the fetus has no risk, or whether it has the same 50 percent risk as the parent. In this way the child can gain information about its risk without the parent at risk learning whether the parent has the Huntington's gene. In one early study at the University of Wales College of Medicine, doctors offered the nondisclosing test to fifty-five couples; the results, say some geneticists, are surprising. Fifty-one of the couples said they would consider aborting a fetus found to have a 50 percent risk. In a report issued by the Wales geneticists in 1987, nine couples had pregnancies that could be tested.[9] In two cases where the DNA couldn't provide a result and in another where the couple changed their minds, no test was performed. In two instances, the geneticists were able to tell the prospective parents that the fetus had inherited safe chromosomes from the unaffected grandparents, and the fetuses were carried to term. But in three cases, the geneticists couldn't "exclude" the risk, and the pregnancies were terminated. And, in one case in which the geneticists couldn't make an accurate reading, the couple decided to abort anyway, perhaps because they were not able to receive a definitive guarantee that the fetus would live a full and healthy life.

By 1989, geneticists around the world had performed several dozen such exclusion tests, although the results mostly have been kept confidential. Michael Hayden in Vancouver describes why in one especially sensitive instance. A woman in her twenties came to Hayden's center seeking prenatal diagnosis. Her maternal grandfather had died of the disease. She didn't want to be tested herself. In fact, the woman's mother didn't know if she had the gene, and also hadn't been tested.

Hayden says that, from available blood samples, the geneticists were able to distinguish between the chromosome from the woman's unaffected maternal grandmother and the one passed to the woman's mother from the affected grandfather. The geneticists didn't know if the chromosome they could see from the grandfather actually contained the faulty gene, or was the chromosome with the healthy gene, they only knew there was a 50 percent chance that it held the HD gene. The test found that the fetus, indeed, had inherited a chromosome from the maternal grandfather. (If it had inherited the maternal grandmother's chromosome, its risk would have been almost zero.) Thus, the fetus had the 50 percent risk of having the gene. The pregnancy was terminated at thirteen weeks' gestation.

Soon afterward, Hayden was contacted by the woman's mother, who was upset that a fetus with 50 percent risk was aborted. The mother asked the geneticist if it wouldn't have been better to first determine if the daughter had the gene. She was concerned how her daughter may feel if, in years to come, she doesn't develop the disease and realizes that she terminated an unaffected fetus. Hayden says, as a result of this case and others, he began to wonder if prenatal testing should be restricted to people who know they carry the gene. In a lengthy paper discussing these and other ethical issues, Hayden and his colleagues decided that the exclusion test is legally and ethically appropriate for some individuals.[10] "But," he says, "that doesn't make the individual cases any less complicated and painful."

Indeed, among geneticists involved in Huntington's disease, there is a quiet, but intense debate over the ethics of aborting any fetus whose disease won't erupt until later in life. Perhaps by then there will be a cure, or at least treatments to mute the disease's symptoms, some say. Others argue, however, that abortion for even the slightest of risks is justified.

"I've had several conversations with people who say, 'Well, with prenatal tests we can wipe out the gene in a generation or two merely by not allowing any fetus at risk to be born,'" says Hayden. "While that may be possible, who am I or anyone else to make that kind of decision for people. Because we have the power to eliminate a disease, who's to say we should use it. Do you want to make the decision?"

In a report of his center's experience with prenatal tests, Hayden found that, of twenty couples who were asked if they wanted to make use of the predictive test, eight said they wouldn't terminate a high-risk pregnancy and six were uncertain. Six of those surveyed said they would terminate an at-risk pregnancy.[11] Hayden also spent hours talking to the couples about their choices and their feelings.

"People argue that a 50-percent at-risk child also has a 50-percent chance of not having the gene and living a full and normal life," Hayden says. "But we have people here who have lived a sort of hell all their lives being at 50 percent risk. They don't want their children to live with the same uncertainty. I think for some people the 50 percent risk is itself an illness they don't want to pass on to their children."

Most geneticists believe Huntington's disease is an extreme example of genetic testing. However, Hayden says—and some others agree—that the Huntington's disease test "will eventually serve as a model for similar programs for other late-onset genetic diseases," such as heart disease, cancer, mental health illnesses, and Alzheimer's disease. "These issues must be addressed before this type of testing becomes widespread," Hayden and his colleagues wrote in 1989. "In our experience, predictive testing is far more complex than had been anticipated."[12]

* * *

Barbara and John Wilson of Round Rock, Texas, had their lives upended suddenly in 1983 by the unexpected invasion of cystic fibrosis (CF), a disease they had known almost nothing about. During a regular medical checkup for their two children, Barbara offhandedly mentioned that one of her two young children, four-year-old Sarah, seemed to have a high number of bowel movements. The doctor at first brushed aside Barbara's concern, and then he changed his mind. "Just like that, he says, 'To be sure, let's check for CF,'" Barbara recalls.

Within days the doctor called Barbara with the results. Sarah had cystic fibrosis. So did three-year-old Richard. "I couldn't believe it, I refused to believe it," says Barbara. "It had to be a fluke, it had to be impossible."

But, as she and John soon were to learn, there is never a warning before cystic fibrosis strikes its first victim in a family. This is true even though CF is the most common inherited disease in America. The first sign that an otherwise healthy couple carry the genetic defect is when the disease surfaces in one of their children. Doctors estimate that about one in twenty white Americans unknowingly harbor the mutant gene, making it the most common of all genetic defects.

Even so, the disease affects only one in about two thousand births because it is a recessive genetic illness. That is, a child must inherit two versions of the mutated gene, one from each parent, for cystic fibrosis to take hold. A child who is born with one aberrant version of the gene and one normal version will live an otherwise healthy life. Like Barbara and John, such people are carriers of the gene—they can pass on their mutant copy to succeeding generations but they won't become ill themselves.

But when two carriers, like the Wilsons, conceive a child, they suddenly are thrown into a sort of genetic casino. A biological roll of the dice determines which, if any, of their offspring will be so unlucky as to receive a copy of each parent's mutated gene. The chances of getting both parents' deleterious gene is one in four for each child. The Wilsons had unknowingly rolled the dice twice, and both times they had come up losers.

Barbara and John Wilson, like the thousands of couples each year who realize their gamble only after they've already lost the bet, were shocked by the magnitude of their misfortune. What were the odds, they wondered, that two silent carriers like themselves would meet, fall in love, marry, and conceive? And, what then were the chances that such a coupling would result in both their children receiving the defective genes?

Barbara and John had held off having children until their teaching careers and finances were stable, but they had always talked of having a big family. "John and I loved children, John would have a house full of them if we could have afforded it," Barbara says.

But after the diagnosis came in, the pediatrician advised the Wilsons not to conceive again because there was no prenatal test for the disease. His admonition devastated the Texas couple. "We didn't dare bring another sick child into the world," Barbara says. "It was then that John and I realized that someday we would be alone and childless when what we wanted, what we had planned and expected, was to be surrounded by our kids."

In 1983, doctors knew little about cystic fibrosis other than its pattern of inheritance. Despite decades of study, cystic fibrosis had remained an enigma. Doctors could tell, by studying the disease's passage through a family, that it was the result of a recessive, mutated, gene. But they couldn't figure out how such a defect could cause so many health problems. They didn't know where the gene was, what it looked like, what protein the healthy gene produced, and what role that protein normally played in the body. Thus, they didn't know how to repair the destruction caused by the defect.

Treating the disease, therefore, was akin to plugging a hole, or more accurately, a series of holes, in a dike. Children with CF often have trouble breathing because a thick mucus clogs the airways of their lungs. The mucus, in turn, becomes a hospitable environment for all sorts of bacteria. As a result, these children often contract lung infections accompanied by bronchitislike coughing attacks. Breathing can become so impaired that a child's life is threatened. The disease also damages the pancreas, causing a multitude of digestive problems that culminate in severe malnutrition. Many kids struggle for years and then succumb to the disease before they are teenagers. Others with less severe forms of the disease can live into their twenties or thirties. Because of new antibiotics, some are living into middle age.

The Wilsons began to experience a double dose of all these problems, as both Sarah and Richard developed typical symptoms. To keep up their strength, the children received as many as thirty-two pills a day, mostly in the form of synthetic enzymes no longer produced by the pancreas. To keep the children's airways free of congestion, the Wilsons would spend twenty minutes every morning and again at night, banging away at their children's chests with a mechanical clapping device designed to loosen mucus from the lung walls. Nonetheless, within the first two years, Sarah and Richard had numerous bronchial infections, requiring heavy doses of antibiotics. Both children attended regular public school, but often had to sit out physical activities, especially when the weather turned hot and muggy. Neither child was physically strong, "and it was a constant battle to keep Richard from losing weight," Barbara says.

Barbara quit her job, and, while many of the medical bills were covered by John's health insurance, other costs, such as nutrition supplements, were beginning to weigh the family down. "We were being drained financially and emotionally," Barbara says. "And the prospects for treatments, or cures, didn't seem very bright."

But the Wilsons did have some good luck, too. Round Rock, a suburb of Austin, is relatively near Houston, where, at Baylor College of Medicine, Arthur Beaudet treated CF and other genetic illnesses. Among Beaudet's principal research projects was devising a prenatal test for CF. His hope, of course, was the eventual discovery of the gene. By 1983, several teams of researchers, including one involving Beaudet, already had begun the arduous task of using riflip probes to locate the gene. When John and Barbara came to him for help in 1985, neither the gene nor a probe for it had yet been uncovered. But Beaudet offered the Wilsons, who desperately wanted to have another child, some hope.

Researchers elsewhere had discovered that an intestinal enzyme usually present in fluid surrounding the fetus was reduced in those with CF. The test had its problems; it had a high false positive rate, which meant that many unaffected fetuses tested as if they had the mutant gene. Nonetheless, that year, Barbara became pregnant. "The doctors offered us a chance to know and we jumped at it," she says. Sixteen weeks into the pregnancy the doctors performed amniocentesis designed to measure the enzyme. But when the results were equivocal, the Wilsons terminated the pregnancy. "I couldn't risk another child having the disease," she says. "It wouldn't be fair."

When, in the fall of 1985, Barbara Wilson heard that a marker for the gene that causes cystic fibrosis had been discovered, she quietly decided "to try to have a child again." Scientists at the Hospital for Sick Children in Toronto and at Collaborative Research, the biotechnology company, used a riflip probe to track the gene to chromosome 7. Barbara Wilson got pregnant again, and in the winter of 1985–86, the Wilsons were among the very first families to use the marker to predict whether their offspring had inherited both versions of the defective gene. This child didn't, and was carried to term.

In the summer of 1989, the Wilsons twice received good news. Their fourth child was born, free of disease, and the researchers in Toronto and at the University of Michigan, performing a "reverse genetics" search, isolated the gene for cystic fibrosis. Within days, scientists began setting up experiments to identify the gene's protein and the method by which it wreaked havoc on young children, a task they estimate would still take years to accomplish. The Wilsons' affected children might live long enough to benefit from this research. "You know we love all our kids," Barbara says. "Sarah and Richard are very dear and special children, very sensitive and bright and everything a parent would want. But I never knowingly would bring another affected child into the world. The [prenatal test] gave us the possibility of having a somewhat normal life and we are thankful for that. Now we hope the discovery of the gene will save our kids."

The discovery of the marker, and then the gene, for CF comprises one of the more important developments to come out of riflip technology. Because the mutated gene for CF is so common, geneticists point to the experiences of the Wilsons and others as a hallmark in the era of predictive medicine. Within four years of the gene marker's discovery, hundreds of couples who had suspended plans for more children were able to conceive again, knowing they would have the chance to terminate an affected fetus.

But as with Huntington's disease, the prenatal test produced a fallout of complications that few geneticists were prepared for. One New Orleans couple, like the Wilsons, learned they were at risk for CF when their first child was born with the disease. After the marker was found the family contacted the geneticists at Baylor College of Medicine and then asked their insurer to cover the costs of the prenatal exam. The insurer told the couple that it would pay for the exam, but also told the family that, if the test showed the child was affected, it had to be aborted, says Susan Fernbach, a genetics counselor at Baylor. The insurer didn't want to cover the huge health costs produced by a CF child, especially since it already was paying the bills for the couple's first child.

The Baylor geneticists called in Philip Reilly, a lawyer specializing in medical ethics in Boston, for advice. "It was the first example I had seen of a third-party effort to force a family into making what for them had to remain a private and personal matter," Reilly says. Several calls to the insurer convinced the company to stay clear of the family's decision. "But the experience was a real jolt," Reilly says. "All of us realized we had just crossed over into a whole new realm of medicine. Ethicists hadn't really seen anything like this before."

The prenatal test also sparked a series of debates among some people about the propriety of terminating a pregnancy involving a disease like CF, in which victims can live long, albeit disabled lives. The prenatal test can only detect whether the mutation exists, it can't determine whether the defect will cause a mild or severe form of the disease. Some children, like the Wilsons', can live for years, while others succumb early on. Brian Scully, a doctor who treats CF patients in New York, complained in an interview in 1986 that couples who abort an affected fetus "aren't giving the child the chance it deserves." He noted that, with new antibiotics and aggressive monitoring, his patients are living "longer and longer."

Scully pointed to a patient of his, Dorothea Ridenour, a woman in her late twenties with CF who called her life, "good, almost normal." She and her husband had a four-year old child. "I've had a job I liked, I cook, shop, take my son to school," she says. But she had problems, too. She was separated from her husband and son for long hospitalizations, and even on good days her husband or her mother spent hours pounding on her chest to break up the congestion.

Scully said that people like Ridenour who make this kind of fight "enrich all of us who work and live with them. I wouldn't ever underestimate their suffering but these are special people whose courage contributes a lot." Scully argued that parents at risk who would abort an affected fetus shouldn't try to have kids at all. "Aborting these kinds of people isn't right," he says.

Indeed, some doctors have refused to provide prenatal exams for some of the genetic diseases for which predictive tests can be performed. In Boston, Thaddeus Dryja, the eye doctor who isolated the gene for the rare eye cancer retinoblastoma, won't use the gene discovery to perform a prenatal test. "Couples have asked to have a prenatal test and the technology exists to do it," says Dryja, who tests newborns at risk to see if they have inherited the gene and must be monitored for the first signs of the eye cancer. "There is a treatment for the disease if caught early," Dryja says. "Someone is going to start offering prenatal tests [for retinoblastoma], but it's not going to be me."

Abortion concerns have blocked some families' efforts to have prenatal testing. Pat Ward, the counselor at Baylor College of Medicine, says one family wasn't able to get tested for hemophilia, a blood disorder caused by a mutated gene on the X chromosome. The disease, which makes it difficult for the blood to coagulate and can make otherwise simple injuries disastrous, affects males who inherit the defective X chromosome from their mothers. Ward says a woman worried that she might carry the defective gene. Her male cousins had the disease, but when she asked her aunt for blood for a DNA analysis, the aunt turned her down. "The aunt felt strongly that she couldn't participate in a study that would result in abortion," says Ward. "She absolutely refused to have her blood or her sons' blood drawn."

The need to address these and other problems certainly will grow as scientists plot succeeding genes onto their rightful spots in the human gene map. Each newly identified gene will bring with it its own complexities, forcing the genetic counselors each time to holler to the bioethicists for help. To date, the counselors have dealt only with issues arising from single gene diseases. Surely, the issues will become even more complicated for diseases—such as cancer, heart ailments, and psychiatric disease—that are polygenic, caused by multiple genes. Predictive tests for these illnesses will be based on genes that merely heighten one's risk of illness. Doctors and their patients then will be faced with information not nearly as clear-cut. Instead, test results will take on the vernacular of the casino, as health risks will be measured by relative odds, as those people with the larger set of certain defects are placed at a higher risk of illness than those with a smaller pool of such genes.

How will we deal with such ambiguous data in forecasting future health? The enthusiasts, such as Mark Skolnick, Ray White, Bert Vogelstein, and Jan Breslow, while cognizant of potential dilemmas, are convinced that the new genetic insights, nonetheless, will lead to a new era in medicine in which preventive measures will rival the breakthroughs of the past century in antibiotics, surgery, and vaccines. Even so, during the intervening time from when a new gene is located by a riflip marker and its presence becomes detectable to the time when the gene is actually isolated, its protein product known and treatments devised, we will have to grapple with the kinds of conundrums described above. Until the cures come, the new era of predictive medicine, filled with promise and problems, will reign.

CHAPTER 14

A Niche in Society

On the evening of November 11, 1989, three prominent medical geneticists huddled in a suite in the Hyatt Regency Hotel in Baltimore's recently renovated waterfront district. The scientists had gathered at a hastily convened caucus to compose a document they hoped would delay a stampede. In previous weeks, geneticists around the country had begun talking seriously about conducting the most ambitious program ever undertaken in genetics: screening the entire U.S. population for the presence of a gene, the gene that causes cystic fibrosis.

The three geneticists were in Baltimore to attend the annual meeting of the American Society for Human Genetics. Thomas Caskey, a genetics researcher and physician at Baylor College of Medicine in Houston, as well as the president of the society, had asked Arthur Beaudet, a colleague at Baylor, and Michael Kaback, chairman of the department of pediatrics at the University of California in San Diego School of Medicine, to meet with him before the society began its sessions the next day. The three men were deeply troubled.

In August, researchers from Toronto and the University of Michigan had discovered the gene for cystic fibrosis. The discovery was a cause for scientific celebration—but it also posed a social dilemma. Unlike muscular dystrophy or Huntington's disease, cystic fibrosis is the result of a defective recessive gene; the disease develops in a child only if he or she has inherited two defective copies of the cystic fibrosis gene, one from each parent. But a recessive disease also means that there are "silent" carriers of the disease, people who carry a single copy of the defective gene but show no ill effects of its presence. Of the three to four thousand so-called inherited diseases caused by a single defective gene, none produces as many silent carriers as cystic fibrosis: One in twenty Caucasians carries a defective version of the gene.

The discovery of the cystic fibrosis gene in August meant that almost immediately it would be possible to test any individual—not just the members of a family with a cystic fibrosis child—to see if they were carriers of the genetic disease. Once the cystic fibrosis genetic test reached the market, there would be an unprecedented flood of demand for it. Young couples would want to know if they had any risk of bearing a cystic fibrosis child. Public health officials would want to screen the population for carriers in an attempt to reduce the incidence of cystic fibrosis. And biotechnology companies would want to exploit a potentially huge market for cystic fibrosis tests. Indeed, within weeks of the gene's discovery in August, two companies near Boston, Collaborative Research and Integrated Genetics, and one in Dallas called GeneScreen began making plans to distribute carrier-detection kits to medical laboratories across the United States. Collaborative even mailed letters to doctors urging them to send in samples of their patients' blood to be tested for the cystic fibrosis gene.

"This is a gold rush," Norman Fost, a geneticist and pediatrician at the University of Wisconsin Medical School in Madison told the *New York Times*.[1] "The potential market for this screening is at a minimum a billion-dollar a year industry."

But Caskey, Beaudet, and Kaback, as well as a handful of other geneticists with years of experience in genetic counseling, feared that counselors at medical centers would be overwhelmed by people who would want to know if they were carriers of the defective gene. For weeks prior to the November meeting, Kaback had warned anyone who would listen of the dangers posed by widespread, uncontrolled carrier screening. He had called for a temporary halt for testing plans.

Kaback, who was to become president-elect of the genetics society that week, felt it was his responsibility to put a brake to the screening fever. Too many questions first had to be answered. For instance, he wondered, should the carrier test be limited to young couples planning to have children or should newborns also be tested? If children are tested, who would explain the meaning of a positive test to them and where would such information be kept? How would the results be kept private? Would silent carriers have to divulge the information to insurers, or to their prospective spouses prior to a marriage?

That November night in Baltimore the three men toiled to construct a statement so strong it would convince everyone involved not to launch ad hoc cystic fibrosis carrier screening programs. At the same time, the geneticists couldn't help but see the irony of their situation. For more than five years molecular geneticists in the United States, Canada, and Great Britain had been in a furious race to find the cystic fibrosis gene. They had given little thought to what would happen if and when the gene was found.

The race got underway with Lap-Chee Tsui (pronounced *Choy*), a thirty-one-year-old molecular biologist who in 1981 joined the laboratories of Manuel Buchwald at Toronto's Hospital for Sick Children as a postdoctoral researcher. Tsui grew up in Hong Kong and had come to North America in 1974 for graduate studies at the University of Pittsburgh. Young and, perhaps, naive, Tsui was eager to take on the cystic fibrosis gene despite the daunting obstacles and the many dead ends researchers had hit in recent years. He brought to the labs a youthful optimism and effervescence, as well as a gift for using the most up-to-date tools of molecular biology in ways that often were overlooked by his colleagues.

Tsui knew the hospital had an invaluable resource. It was a regional treatment center for those families in Ontario who were stricken with cystic fibrosis. In its computers were the names of many families having at least one living child with the disease. Tsui knew that the DNA from those families could be vital to a gene search—if there were adequate genetic markers to light the way. Of course, the 1980 paper by Botstein, Skolnick, White, and Davis provided the match Tsui and other researchers were looking for.

"Everyone read Botstein's classic paper," Tsui told *Science* magazine's Leslie Roberts several years later. "I read that paper and thought, 'Geez, this is so simple.'"[2]

But simple, it was not.

Tsui understood from reading Mark Skolnick's segment of Botstein et al. that he would significantly raise the odds of finding a marker closely linked to the cystic fibrosis gene if he could locate very rare families who had two or more living members who had the disease. Between 1982 and 1984, Tsui, Buchwald, and their colleagues, along with help from the Cystic Fibrosis Foundation in the United States and Canada, contacted dozens of families, finally settling on fifty in which at least two children had the disease. Of those fifty, two separate families, remarkably, were found to have five siblings, each affected by the disease. Another eight families had more than three affected children still alive.

In London, Robert Williamson and Kay Davies, who were instrumental in helping to find the muscular dystrophy gene, were collecting cystic fibrosis families as well. Arthur Beaudet in Houston also treated numerous families haunted by cystic fibrosis, and Ray White had made contact with cystic fibrosis families in the Utah area. At least three other research groups in the United States and Europe had decided to focus on finding the cystic fibrosis gene. They, too, were collecting families.

At the time, no one knew which chromosome contained the cystic fibrosis gene, so there was no choice but to blindly start testing DNA from the families with one riflip probe after another, hoping against hope that one of the probes might stick to a riflip near the cystic fibrosis gene. Realizing he had little time to waste, Tsui in 1984 quickly set about testing his collection of families' DNA with riflip probes selected randomly from all over the genome. By late 1984, just when Tsui was running out of probes to test, he and Buchwald were approached by Collaborative Research, the Boston-area biotech firm whose scientific advisory board included Botstein and whose scientific staff included Botstein's then wife, Helen Donis-Keller. Botstein had convinced the company to focus on developing probes for the riflips that he believed were scattered throughout the human genome. Consequently, Collaborative had developed more than two hundred riflip probes and was willing to let Tsui use them in his hunt for the cystic fibrosis gene.

"Everyone wanted Collaborative's probes," says Buchwald. "We couldn't believe our good fortune."

In 1985, while Tsui was attending an international gene mapping workshop in Helsinki, his colleagues in Toronto found that one of the Collaborative probes was weakly linked to the cystic fibrosis gene. At the same time, a team of Danish scientists said they had a probe that appeared to be linked to the gene. The race to the gene was suddenly underway in earnest.

"It was a nerve-racking time. I couldn't sleep," Buchwald recalls.

Within a few weeks, Collaborative had determined that the probe used by the Canadians was sticking to a riflip on chromosome 7, but the company asked the Canadian scientists to keep the location quiet until Collaborative could establish its patent position. Nevertheless, rumors circulated that the cystic fibrosis gene lay somewhere on chromosome 7. Almost immediately, White in Utah and Williamson in London focused their search in that area, and in a few weeks each of them found riflips located even closer to the gene than Tsui's. The three teams then rushed their findings into print, all of them appearing in the same issue of *Nature* in late November.[3-6]

The probes found by White and Williamson flanked the gene like bookends, providing an extraordinarily accurate set of markers that could determine if a fetus had inherited the cystic fibrosis gene. The probes were useful for prenatal testing, but only if DNA was available from other members of a family known to have the defective gene. The probes were of little use in detecting silent carriers of the defective gene among the general population, and few at the time bothered to think about carrier screening and its implications.

Week after week for more than a year, the Toronto lab tested new riflips—250 in all—until they found two that were mapped to spots between the White and Williamson markers and hence were much closer to the cystic fibrosis gene. At that point, Tsui decided he needed help and formed a collaboration with Francis Collins at the University of Michigan, who had developed a method of traveling along stretches of DNA called "gene jumping." Just as the two teams embarked on their march toward the gene in the spring of 1987, Williamson in London reported that he'd found a sequence of DNA he believed was the gene. The spirits of Tsui, Buchwald, and, now, Collins, plummeted.

By the fall, however, it was shown that Williamson didn't have the gene, but instead, a marker that linked so tightly with the gene that it was within several thousand base pairs of it. But Williamson's initial report cleared out the competition; White, Beaudet, and others dropped their cystic fibrosis projects. Only Tsui and Collins hung on.[7]

By this time, researchers believed the cystic fibrosis gene encoded for a protein similar to proteins made by sweat glands. Tsui and Collins began fishing out segments of DNA that had the earmarks of a gene and compared them to genes known to produce sweat gland proteins in the lungs and pancreas. After a year of experiments, they found a DNA sequence they believed was the starting end of the gene. "To finish the [gene's] cloning we spent night and day with lots of people," Tsui says. "We could only get bits and pieces and then had to fit everything together."[8]

Finally, in the early summer of 1989 they found a long DNA fragment that looked as if it might be the gene. By testing it in people with the disease, they found that about 70 percent of people afflicted with cystic fibrosis had the same piece of mutated DNA. It was the cystic fibrosis gene. The Toronto–Ann Arbor scientists quickly found that the mutant gene normally produced a protein whose structure resembled proteins anchored in cell membranes that help carry an energy-producing chemical into the cell to fuel its various functions.

The discovery of the gene was widely hailed. "The long march to the cystic fibrosis gene was obviously arduous. But the successful procedures worked out by Tsui, Collins, and their colleagues for isolating the gene should also be applicable to the identification of the gene causing other genetic diseases," proclaimed Louis Kunkel in *Science*.[9]

In San Diego, Michael Kaback, the medical geneticist, was overjoyed. But within a few days, "I got worried," he says. With the mutant gene now known, many people would mistakenly believe it would be possible to test anyone for the gene, including the millions of silent carriers of cystic fibrosis. And Kaback had firsthand experience in carrier testing of large populations. In the mid-1970s, he had developed the national program to test the population for carriers of the Tay-Sachs gene, a fatal illness also caused by the inheritance of two otherwise silent recessive genes. Explaining the meaning of being a carrier to people "was very hard," says Kaback. It requires that people be patiently counseled by a professional with a lot of time. Most people didn't know what Tay-Sachs was, and most don't know what cystic fibrosis is, either. But the number of silent Tay-Sachs carriers was a fraction of those carrying the cystic fibrosis gene. "I knew there weren't nearly enough genetics counselors to handle the load of people who would need attention," he says

Kaback's concern was heightened when Collaborative and the other companies began setting up plans to commercialize a gene detection kit. In a letter to physicians, Collaborative strongly implied that obstetricians might be liable to malpractice suits if they didn't test prospective couples who later conceive a child affected by the disease.

Kaback also believed the test wouldn't be as accurate as needed. Since the gene defect identified by Tsui and Collins would be found in just 70 percent of cystic fibrosis patients, that meant that a test based on the gene would only detect 70 percent of the mutated cystic fibrosis genes. It also meant that 30 percent of the carriers would be told mistakenly that they were free of the gene.

Outside Boston, Philip Reilly, a physician and lawyer specializing in bioethics, wrote a letter to Tom Caskey echoing Kaback's concerns.

Reilly, director of the Shriver Center for Mental Retardation, knew that the screening could warn tens of thousands of people of their risk of giving birth to an affected child. Once warned, these silent carriers of the gene might want to make certain they don't marry one another. If they married anyway, they would be forewarned of the risk and, if they chose, they could undergo prenatal testing to avoid giving birth to an affected child.

But like Kaback, Reilly knew from experience in other small-scale genetic testing programs that most Americans are unschooled in genetics. The test results, they feared, would be misunderstood and, perhaps, misused.

"There is plenty of reason to believe that carriers, if identified, will be stigmatized, by their employers or insurance carriers, as being sick or as being at risk of getting sick even though that's preposterous," Reilly says. "There's concern that even many doctors don't really know what it means to be a cystic fibrosis carrier. The technology has gotten way ahead of us."

But, says Reilly, "it's inevitable that carrier testing is going to happen and when that happens we'll be testing millions of people. What this means is that we better get a handle on what kind of information we'll be producing, who will use it, and keep it from being mishandled, misunderstood, or poorly protected. And while some people still say we have time to deal with the broader issues beyond cystic fibrosis, scientists who see how quickly advances are coming know we ought to begin thinking about this now."

Thus, at the urging of Reilly and Kaback, Caskey, the genetics' society president, sought a strong policy statement to temporarily hold off any testing. At the sessions held in November in Baltimore, the geneticists debated the issue, finally agreeing to a ban on testing until the issue could be further hashed out in the spring.

"Eventually, probably by the end of [1990], this test will be out and it's going to be very valuable, but we're just not ready right now," Reilly said. "Our best efforts to conduct the test in a controlled way are going to surely be overwhelmed by demand. Just wait until the first article appears in a large-circulation woman's magazine. Every couple in America thinking of having a child is going to want to be tested. Then what?"

Spurred by the worries raised at the genetics society meeting, federal health officials convened a special workshop four months later, in March 1990, to decide the issue. In a crowded auditorium at the National Institutes of Health in Bethesda, Maryland, a special panel of genetics experts including Reilly, Kaback, Beaudet, Tsui, and Collins heatedly debated whether to lift the informal ban on unrestricted screening. After arguing for two days, the twenty-three-member panel decided that the United States medical system and the public were unprepared to deal with widespread carrier testing.

The principal reason for blocking a screening program was that in preceding months Tsui, Collins, and other researchers had hit an unexpected glitch. By March, six months after the gene's discovery was reported, the scientists still hadn't found a second major mutation in the gene. Thus, the test could detect the defect in about 75 percent of those people who carried the gene, but it couldn't detect the other 25 percent of the carriers.

Those attending the workshop were baffled. On one hand, they acknowledged that even the existing, but imperfect, test, if used widely, would pick out many people with the common defect. These people, if tested before they married or conceived a child, would be warned of their risk of having a child with cystic fibrosis. But many other people would be told they weren't carriers, when, in fact, they were.

"My fear is that this kind of multiple defect problem is going to be very common in other [disease] genes as we uncover them," said Haig Kazazian, a respected geneticist at the Johns Hopkins University School of Medicine, who cochaired the government workshop with Arthur Beaudet. "We thought that once we found a gene, boom, we would be able to test the population for the problem. But if the cystic fibrosis gene is an accurate model, and I think it is, each gene is going to pose similar headaches."

Indeed, many at the workshop echoed Michael Kaback's broader concerns about what the group would finally recommend. "You know, this is the first gene that is widely carried," he told the workshop. "If we botch this one, we're going to have big problems in the future."

Ultimately, the group decided to enter the era of widespread screening gingerly, recommending that the federal government first support broad educational programs for doctors and the public, and to fund pilot screening programs before widespread screening was approved. But some attending the meeting were dubious that the workshop's panelists could stop the demand for carrier testing. Many of the geneticists said they already were receiving inquiries from couples who had read newspaper articles about the availability of a test.

Officials from Collaborative Research and Integrated Genetics acknowledged that they were conducting carrier tests for doctors who requested them, generally for patients who knew that some relative had cystic fibrosis, where there was a higher-than-normal likelihood that the gene was being carried. "CF testing is happening and it's going to happen no matter what these folks here say," said Gerald Vovis, a Collaborative official.

Thus, the geneticists were beginning to encounter the real-life implications that society will face as the human genome is mapped. As each new gene is uncovered, new genetic tests will become cheaply and easily available. And the demand for genetic testing will be impossible to resist, even if the information produced by the tests is misunderstood or falls into the hands of those who may misuse it. The hundreds of thousands of people who have had parents or siblings die of cancer, heart disease, diabetes, or a multitude of other diseases may well seek these genetic tests. Physicians may encourage genetic tests to confirm their own suspicions of a genetic disease, to uncover a patient's future disease risk, or to avoid the risk of a future malpractice suit for failing to prescribe a genetic test.

And parties other than family members and their doctors are going to demand genetic tests. An internal report by the United States insurance industry in 1989 predicts that "by the year 2000, genetic testing may be standard within the medical community. Much the same way that cholesterol and blood sugar tests are now a part of routine examinations, genetic tests may be carried out to test for predispositions to such common diseases as cancer, heart disease and hypertension."[10]

In October 1989, just one month before cystic fibrosis testing was being debated in Baltimore, the journal *Science* convened a meeting of gene mappers and sequencers, which it called Human Genome I, to hash out strategies for identifying all the human genes. Those present heard the leaders of the formal gene mapping enterprises make some startling predictions.

Charles Cantor, director of the United States Department of Energy's DNA sequencing project, told reporters covering the meeting, "You can probably sense our optimism that we really are going to find and identify all the hundred thousand genes that make up the set of instructions that determine why we're human."[11]

If that weren't audacious enough, James Watson, codiscoverer of the structure of DNA, almost facetiously projected the precise day by which the human genome will be mapped and sequenced: September 30, 2005. Watson, as head of the National Institutes of Health's mapping segment of the genome project, had chosen October 1, 1990, as the date when mapping was to commence—it was the beginning of the United States' federal 1991 budget year when the genome project was to receive its first funds. "We've said [the project will take] 15 years and so I guess we have to declare a date when the 15 years starts," he said. "We really should get it done in 15 years, not in 20."[12]

A hint of just how widespread testing for genetic diseases will become as the genes are mapped is offered by Johns Hopkins geneticist Neil A. Holtzman in his recent in-depth study of predicting genetic risks, titled *Proceed with Caution:* He foresees 2.4 million women each year undergoing prenatal tests for chromosome abnormalities and almost 2.8 million a year being tested to see if they are carriers of the genes for the four major inherited disorders—cystic fibrosis (750,000), sickle cell anemia (152,000), hemophilia (940,000), and muscular dystrophy (940,000).[13]

These astonishing estimates pale beside the potential annual volume of tests for genetic susceptibilities: 16.2 million a year, including 3.8 million for a susceptibility to diabetes and 3.4 million for coronary artery disease (the remainder involving tests for susceptibility to cancer, manic-depression, and Alzheimer's disease). Even so, "The estimates are conservatively based—that is, they tend to minimize the number screened for any condition in any one year," Holtzman says.[14]

There is, of course, convincing evidence that this kind of screening for defective genes can lead to a dramatic decline in a disease incidence. One recent and clear-cut example involves the Hasidic Jews of Brooklyn. The Hasidim, like all Ashkenazi Jews of Eastern European origin, are at especially high risk of developing Tay-Sachs disease. The disease, like cystic fibrosis, is caused by a recessive gene—the two silent carriers conceive a child who inherits each parent's copy of the Tay-Sachs gene. The consequences of this disorder are fatal; the child gradually becomes paralyzed and dies before age five.

A decade ago, children with Tay-Sachs disease were common among the Brooklyn Hasidim because of their tendency to marry within the group and because of a religious ban on contraception. Prenatal diagnosis for the disorder, even though available, was out of the question since the Hasidim forbid abortion. (Why diagnose a fetus if an affected one won't be aborted?) The local Kingsbrook Jewish Medical Center had an average of thirteen Tay-Sachs-afflicted children in the hospital at any one time.

In 1984, the medical center's neurologists, a local Hasidic organization called Dor Yeshorim, and the United Jewish Appeal–Federation of New York, began a program in the yeshivas and high schools to encourage the students to be tested for Tay-Sachs carrier status. To avoid stigmatizing the students, the testers promised complete confidentiality. The results, with numbers substituted for names, were stored in a Dor Yeshorim computer.

By 1989, the program was testing three thousand high school students a year. The number of Tay-Sachs children in the Kingsbrook center at any one time had dropped to two or three.

The key to this success was the matchmaker, the UJA–Federation says. "Most Hasidic marriages are arranged, and the marriage brokers protect the identity of persons found to carry the Tay-Sachs gene. In Dor Yeshorim's system, the marriage brokers run their clients' codes through the computer and if two potential partners have the gene, the match is never made," the federation explains.[15]

Widespread genetic screening could, of course, produce considerable benefits to both the individual and society. Holtzman of Johns Hopkins estimates that, of nearly 2.8 million women who could be screened each year for the four major genetic diseases, about 50,000 would discover they are carriers of a defective gene and at risk of bearing affected children. These carriers can now choose to become pregnant or not in a much more informed manner than previously, and if they do become pregnant they have the option of undergoing a prenatal diagnosis to detect whether an offspring is affected by the disease.

Detecting a genetic susceptibility, particularly early in life, also promises benefits to those who inherit the susceptibility. Scientists who are chasing down the genes involved in cancer believe they will soon have tests to identify who carries a higher-than-normal risk of developing the disease. Researchers at Johns Hopkins, under the leadership of Bert Vogelstein and others chasing after tumor suppressor genes, expect they will produce several gene candidates that could be detected in prenatal tests. A child born with a damaged cancer suppressor gene might be schooled from an early age to avoid diets high in fat, or be warned of the lethal potential of ever taking up cigarettes. Such a child could enter lifelong periodic screening programs to detect colon, lung or breast cancer, diseases best treated if diagnosed early.

Holtzman estimates that of the 16.2 million children and young adults who might be tested annually for genetic susceptibilities, 810,000 would discover they may be predisposed to a major late-onset disease that they might be able to avert by taking some preventive action. Can anyone dispute the personal and public health benefits of such programs if instituted on a national or international basis?

But there is, of course, a price to be paid for this newfound knowledge. The elucidation of the human genome will help identify an individual's genetic profile, one's personal collection of genes. And, once that happens, every person's individual set of inherited characteristics, biological tendencies, weaknesses and strengths, will be revealed. Even before the complete genomic map is drawn, the broad outlines of a personal genetic profile could be constructed based on those genes already mapped and identified. By use of a simple blood test, it will be possible to see an individual's private constellation of known genes, and, thereby, reveal the biological foundation upon which is built a person's particular health, personality, and physical and mental talents.

Throughout the 1980s, amidst the excitement over the new technology and the exuberantly competitive races among scientists to map specific genes, few researchers found the time to examine the broader social implications of breaking through into the new genetic frontier. One of those few perceptive scientists who stood back and scanned the scientific horizon for the presence of storm clouds was Arno G. Motulsky, the pioneering medical geneticist at the University of Washington in Seattle. In 1983, Motulsky was the first to publish a paper alerting scientists to the ominous side to the technology described in Botstein et al.[16]

Motulsky declared, "As public bodies assume a more direct role in the health system in many countries, confidentiality may become eroded and genetic information may be used by social and health planners to assign individuals their niche in society."[17]

In 1984, Marc Lappé, then at the University of California School of Public Health in Berkeley, echoed Motulsky in a strongly worded warning published in the *Hastings Center Report*, the journal of a world-renowned medical ethics institute in Briarcliff Manor, New York.[18] At the time, the Huntington's disease test was the only genetic test to have resulted from the new technology. But Lappé wrote that he expected others to follow and that such tests soon might "become tools of great social power and potential coercion."

Based on studies he and other ethicists had conducted over the years in analyzing prior uses of genetic tests, Lappé was the first to caution that a test result that provides an adverse medical diagnosis "can set in motion a sequence of personal reassessments that leads to feelings of reduced self-worth and autonomy." But, as importantly, he warned, the same findings can be used by employers, insurers or government agencies, in ways deemed as in the interests of the public welfare, but which, nonetheless, will infringe upon individual freedoms.

"When we can know that a person is at risk for a major disabling or costly condition long before he or she manifests symptoms (for example, when we do a prenatal test for hemophilia)," Lappé wrote, "does society have a right to invoke prepayment, taxation, or other schemes to offset the inevitable cost of the disease? The obvious rejoinder is that such a scheme is regressive, and has a disproportionate and hence coercive impact on the poor. But this ignores the overwhelming social reality that we are reaching a point where virtually no one can afford to pay for the longterm care of chronically diseased individuals.

"The social reality threatens to drive medicine away from its largely humanitarian base, and toward the use of predictive tests that allow cost savings no matter how discriminatory or prejudicial the process."[18]

At the time, the concern expressed by Motulsky and Lappé that genetic tests would be used to label people according to their inherited traits was scoffed at or ignored by most gene hunters. Even scientists optimistic about the gene mapping prospects back then believed it would be many years before enough genes would be identified to cause much of an ethical ruckus. But by the decade's end, Motulsky and Lappe were vindicated by a series of events that, taken together, began to force the scientists to see that the power of their new genetics is spreading far beyond the confines of medicine, and the prevention and treatment of disease.

In June 1989 a remarkable report was issued quietly in the United States. The report, compiled by the American Council of Life Insurance, says that, "As genetic testing is perfected, society will be forced to confront important issues that have never before been of such widespread concern. Profound ethical questions will be posed concerning the practice of medicine, procreation, employment, privacy, individual versus societal rights, confidentiality, 'the right to know' and the 'right not to know.'"[19]

This forty-seven-page report, titled "The Potential Role of Genetic Testing in Risk Classification," was circulated only to the Council's member companies, the major providers of health and life insurance in the United States. It was designed to prepare the member companies for the day when large numbers of people know their genetic profiles. "We think genetic testing will be a topic of extraordinary importance" among insurers, says Robert Pokorski, an author of the report. He believes that, when people are apprised of their genetic predispositions to disease, insurers will need to have access to the same information. Otherwise, insurers, in addition to the population covered by insurance plans, will be hurt as those people who know themselves to be at high risk of developing disease begin loading up on health and life insurance.

Three months later, across the Atlantic, a group of well-meaning doctors in Great Britain's Royal College of Physicians attacked their country's National Health Service for failing to provide prenatal diagnoses for all British women at risk for having children with spina bifida or Down's syndrome.[20]

The physicians called for nationwide screening and prenatal diagnosis that would become an intrinsic part of the health service's maternal health care program. They estimated that this affirmative step could prevent the birth of two thousand severely handicapped infants a year.

The Royal College declared that "if the costs of the whole programme were aggregated, it is cheaper to screen and counsel the whole population than it is to treat affected children who would otherwise be born to unprepared parents." The annual cost of treating an afflicted child was estimated at ten thousand pounds (approximately fifteen thousand dollars); the cost of the prenatal test only a few pounds. Of course, the college added, a "code of practice" would have to be drawn up to deal with the inevitable ethical problems associated with a program requiring mandatory genetic inspection of prospective newborns.[21] To a handful of watchful ethicists, the screening program looked like the first opportunity for the health service to collect genetic material from thousands of newly born infants.

But what happens if large numbers of people are identified as possessing defective genes before they or their offspring manifest disease? Medical ethicists Kathleen Nolan and Sarah Swenson of the Hastings Center point to the possibility of changing the concept of who is normal and who is not. "For example," they note, "are asymptomatic young adults who carry genes associated with the later development of bipolar disease [manic-depression] 'diseased' even though not ill?"[22]

They add: "The ability to identify and treat many common conditions risks establishing a populace of 'diseased' individuals and increasing the medicalization of daily life. Failure to adhere to a suggested therapeutic regimen may then result in individual guilt and self-recrimination, or worse, a 'blaming of the victim' by others through effects on insurance premiums or more subtle forms of stigmatization."

Nolan and Swenson note, for example, that detection of genes predisposing people to abnormally high cholesterol levels is possible. Thus, "a zealous clinician might be tempted to argue that failure to maintain an identified child on a low-cholesterol diet constitutes a new category of parental medical neglect."

Genetic labeling has already occurred to a small degree. In the 1970s, following development of a test to detect the gene for sickle-cell anemia, a screening effort was launched to detect carriers of the gene, the so-called sickle-cell trait. "Carriers of the trait who never developed any clinical problems were considered as mildly affected by the public or even by some physicians who were unaware of the harmlessness of the carrier state," Motulsky noted in his 1983 paper.[23]

The U.S. Air Force Academy went so far as to ban carriers of the sickle-cell trait from its ranks in the mistaken belief that, since the gene was for the blood's oxygen-carrying hemoglobin, the carriers wouldn't be able to function at high altitudes. The ban was subsequently lifted following public and medical protests. (The ban and the carrier screening program behind it were halted following protests of racism, sickle-cell anemia being primarily a disorder of blacks.)

Holtzman recalls one instance of genetic labeling in one of his young PKU patients. "With permission of the child's parents, I communicated to their son's school that the child had to have a special diet because he had PKU. Although it should have been apparent to his teachers that the child was functioning normally, he was placed in classes for the learning disabled. The school authorities knew PKU was associated with mental retardation but not that early dietary treatment could prevent it."[24]

Genetic labeling or stigmatization is not necessarily a problem if the results of genetic tests are under control of the individual, who may then decide who should and who shouldn't know his genetic affirmatives. But there is good reason to believe that few individuals will be able to keep their genetic profiles private.

The 1989 insurance industry report of the American Council on Life Insurance made it clear that their industry will demand access to genetic profiles of those seeking life and disability insurance. Their business, the report notes, is based entirely on "risk assessment," that is, determining an individual's risk of future disability or early death. Premiums are based on "the principle of equity: Policyholders are charged equal premiums for equal risks."[25]

The report then adds: "If insurers were unable to use genetic tests during the underwriting process because 'risks should only be classified on the basis of factors that people can control,' then equity would give way to equality (equal premiums regardless of risk) and private insurance as it is known today might well cease to exist."

What the industry fears, of course, and perhaps rightly so, is a dramatic skewing of their own risks. "If test results were unavailable to insurers, applicants who already knew from tests performed by their own physician that they were predisposed to illness or early death could buy large amounts of insurance coverage. Moreover, they would be doing so at rates that do not properly reflect their known risk. If large numbers of such applicants bought insurance, or if large amounts of insurance were purchased, the ensuing claims would markedly exceed projected losses." At the same time, the report notes later, "if an individual underwent extensive genetic testing and all the results were negative, that individual might be inclined to seek less insurance than might otherwise be the case." In other words, without knowing insurance buyers' genetic risks, the industry could easily find itself unwittingly insuring an unusually large number of high-risk individuals and fewer low-risk individuals—and insuring them all at the wrong premium.

At first, the insurers aren't likely to demand genetic tests of their clients, since the tests coming into use in the next decade will probably involve less common diseases. But the insurers may well ask applicants for results of certain genetic tests.

"As technology is perfected and/or genetic testing within the general population becomes common, insurers may eventually consider ordering genetic tests," the report declares.

Employers also may demand to see or even to require genetic tests of their workers and/or job applicants. It is quite likely that as genetic tests uncover an increasingly wide spectrum of genetic susceptibilities, at least some people will be excluded from certain jobs. A young person with a susceptibility to schizophrenia, manic-depression, or other mental illness may well be excluded from any job involving firearms, such as a policeman. Corporate managements well may want to know if a candidate for promotion to a key management job is susceptible to alcoholism, mental illness, or Alzheimer's disease.

"An apparently healthy airline pilot at risk for one of the gene-influenced forms of heart disease could suffer a heart attack in flight, endangering all on board," Holtzman notes. "Some persons with retinitis pigmentosa insidiously lose their sight in adult life, beginning with night blindness and resulting in their impaired ability to perform jobs that require visual acuity."[26]

Employers may face a dilemma. If they exclude some workers from certain jobs because of their genetic susceptibilities, they could be charged with discriminating against the worker rather than trying to make the job safe. This is particularly true if, as seems likely, there are genetic aberrations that make some people more susceptible to the harmful effects of chemicals or other substances used in the workplace. On the other hand, if the employer fails to determine an employee's genetic susceptibilities to workplace hazards, and the worker is subsequently injured, the employer might well be held liable for negligence.

The problem is limited at the moment because science has yet to detect many genetic susceptibilities affected by workplace environments. But most researchers expect these genes to be found. Some already are known. Tests can now detect a condition that can make some workers especially vulnerable to chronic lung disease, such as emphysema, a severe breathing problem. The condition is the result of a defect in the body's production of an enzyme called alpha-one-antitrypsin, or AAT. The enzyme is crucial to a defense mechanism used by the lungs to ward off damage from dust, smoke or other irritants inhaled by an individual.

The gene defect isn't common, about one to two percent of people of northern European descent carry one copy of a damaged gene for the enzyme. Two defective copies must be inherited for the enzyme to become inactivated. Thus, about twenty to forty thousand people in the United States carry two copies of the gene, and consequently are at high risk of developing chronic lung disease, if exposed to a dusty workplace or if they smoke.[27]

Marc Lappé, who by decade's end had moved to the University of Illinois at Chicago, where he studied the issue of susceptibility to the lung disease, says several countries have considered screening newborns for the presence of the defective AAT gene. In Sweden a newborn screening program was begun and then terminated when public health officials realized that carriers of a single defective gene were at no special health risk. But, Lappé says, there is ample reason to identify those people with both copies of the defective gene.

"The ethical issue here centers on what governments and employers will do once they know someone ought not work in a dusty or smoky factory," he says. "If what they choose is to ban the person, then we have a serious problem of genetic-based discrimination. A more enlightened, but understandably more expensive, approach would be to clean up the factory."

Lappé says that Swedish health officials are pressing industries in Sweden to sanitize the workplace, but "the issue is still up for grabs in the United States and elsewhere." Even so, to date there are no screening programs to pick out AAT gene defects among workers. But Lappé expects such programs to be instituted sometime soon, another indication that population screening is on the near horizon.

By 1989, several ethicists, including Lappé, were becoming especially worried that the ethical issues posed by a few known genes would soon be compounded. At the root of their concern were the brash expectations of the organizers of the formal genome mapping projects that all human genes would soon become known.

Dorothy Nelkin, a sociologist at New York University, and Laurence Tancredi, director of the health law program at the University of Texas Health Science Center in Houston, capture these worries in a book published in late 1989, called *Dangerous Diagnostics*.[28] Nelkin and Tancredi are especially worried that employers, increasingly frightened by their skyrocketing costs of health, life, and disability insurance, will use the gene discoveries coming out of the mapping enterprises to create a class of biologically unemployable people.

These people will be denied employment because they will be found, through routine genetic screenings, to be at high risk of developing a genetic illness, or highly predisposed to a common illness such as heart disease. Thus denied access to jobs, these people will become the "biological underclass," a new category of citizens destined for the lowest levels of the economic strata.

"Could not genetic diagnostic techniques create a growing class of unemployables, not on the basis of existing symptoms but on the anticipation of possible future symptoms?" the authors asked. "In a competitive economic environment, industry must try to select the 'best' employees on the basis of both potential productivity and future health. If future medical risk becomes a criterion, individuals could find themselves on genetic blacklists, classified as unfit for work."[29]

Already some union leaders see the problem as a distinct possibility. Anthony Mazzocchi, an official of the Oil, Chemical and Atomic Workers Union, told *Wall Street Journal* reporter Alan Otten that identifying susceptibilities will constitute "a genetic scarlet letter that will follow [the worker] the rest of his life."[30]

Suddenly, the bioethicists' vernacular was being filled with new phrases defining the fear attached to exposing one's genetic individuality. At a conference held outside Washington, D.C., in late November 1989, Lappé told a gathering of health reporters that he worries that insurers and employers would soon become "genetic police." Nelkin and Tancredi spoke of "biological discrimination."

Tancredi says genetic tests, like many other medical tests, take on an especially pernicious power because the results are often viewed as clear-cut when they aren't. Few people understand that a genetic reading, except in the case of a single-gene disease, is an imprecise measurement. Indeed, even the presence in a fetus of a single-gene defect such as cystic fibrosis provides only an equivocal prediction of the child's future health. That's because the disease is variable, and some children respond remarkably well to therapy. But those variables cannot be measured by merely reading the results of a genetic test. The test says the gene is defective; it doesn't say how well the child, his parents, or his doctors will respond to it.

In fact, the gene tests tell nothing about what an individual has done to thwart, or, at least, mute the effect of a "bad" gene. The genetic profile may someday show that an individual has inherited a predisposing gene for cancer, but it won't supply any information about whether the individual countered the gene's deleterious action through any number of "healthy" behaviors. Inheriting an altered gene on a particular chromosome may soon indicate that the individual is at risk of developing, say, lung cancer. But if that individual never smoked, the altered gene could be harmless. The genetic profile would tell none of this.

Nonetheless, society is enthralled by the illusion that test results are incontrovertible, Tancredi says. People increasingly want ways to evaluate one another; employers want healthy, productive workers, insurers want to take only reasonable risks. Genetic tests will become additional data from which individuals and institutions can make appraisals.

Some social scientists are beginning to track how this obsession for testing is being confounded by the new genetic information. One of these scientists is Paul Billings of Harvard University School of Medicine, who in 1989 placed advertisements for anecdotes from people who felt they were unfairly discriminated against because of the disclosure of genetic information. One case involved a man who knew he carried a gene for a particularly exotic-sounding genetic disease called Charcot-Marie-Tooth, or CMT, named after the three researchers who identified it.

CMT is a nerve disorder that manifests itself late in life as a weakening of the limbs, mostly due to an atrophying of leg muscles. The disease is highly variable; it can strike quickly or take years to develop, and only rarely leaves its victims severely disabled. Most people with the disease never even know they have it. Nonetheless, the man with CMT told Billings that his car insurance policy was canceled when he advised the insurer, in an answer to a general medical questionnaire, that he carried the gene. The man wasn't sick, nor was he exhibiting any symptoms of CMT, Billings says.

"Insurability should be based on objective clinical findings, not on genetic tests," says Billings. "Say someone is disabled only when he is disabled, not when you measure a trait in his DNA that might eventually make him disabled."[31]

As it becomes certain that the various gene mapping enterprises will produce, in fact, a deluge of genetic data—much of which will at first be difficult to understand or evaluate—the pioneering gene-searching scientists find themselves drawn into increasingly contentious debates. David Botstein, who in 1988 joined Genentech, the South San Francisco biotech company, and became an advisor to the United States' Human Genome Project, found himself arguing over the relative importance of genes and environment, a revival of the old nature-versus-nurture debate.

"The brain is hardware," Botstein told the *New Republic* magazine in April 1989, "and education—in fact, arguably all experience—is the software. Now, the question is, can I increase the ability of an IBM-PC to do word processing by giving it the brains of a Cray?"[32]

By decade's end, Nancy Wexler perhaps had gained the most insights into the problems caused by predicting the future through her experience with Huntington's disease. Early in 1989, Wexler was named to head an advisory board to the Human Genome Project dealing specifically with expected ethical, legal, and social issues. A report issued by the advisory group in December 1989 strongly reflected Wexler's personal view that much regulation and education will be needed if the public is to understand exactly what a genetic test result really means.[33]

The report noted that the genome mapping goal, of course, was to provide improved "treatment, prevention and ultimately cure. [But] the interim phase, before adequate treatment is available is the one in which the most deleterious consequences can occur, such as discrimination against gene carriers, loss of employment or insurance, stigmatization, untoward psychological reactions and attention. Once effective treatment is available for an illness, most of these problems disappear. As the fruits of the Human Genome initiative are realized, there will be an increased need for improved professional and public education to take advantage of the information gained."

Inevitably, the mapping of the genome will raise the question of whether a new eugenics program is in the making. An inkling of this debate could be found in the pages of *Science* in late 1989. The fracas was set off by Daniel E. Koshland's editorial in the journal's October 13 issue.[34]

In the editorial Koshland admonishes those among his scientific peers who are raising doubts about the wisdom of the various genome projects. Koshland calls them the "sky is falling" group. "Sequencing the human genome puts us on the threshold of great new benefits and some real avoidable risks," he writes. "There are immoralities of commission that we must avoid. But there is also the immorality of omission—the failure to apply a great new technology to aid the poor, the infirm, and the underprivileged. We must step boldly and confidently across the threshold."

Four weeks later the journal published several letters responding to Koshland's clarion call. S.E. Luria of MIT's department of biology wrote that the genome project "has been promoted without public discussion by a small coterie of power-seeking enthusiasts."[35] Luria said that Koshland himself hinted of "a eugenic program targeted to the 'poor, the infirm, the underprivileged.' Are they to be transformed (or perhaps altogether eliminated) by eugenic applications of the new technology?"

Wrote Luria: "Will the Nazi program to eradicate Jewish or otherwise 'inferior' genes by mass murder be transformed here into a kinder, gentler program to 'perfect' human individuals by 'correcting' their genomes in conformity, perhaps, to an ideal 'white, Judeo-Christian, economically successful' genotype?"

Luria, it seems, was not alone. Among the other letters expressing similar concerns was one written by Ari Berkowitz of the department of biology at Washington University in St. Louis.[36] Berkowitz wondered if the availability of genome data could "mean that couples will be able to choose characteristics they would like their child to have? Is this something we want to come about?"

He added: "What about the much more likely possibility that the power to alter human genes will encourage well-meaning researchers and statesmen to create human beings with characteristics they see as beneficial, something along the lines of Aldous Huxley's *Brave New World?* How would the cost of dehumanization inherent in fabrication of people compare with the benefits of eradicating certain disease?"

Thus, as the various genome projects were about to get underway, scientists were beginning to face the dilemmas posed by their discoveries. Yet, even as they debated the consequences of gene mapping, scientists elsewhere already were conducting experiments designed to give prospective parents even greater control over the genetic fate of their future children.

CHAPTER 15

Choices

In the fall of 1989 two British molecular biologists offered readers of the *Lancet* a glimpse into the future of prenatal genetic diagnosis.[1] In a "Preliminary Communication," Cathy Holding and Marilyn Monk of the Medical Research Council's Mammalian Development Unit at University College, London, described an experiment with a strain of inbred laboratory mice that lacked a gene for beta-hemoglobin, one of the forms of hemoglobin used by red blood cells to carry oxygen. The absence of this gene makes these lab animals ideal models for studying a severe inherited human anemia known as beta-thalassemia.

In the experiment, a male and a female mouse, each lacking the beta-hemoglobin gene, were mated. Three days later, before the newly fertilized egg had become implanted in the womb, it was washed out of the oviduct of the impregnated female. At this stage, the fertilized egg had divided three times, producing a microscopic clump of eight cells.

The biologists then separated one cell from the eight-cell clump and extracted its DNA. At this point the two biologists brought into play a new technology known as a polymerase chain reaction. They snipped out the segment of DNA where the beta-hemoglobin gene is normally located, dropped the snippet into an appropriate chemical broth, and added a combination of enzymes. The enzymes triggered a rapid chemical copying of the DNA, producing billions of copies of the original DNA within a couple of hours. This rapid chemical copying produced enough DNA for a probe to determine whether the beta-hemoglobin gene was absent (which, of course, it was, since neither parent had the gene).

The exercise was repeated with fertilized eggs from twenty-five additional mouse matings and only once did the technique fail to detect the absence of the beta-hemoglobin gene. Since the DNA from each isolated cell was an exact duplicate of the DNA in the remaining seven cells, each success was tantamount to a genetic diagnosis of the remaining seven cells. Had they wished, Holding and Monk could have returned the remaining seven cells to the female mouse. The seven-cell clump would become implanted in the womb, producing a pregnancy (and eventually a baby mouse) that lacked the beta-hemoglobin gene.

The experiment wasn't an attempt to improve the prenatal diagnosis of beta-thalassemia per se. Instead, it was an attempt to find out if the new polymerase chain reaction technology permitted the diagnosis of a single gene defect using only the DNA from a single cell. Obviously, it did.

"In theory, any disease for which the defect is known at the level of the DNA base sequence is amenable to this approach," the two biologists declared.

"Preimplantation diagnosis of genetic disease provides an alternative to the therapeutic abortion offered to couples at risk of producing children with severe inherited disorders," Holding and Monk asserted. "Preimplantation diagnosis could allow identification of normal and mutant embryos . . . and the replacement in the mother of only those embryos shown to be free of the defect."

The experiment introduces an entirely new dimension into the concept of prenatal genetic diagnosis, that of making a genetic diagnosis *before* pregnancy, thereby circumventing the question of abortion. Such "preimplantation diagnosis" holds staggering implications for the use of the gene discoveries that are destined to come out of the mapping of the human genome. As prenatal genetic diagnosis becomes simpler and easier, the temptation will arise to use it for less severe genetic aberrations. It appears highly likely that young couples, possibly those in the next generation, will be able to make choices about the genetic traits of their children that would astonish today's generation. As the genetic secrets of stature are uncovered, for example, couples would be able, if they desired, to select the height of their children within certain limits. As the gene mapping proceeds, other traits affecting intelligence, athletic or musical ability, even personality could become matters of parental choice.

At present, prenatal genetic diagnosis is limited by both technology and societal pressures. The most commonly used prenatal diagnostic technique is amniocentesis. At about the sixteenth week of pregnancy a sample of the amniotic fluid is taken from the placental sac by a hollow-bore needle. The fetal cells floating in the fluid are extracted and cultured in the laboratory for a week or two until enough cells are available for a cytogeneticist to spot chromosome abnormalities, and enough DNA is available for the molecular geneticists to detect defective genes.

The genetic diagnosis by amniocentesis is usually completed by the eighteenth week of pregnancy, which is dangerously close to the time when an abortion (should the diagnosis detect a defect the parents are unwilling to accept) becomes too risky to attempt.

A newer technique greatly reduces the risk associated with abortion by making the diagnosis several weeks earlier in the pregnancy. This new technique, chorionic villus sampling, or CVS, involves snipping out a small sample of the newly forming placenta at about the eighth week of pregnancy and using its cells for a genetic diagnosis.

Neither procedure is recommended unless the woman is willing to undergo an abortion should the genetic diagnosis indicate a defective fetus. Thus, prenatal genetic diagnoses are, for all practical purposes, rarely used by those who have moral objections to abortion. Prenatal diagnoses also are denied many poor women whose access to abortion is restricted by government policies that prevent the use of tax monies for abortion.

Even for those couples who may accept abortion, there are some limitations. Most health insurers will cover the one-thousand-dollar-or-more cost of a prenatal diagnosis only if the woman is at a higher-than-normal risk of bearing a severely affected child, specifically a woman over thirty-five, the age at which the risk of bearing a child with Down's syndrome begins to rise, or a woman who has either already borne one child with a severe genetic defect or who, by virtue of a family history, is likely to be a carrier of a severe genetic defect.

The Holding-Monk experiment suggests that many of these limitations on prenatal genetic diagnoses may quickly fall aside. Particularly significant is Holding and Monk's use of the term "preimplantation diagnosis." *Webster's Third New International Dictionary* defines a human embryo as "the developing human individual *from the time of implantation* to the end of the eighth week after conception." (Emphasis added.) The clump of eight fertilized cells used in the experiment was extracted from the mouse while it was still in the oviduct and before it had become implanted in the womb. Hence, the eight-cell clump couldn't be defined technically as an embryo but rather, as some researchers would have it, a "pre-embryo."

Holding and Monk's use of the term "pre-embryo" quickly brought charges that it was a subterfuge. In a letter to the *Lancet*, an English physician cited the rebuttal of a member of a British committee that investigated the use of embryonic tissues for research: "Those who are introducing 'pre-embryo' into the vocabulary know full well the research is indeed contentious and that fundamental issues have yet to be resolved . . . they are themselves manipulating words to polarise an ethical discussion."[2]

There is, however, ample precedent for regarding a four-cell or eight-cell human "pre-embryo" as different from a human embryo. That precedent comes from the in vitro, or test-tube, fertilization procedure that is offered infertile couples by clinics throughout the industrialized world. The in vitro fertilization procedure involves first using a pharmaceutical to induce multiple ovulation in the woman. Six to eight eggs are then removed from the woman and fertilized in the test tube with the potential father's sperm. The fertilized eggs are allowed to divide two to three times to the four-to-eight-cell stage. One or two of these "pre-embryos" are then returned to the woman in hopes that a pregnancy will ensue. The unused "pre-embryos" are frozen and stored.

The reason for obtaining and fertilizing several eggs is that the chances of becoming pregnant by this method can range from as low as 20 percent in older women to 60 percent in women under age thirty. Thus, if the first attempt fails to result in pregnancy, a second set of "pre-embryos" can be thawed and used, avoiding the necessity of having to obtain a fresh supply of eggs from the woman. If a pregnancy does occur and is brought to term, the unused frozen "pre-embryos" often are discarded with the parents' permission, a practice that so far hasn't provoked opposition from antiabortion groups.

Preimplantation or "pre-embryo" genetic diagnosis, as suggested by Holding and Monk, is directly adaptable to in vitro fertilization. Each fertilization of an egg in the test tube is a separate conception and each "pre-embryo" has its own distinctive genetic makeup. A defective gene from one of the parents may be inherited by some of the "pre-embryos" but not others. Since removal of only one of a "pre-embryo's" eight cells doesn't impair the ability of the remaining seven cells to develop into a fully formed fetus, it is possible to test each of the pre-embryos for the defective gene and choose only those free of the defect for return to the would-be mother. Those "pre-embryos" found defective could be discarded. Disposing of a few microscopic clumps of cells from a test tube may be morally easier and less objectionable to many young couples than an abortion at the eighth week of pregnancy.

If some couples or physicians find it morally objectionable to discard four-cell or eight-cell "pre-embryos," an experiment in Chicago in 1989 poses a way around it. At the annual meeting of the American Society of Human Genetics in Baltimore in November 1989, researchers from the Illinois Masonic Medical Center in Chicago reported they had made a genetic diagnosis in an immature human oocyte or egg before it was fertilized.[3]

Yuri Verlinsky and his colleagues reported they had flushed out eight immature eggs from a woman who had an enzyme deficiency caused by a defective gene. The immature eggs were just at the stage where the normal twenty-three pairs of chromosomes were being separated to give the egg a set of twenty-three single chromosomes. The chromosomes being discarded were still in the eggs. The researchers were able to extract the about-to-be-discarded chromosomes and subject their DNA to the new polymerase chain reaction. The scientists could then determine which eggs had retained the defective gene and which had discarded it. The eggs that were free of the defective gene were then fertilized in the test tube with the husband's sperm and reimplanted in the woman.

There seems little doubt that an unfertilized human oocyte or egg fails to qualify as a human embryo and, hence, the technique promises to circumvent any moral objections to the preimplantation diagnosis of genetic defects, at least for those defects carried by the mother.

At the moment, the concept of pre-embryo selection is so novel that any speculation on its potential use seems farfetched. But technology has a way of changing social and moral precepts. A generation ago amniocentesis was a novel research technique that few thought would be widely used considering its implied acceptance of abortion, which, before 1973, was illegal in the United States. Today, more than three hundred thousand American women a year undergo amniocentesis for prenatal diagnoses.

The first clients for pre-embryo selection undoubtedly would be couples who have a high risk of a transmitting a defective gene. These would include couples in whom the wife is a carrier of the muscular dystrophy gene or in whom both the husband and the wife are known carriers of the cystic fibrosis gene. The DNA probes to diagnose both of these disorders are already available. As early as 1986, when news broke that DNA markers had been found that made it possible to diagnose Huntington's disease, muscular dystrophy, and cystic fibrosis prenatally, some in vitro fertilization clinics said they would offer their clients genetic tests for these disorders when such tests became available. None of the clinics have publicly disclosed whether they have actually performed such tests on pre-embryos.

If in vitro fertilization clinics begin providing pre-embryo selection for genetic defects on a wide scale, as appears possible, there is no reason to believe they will limit their services just to the detection of genetic disease. As explained in earlier chapters, it soon will be possible to determine if an eight-week-old fetus, or a test-tube pre-embryo, carries the genes that would render the future human susceptible to cancer, heart disease, diabetes, manic-depression, schizophrenia, or any of a host of other disorders. If the tests exist and are readily available, many young couples will be tempted to use them. "Some prospective parents demand assurances of fetal well-being that require otherwise unnecessary testing," note medical ethicists Kathleen Nolan and Sarah Swenson of the Hastings Center.[4] And, they add, "Sophisticated genetic scrutiny does seem a natural response to heightened parental hopes for and expectations of 'perfect babies'..."

It already is technically possible to select a pre-embryo for sex.[5] Fifteen months before Holding and Monk reported their experiment, an in vitro fertilization research team in Edinburgh, Scotland, reported they had taken single cells from donated human male pre-embryos at the two-to-eight-cell stage and, with the aid of the polymerase chain reaction, had been able to detect the Y chromosome. The test confirmed that the pre-embryos with the Y chromosome were male and those without the Y chromosome were female. The Edinburgh experiment, in other words, showed that the sex of pre-embryos could be determined. (The Edinburgh experiment detected a chromosome; the Holding-Monk experiment fifteen months later marked an advance in that it detected a single gene.)

Would a young couple go so far as to use in vitro fertilization and pre-embryo selection just to select the sex of their future child? In 1987, the physicians' news magazine *Medical World News* surveyed prenatal diagnostic clinics, asking them if they were encountering couples seeking amniocentesis or CVS solely to determine the sex of the fetus.[6] Eight of the nine clinics checked said they did occasionally get such requests. The clinic at the University of California, Los Angeles, said it received about fifty such requests a year.

About 60 percent of the geneticists surveyed said that, if a couple requested a sex determination of the fetus, they would either do the test or refer the couple to a clinic that would. A clinic in Fairfax, Virginia, estimated that eighty of the two thousand CVS procedures it had performed up to that time had been done solely for sex determination. Interestingly, the Fairfax clinic was one that offered both prenatal testing and in vitro fertilization services.

What the couples intended to do once they learned the sex of the fetus wasn't known, or at least wasn't reported by the magazine. The assumption, however, is that some couples were willing to undergo an abortion at the tenth or even the eighteenth week of pregnancy if the sex of the fetus wasn't what they desired. If some couples today are willing to undergo an abortion just for the sake of selecting the sex of their child, it isn't difficult to imagine that many more would be willing to use pre-embryo sex selection if it were available. It would be far less traumatic, considerably cheaper, quicker, and, to many couples and physicians, less objectionable morally than determining the sex of a fetus at the tenth week of pregnancy and dealing then with the issue of whether an abortion is justified.

Pre-embryo selection for genetic disorders, genetic susceptibilities, and sex will be available to young couples within a decade or two. Some young couples will use it, particularly if pre-embryo diagnosis (or oocyte diagnosis) circumvents the emotionally charged issue of abortion.

The mapping of the human genome will rapidly broaden the spectrum of genetic disorders and inherited traits that will be detectable in a pre-embryo. Thus, in a generation or two, young couples, should they so desire, will be able to make choices about the nature of their children that today seem to lie in the realm of science fiction.

As the notion of pre-embryo selection takes hold among doctors and researchers as a viable medical tool, scientists already are conjuring up all sorts of fantastic-sounding schemes. One was prompted by the story, widely reported in early 1990, of a California couple in their mid-forties, who decided to have another child in the hopes of treating their seventeen-year-old daughter who was dying of leukemia. The only way to treat the daughter's leukemia, a cancer of the white blood cells produced by the bone marrow, was by killing off the girl's diseased marrow with radiation and replacing it with healthy marrow donated from someone with a matching HLA blood type. As the family had been unable to find a match after an exhaustive search, they conceived a child, hoping its blood type would be identical to their teenage daughter's. It was. The family hoped that when the new baby was six months old it could donate enough marrow to save its sister's life.

The news reports set off a firestorm as ethicists argued whether it was morally right for a couple to have one child in order to treat the disease of another. But the news reports got some molecular biologists thinking. The California couple had to hope the child they conceived would be a compatible match. Some of the scientists said one way around this problem would be for the sick child's parents to conceive another child in vitro. Scientists, using pre-embryo selection techniques, would then choose the pre-embryo with the matching HLA and deposit only that pre-embryo into the mother's womb. "It sounds unbelievable but it certainly sounds feasible," says David Housman of MIT, who was among those scientists to hear of the notion.

In April 1990, the technique received its first publicity when scientists in Great Britain startled the world by reporting the first human use of pre-embryo selection. Researchers at Hammersmith Hospital in London said they used the technique to identify the sex of five eight-cell embryos fertilized in vitro. All five prospective mothers were at risk of passing along to their sons an X-linked disease, such as hemophilia or muscular dystrophy. The researchers said they used polymerase chain reactions to detect the Y chromosome in one cell that was removed from the eight-cell embryos, thus determining which of the embryos was male and at risk of inheriting the X-linked disease. The researchers then implanted into the mothers' wombs only those pre-embryo cells known to be female. Three of the five women then became pregnant.

Although the scientists used the technique to screen out the X-linked diseases, they said that it could be used for other diseases, too. "We are hopeful that we can use the same approach for the specific diagnosis of other relatively frequently inherited diseases," Robert Winston, who led the research, told *The Times of London*.

It should be noted at this point, however, that there is no basis for fears that some kind of superhuman can be created by genetic selection, or that some dictatorial government will be able to revive the Nazi eugenics program and produce a superior race. Creation of a new human would still require the merging of two genomes, male and female, even if it takes place in the test tube instead of the uterus. The resulting embryo cannot inherit genes that aren't possessed by its parents; the genetic selection is limited to column A and column B, so to speak. Even pre-embryo genetic selection cannot change the fact that a blond, blue-eyed child cannot be born of parents who possess only the genes for black hair and brown eyes.

Moreover, there is a limit on the number of traits a young couple could select for their future child. If one parent possesses a dominant gene, half the pre-embryos likely will possess it. If the parents have two dominant traits they want to see in their future child, only one in every four pre-embryos will possess both dominant genes. The more dominant traits the couple wants, the less likely they are to find a pre-embryo that has the right combination of dominant genes. If the trait is recessive, the odds are even lower: one in four pre-embryos will possess the right two-gene combination for a single recessive trait. The odds of a pre-embryo having two desirable recessive traits jump to one-in-sixteen. For three recessive traits, the odds become one-in-sixty-four. And these are only the odds. If eight eggs are fertilized in the test tube at the same time, it is possible that none of them will possess the desired combination of genes—or that all of them will.

The idea that one could produce a superhuman—or even another Einstein, Mozart, or an athlete comparable to, say, Pele, the soccer star—by genetic selection becomes preposterous. The odds against getting hundreds or even thousands of genes together in the right combination in a single pre-embryo are more than astronomical, assuming that a male and female who could contribute the right genes could be found (and assuming that in some distant century, it's known what combination of genes underlies such talents). Moreover, genes alone don't make geniuses. Einstein, Mozart, or Pele might never have expressed or exhibited their talents had their particular genetic endowments been born into an entirely different social environment.

Nevertheless, young would-be parents in future generations will have the power to select at least some of the traits of their children. A musically oriented mother might want to make sure her child inherits her gene for "perfect pitch," for example. An athletically talented father might want to make sure his son is tall enough to participate in sports or, if the mother's family tends to be relatively short, the father, at the least, might want to make sure the paternal genes for stature will be expressed in his son.

Selection of progeny for physical traits could have some unusual consequences. Take stature, for example. Today, an adult height of five feet, four inches, is considered short for a man. Many parents, if they had a choice, would prefer to have their sons grow to a height of at least five feet, eight inches. If young couples, resorting to pre-embryo selection, began to eliminate males of five-foot-four stature, it would not be long before five feet, six inches, would be considered undesirably short. At some point, of course, selection for stature would become rather ludicrous. Nevertheless, this illustrates how genetic selection could be subject to fashion.

There are hints in the scientific and medical journals, however, that the young parents' choices will extend beyond physical traits; they may also be able to make decisions regarding certain mental traits and perhaps even decisions that bear on their future child's personality and behavior.

Take, for example, the implications of research now underway into a genetic disorder known as the fragile X syndrome. The name stems from the fact that, in those who are affected, one tip of the X chromosome is subject to breaking off. In about 90 percent of boys with the fragile X chromosome, the defect produces moderate to severe mental retardation accompanied by unpredictable impulses to go into tantrums, a behavior that, far more than most other forms of mental retardation, exhausts those parents who choose to care for their afflicted sons.

The effect of the fragile X defect in girls, who have a second "back-up" X chromosome, is markedly different. A few are retarded, some only mildly so. About half are mentally normal. Some of the remaining girls are completely normal and can learn to read and write as well as anyone else. But they do have some unusual learning disabilities.[7] Most of these disabilities are not incapacitating. One, for example, is extreme difficulty in learning left from right. But one learning disability does present a formidable problem for the young girls. This is an inability to understand the concept of a number. A number is an abstract symbol and these girls find it almost impossible to relate an abstract symbol to the real world. They simply cannot do arithmetic. Nor can they grasp what a coin or a dollar bill represents. These particular girls, despite a normal intelligence, often are put in special remedial classes and have an extremely difficult time finishing high school. They are unlikely to hold any job where it is necessary to handle money.

A prenatal test can detect the broken X chromosome in women who've either borne one fragile X child or who come from families where the defect is known to occur. Many couples have chosen to abort a pregnancy if the test shows the fetus has the defect and is male; the couples know from experience the almost unbearable burden imposed on a family by a fragile X retarded boy. But if the prenatal test shows the defect is in a female fetus, the question of abortion takes on an entirely different character. The odds are much smaller that the girl will be severely retarded. Instead, she is likely to be on the borderline between retarded and normal, meaning she will spend most of her school days in special classes. Or there is a good chance that the girl may suffer only the inability to learn arithmetic. Does one avoid the birth of a child just because she may have difficulty getting through school and may end up economically dependent on her parents and her siblings the rest of her life? So far as is known, most couples have chosen to bear a fragile X girl, often because they are already acquainted with such a girl in the family, perhaps their first daughter or a niece, and can't accept the idea of aborting someone who may be similar to a child they already love.

There is an intense effort to identify the several genes that are damaged by the breaking of the X chromosome. Identifying the genes would provide a new insight into a major cause of mental retardation. Uncovering the damaged genes could permit concerned couples to make a more informed choice about whether to bear a female child with the defect. By examining which genes are damaged by the breaking of the X chromosome in a particular female fetus, it should be possible to determine if the girl will be normal, learning disabled, or severely retarded.

This excursion into the genes of the fragile X syndrome, however, has a much broader implication. Implicit in uncovering the damaged genes in the syndrome is the revelation of the normal genes. Somewhere among these genes is a gene (or genes) that, when damaged, renders the girls unable to deal with numbers. This gene, in its normal state, must have something to do, directly or indirectly, with the ability to comprehend and use abstract symbols.

The ability to understand abstract symbols happens to be one of the traits measured by IQ tests. The fragile X syndrome research, then, suggests that the gene mappers eventually will uncover this and many of the other genes involved in the traits which IQ tests are designed to measure. Whatever else one may think of IQ tests, the traits they measure are those that are required to succeed in Western industrialized society. It is not difficult to imagine that, if a young, well-educated couple were selecting among eight test-tube pre-embryos, they might choose the one most likely to score highest on IQ tests.

A recent study in Sweden suggests that other attributes that play a role in success might well have genetic underpinnings and thus could be selected for. Geneticists discovered a few years ago that 5 to 10 percent of the population have a genetic defect that inhibits the action of an enzyme involved in metabolizing tranquilizers. Those who possess this defect are more likely to suffer side effects of the drugs than most people. They are known, in the jargon, as "poor hydroxylators."

In early 1989 a Swedish team studying this genetic defect developed the impression that poor hydroxylators have a personality distinct from normal hydroxylators.[8] Intrigued, they gave personality tests to 51 "poor hydroxylators" and 102 normal or "extensive hydroxylators." Both groups were about average on most of the tests, but in one there was a difference. The poor hydroxylators showed higher scores on vitality, alertness, efficiency, and "ease of decision making." The genetic defect that makes a few people poor metabolizers of tranquilizers also affects some internal chemical of importance to the central nervous system which, in turn, affects how easily one makes decisions.

Extending this a bit further, the genes influencing personality and behavior also are likely to be uncovered at some point in the exploration of the human genome. No such behavioral genes have been found yet but there is growing evidence that they exist. These genes, it must be emphasized, do not determine personality or behavior; there seems little question that the social and psychological environments that envelop a child in his or her formative years are the most powerful influences on personality and future behavior. But the genes can influence how a child reacts to the environment.

Take, for example, a study reported in early 1988 by psychologists at Harvard and Yale universities on the biological bases of shyness.[9] In their published report, the psychologists noted that the reaction of young children suddenly exposed to an unfamiliar situation varies between two extremes. "About 10% to 15% of healthy 2- and 3-year-old children consistently become quiet, vigilant, and affectively subdued in such [unfamiliar] contexts for periods ranging from five to 30 minutes. An equal proportion is typically spontaneous, as if the distinction between familiar and unfamiliar were of minimal psychological consequence."

To find out if this tendency to extreme shyness had a genetic underpinning, the psychologists studied a group of two-year-old children and picked out those who displayed extremes of both behaviors. They periodically observed the children's behavior over the next five and a half years, seeing how they reacted in the playroom to unfamiliar situations. They also measured each child's heart rate, eye dilation, and output of stress hormones at times they were exposed to the unfamiliar situations.

The children who were most inhibited at age 2 were most likely at age 7.5 years to have the highest heart rates and to have unusual fears, such as fears stirred by violence on television, worries about being kidnapped, or going to the bathroom alone at night. The children who were uninhibited and outgoing at age 2 had the lowest heart rates and almost no unusual fears at age 7.5 years. The reactions of the children's eye pupils, their muscle tensions, and hormone production (measured in the urine) also correlated with their behavior.

The study led the psychologists to speculate that the inhibited children, at least those with the extreme form of shyness, had been born with a brain and nervous system that had a low threshold for arousal when exposed to unexpected changes or novel events. In other words, the reaction of the nervous system that produces anxiety and fear is more easily triggered in these children than in the outgoing children.

Children born with this low threshold were more likely to be affected or more severely affected by stresses early in life, leading to their shy, inhibited behavior as they grew. Such stresses might be hospitalization, death of a parent, quarrelling between the parents, or mental illness in a family member. "An older sibling who unexpectedly seizes a toy, teases or yells at an infant who has a low threshold for limbic arousal might provide the chronic stress necessary to transform the temperamental quality into the profile we call behavioral inhibition," the psychologists suggested.

In his prescient 1983 paper in *Science* on the social and medical impact of the new genetic discoveries, medical geneticist Arno G. Motulsky warned, "As long as such [genetic] knowledge only concerns genes affecting variables of physical health and as long as testing remains voluntary, society may be able to cope. But when we learn more about the genetics of personality and mental traits, new problems will arise."

Motulsky didn't speculate on what these problems might be. But the possibility that—in one, two, or perhaps three generations—young couples will have the power to select their children for at least some mental and behavioral traits, as well as for physical health, does suggest some disturbing scenarios.

The passions devoted today to the question of abortion could turn in the future to questions arising from pre-embryo genetic selection. Will the use of such selection be limited only to genetic diseases or will it be broadened to include testing for genetic susceptibilities? Should use of the procedure be banned for detecting normal mental and physical traits? Is discarding an eight-cell "pre-embryo" tantamount to abortion?

One scenario lends reality to the "hereditary meritocracy" postulated in the early 1970s by Harvard psychologist R. J. Herrnstein.[10] A meritocracy is a society in which everyone rises to the position and status suited to his or her ability. In this hypothetical society there are no barriers such as race, religion, politics, or nepotism preventing a person from seeking the role in society he is best able to perform. If IQ scores are, as they are claimed to be, a measure of important traits needed to succeed in society, then in a meritocracy those with the higher IQs will rise to the higher strata while those with the lower IQs will occupy the lower strata.

At the time Herrnstein described the meritocracy, a furious debate had broken out over whether differences in IQ scores among individuals and among groups were due to genetic factors or cultural factors. The psychologist showed what would happen in a meritocracy if these IQ differences were, in fact, due largely to heredity. People tend to mate with others in their own social class, a phenomenon psychologists call "assortive mating." Thus, in a meritocracy, the high-IQ individuals in the upper strata are likely to marry each other and bear children, passing along to their progeny those genes that ensure success, i.e., the genes for the traits measured by the IQ tests. Those in the lower strata with the lower IQs similarly will mate assortively, passing their low-IQ genes to their progeny. After a few generations, the society will become rigidly stratified and people literally will be born to their status in society.

In the real world, of course, this "hereditary meritocracy" is overridden by artificial social barriers. Children of the rich inherit their wealth and societal position even if they don't inherit a high IQ. And in the Western democracies, at least, those with a high IQ born in a lowly stratum can fight their way to the top—or if the social barriers are too strong, remain in the lower levels of society to pass their high IQ genes to their children. Either way there is a constant mixing of the gene pool that prevents the society from becoming stratified by inherited differences in intelligence.

But there are reasons to believe that, if future generations are able to select, even to a modest degree, the genetically determined traits of their children, the society may begin to experience some of the consequences of a "hereditary meritocracy." Today, birth control and prenatal diagnosis by amniocentesis or CVS are most commonly used by couples in the middle and upper socioeconomic classes. These couples have both the means and the education to take advantage of the technology and seem to be far less concerned about the morality of abortion or constrained by its cost than those further down on the social and economic ladders.

Moreover, young, well-educated, middle-class couples of today comprise the first generation in history to grow up with choices in childbearing via reliable birth control and prenatal diagnosis. The availability of childbearing choices, even limited ones, seems to have instilled a subtle but significantly different view of childbearing in this generation from that held by previous generations. Where expectant parents in the past might have hoped and prayed that their child would be healthy and endowed to succeed in the world, the new generation seems to *expect* to have such a child, a "perfect baby." They seem willing to go to some lengths, i.e. abortion, to avoid having a defective child or, in some cases, to avoid having a child of the undesired sex.

The advent of tests for genetic disease and genetic susceptibilities to future disease will, without question, strengthen this desire and expectation for a "perfect baby" among affluent middle-class couples.

By contrast, the use of birth control, prenatal diagnosis, and abortion is less common, and in some places quite rare, among the poor and economically disadvantaged. Those who are struggling to survive in a hostile world often have neither the energy nor the education to concern themselves with such luxurious concepts as bearing a "perfect baby." Even if the poor are aware of prenatal diagnoses, society, by refusing to approve public monies for abortion, effectively prevents them from using such technologies.

There is no reason to believe that this discrimination will disappear in the future. When the ability to select pre-embryos for desirable mental traits as well as freedom from genetic diseases and susceptibilities becomes widely available, it will be utilized most widely by the middle and upper socioeconomic classes and used least by the poor and disadvantaged. The young, well-educated couples of the future likely will put a premium on bearing a child who will do well in school and in adult life, that is, a child who will score high on IQ tests.

It wouldn't take many generations of this discriminatory genetic selection to produce an ever-widening gap between the upper and lower strata of a society. Herrnstein's description of a society in which a butcher's son has little opportunity to be anything but a butcher and an executive's child is born to be an executive may not be as unreal as it might seem today.

The democratic alternative would be for the society to step in at the very beginning and make pre-embryo selection (and its implied use of birth control) available to the poor at public expense. But this raises other specters. The current experience with the failure of reliable birth control techniques to curb unplanned pregnancies among poor teenage mothers shows that it is not enough simply to make the technology available. An extensive education program among the young and disadvantaged would be required to make sure they are aware of the power to select their offspring for genetic health and for certain physical and mental traits. Inherent in such an education program is a description of which genetic traits are desirable and which are undesirable.

Moreover, if public monies are used to pay for, say, pre-embryo selection for the poor, it is inevitable that attempts will be made to dictate which genetic tests can be done and which will be prohibited. In a democracy, the prohibitions for the poor should also be applied to the well-to-do. In short, the temptation for the state to step in and begin dictating the genetic traits of future generations is high.

It is easy to dismiss such matters as being either farfetched or of distant concern. But science advances without regard to whether society is prepared to cope with its discoveries. In 1978, when the germ of the breakthrough in genetic mapping was sown at Alta, few in that ski lodge could have predicted that within a dozen years two young British molecular biologists would make a genetic diagnosis of an eight-cell human embryo, that Nancy Wexler would have to agonize over submitting to a diagnosis for a genetic disease for which there is no treatment, that a debate would erupt over whether virtually the entire population should be tested for the cystic fibrosis gene, or that insurance companies would be seriously considering examing an individual's genetic profile. Yet society now faces such issues. Perhaps it's time to consider the ultimate implications of exploring the human genome.

NOTES

much of the material in the preceding pages was gathered in the process of reporting and writing news articles for a daily newspaper. Such "research," involving interviews and coverage of press conferences and scientific meetings, over a space of five years and in a multitude of locales, is hardly conducive to meticulous record-keeping. Consequently, a written source cannot be cited for every quotation and fact, although, where possible, the original source is indicated in the text. In the course of news reporting, however, the authors did consult a large number of scientific and general articles and books, which subsequently were used in writing this book. Whenever possible and appropriate, these sources are noted below. Generally, but not always, the notes for the articles follow the style of scientific journals: Where an article had three or more authors, as is often the case, only the name of the first author is cited, followed by the title of the article, the name of the scientific journal in which it appeared (with the volume number and page number[s] separated by a colon), and finally the year of publication. More than one author is cited in a note when those authors are mentioned in the text. The authors have not provided a separate bibliography, since all relevant sources have been identified in the notes.

Introduction: Breakthrough

1. There is, also, the promise that someday doctors will be able to repair genetic-based illnesses by replacing defective genes in the body with healthy ones. There are a handful of scientific pioneers now developing techniques called "gene therapy" and initial experiments in some laboratory animals suggest that, indeed, broken or missing genes can be replaced. While early efforts at gene therapy have received some high-profile attention by the mass media, even the most optimistic of the scientists believe that widespread use of this tool in treating medical problems is still many years off, if it is possible at all. This book will not discuss gene therapy, but focuses, instead, on the discovery of the genes which someday might be candidates for such treatment. Those who may want to explore the prospects of gene therapy are invited to read *Human Gene Therapy*, Eve K. Nichols, Institute of Medicine, National Academy of Sciences (Cambridge, Massachusetts: Harvard University Press, 1988).

2. Gregor Mendel, "Versuche über Planzenhybriden," *Verh. naturforsch. Verein in Brunn*, vol. IV, 1866.

Chapter 1: From Curse to Crusade

1. George Huntington, "On Chorea," *Medical and Surgical Reporter* 26:317–21, 1872.
2. Ibid.
3. Ibid.
4. Mendel, "Versuche über Planzenhybriden."
5. Nancy Wexler, "50/50: Genetic Roulette."
6. Ibid.
7. Ibid.
8. Milton Wexler, "The Structural Problem in Schizophrenia: Therapeutic Implications," *The International Journal of Psycho-Analysis* 32:157–66, 1951.
9. Nancy Wexler, "50/50: Genetic Roulette."

Chapter 2: Alta

1. Corwin Q. Edwards, Mark H. Skolnick, and James P. Kushner, "Hereditary Hemochromatosis: Contributions of Genetic Analyses," *Progress in Hematology*, ed. E. B. Brown (New York: Grune & Stratton), 43–71, 1981.
2. George E. Cartwright et al., "Hereditary Hemochromatosis," *New England Journal of Medicine* 301:175–79, 1979.

3. David Botstein, Raymond L. White, Mark Skolnick, and Ronald W. Davis, "Construction of a Genetic Linkage Map in Man Using Restriction Fragment Polymorphisms," *American Journal of Human Genetics* 32:314–31, 1980.
4. Ibid., 314.
5. Ibid., 328.

Chapter 3: Worcester

1. Ellen Solomon and Walter F. Bodmer, "Evolution of Sickle Variant Gene," *Lancet* II:923, 1979.
2. Arlene R. Wyman and Ray White, "A Highly Polymorphic Locus in Human DNA," *Proceedings of the National Academy of Sciences* 77:6754–58, 1980.

Chapter 4: El Mal

1. Nancy Wexler, "Genetic Jeopardy and the New Clairvoyance," *Progress in Medical Genetics*, vol. 6, ed. A. Bearn, B. Childs, and A. Motulsky (New York: Praeger Press, 1985).
2. Americo Negrette, *Corea de Huntington*, 2d. ed. (Maracaibo, Venezuela: La Editorial Universitaria, La Universidad del Zulia, 1962).
3. Nancy Wexler, "50/50: Genetic Roulette."
4. Ibid.
5. Ibid.
6. Roswell Eldridge, "Gene Mapping: It's a Whole New World for Medicine," *Medical Tribune*, p. 73, 8 May 1985.
7. James Gusella, Nancy Wexler, P. Michael Conneally et al., "A Polymorphic DNA Marker Genetically-Linked to Huntington's Disease," *Nature* 306:234–38, 1983.
8. Lawrence K. Altman, "Researchers Report Genetic Test Detects Huntington's Disease," *New York Times*, p. 1, 9 Nov. 1983.

Chapter 5: Bruce Bryer and Reverse Genetics

1. Muscular Dystrophy Association of New York, "The Legacy of Bruce Bryer," *MDA NewsMagazine* (Spring 1987), 11.
2. Christine Verellen-Dumoulin et al. (including Ronald G. Worton), "Expression of an X-linked Muscular Dystrophy in a Female due to Translocation Involving Xp21 and Non-Random Inactivation of the Normal X Chromosome," *Human Genetics* 67:115–19, 1984.
3. Louis M. Kunkel et al., "Specific Cloning of DNA Fragments Absent

GENOME / 326

from the DNA of a Male Patient with an X Chromosome Deletion," *Proceedings of the National Academy of Sciences* 82:4778–82, 1985.

4. Anthony P. Monaco et al. (including Louis M. Kunkel), "Detection of Deletions Spanning the Duchenne Muscular Dystrophy Locus Using a Tightly Linked DNA Segment," *Nature* 316:842–45, 1985.

5. Anthony P. Monaco et al., "Isolation of Candidate cDNAs for Portions of the Duchenne Muscular Dystrophy Gene," *Nature* 323:646–50, 1986.

6. Ronald G. Worton et al., "Duchenne Muscular Dystrophy Involving Translocation of the dmd Gene Next to Ribosomal RNA Genes," *Science* 224:1447–49, 1984.

7. Peter N. Ray et al. (including Ronald G. Worton), "Cloning of the Breakpoint of an X;21 Translocation Associated with Duchenne Muscular Dystrophy," *Nature* 318:672–75, 1985.

Chapter 6: Dystrophin

1. Louis M. Kunkel et al., "Analysis of Deletions in DNA from Patients with Becker and Duchenne Muscular Dystrophy," *Nature* 322:73–77, 1986.

2. Anthony P. Monaco and Louis M. Kunkel, "A Giant Locus for the Duchenne and Becker Muscular Dystrophy Gene," *Trends in Genetics*, vol. 3, no. 2, p. 35, February 1987.

3. Anthony P. Monaco et al., "Isolation of Candidate cDNAs for Portions of the Duchenne Muscular Dystrophy Gene," *Nature* 323:646–50, 1986.

4. Ibid.

5. Ibid.

6. Arthur H. M. Burghes et al. (including Ronald G. Worton), "A cDNA Clone from the Duchenne/Becker Muscular Dystrophy Gene," *Nature* 328:434–37, 1987.

7. Michel Koenig et al., "Complete Cloning of the Duchenne Muscular Dystrophy (DMD) cDNA and Preliminary Genomic Organizations of the DMD Gene in Normal and Affected Individuals," *Cell* 50:509–17, 1987.

8. Burghes et al., "A cDNA Clone," 434–37.

9. Eric P. Hoffman et al. "Conservation of the Duchenne Muscular Dystrophy Gene in Mice and Humans," *Science* 238:347–50, 1987.

10. Eric P. Hoffman, Robert H. Brown, Jr., and Louis M. Kunkel, "Dystrophin: The Protein Product of the Duchenne Muscular Dystrophy Locus," *Cell* 51:919–28, 1987.

11. Ibid., 925.

12. Eric P. Hoffman et al. (including Louis M. Kunkel), "Characterization of Dystrophin in Muscle-Biopsy Specimens from Patients with Duchenne's or Becker's Muscular Dystrophy," *New England Journal of Medicine* 318:1363–68, 1988.

13. Ibid., 1364.

14. Lewis P. Rowland, "Dystrophin: A Triumph of Reverse Genetics and

the End of the Beginning," *New England Journal of Medicine* 318:1392–94, 1988.

15. T. A. Partridge et al., "Conversion of mdx Myofibres from Dystrophin-Negative to -Positive by Injection of Normal Myoblasts," *Nature* 337:176–79, 1989.

16. Rowland, "Dystrophin," 1392–94.

17. "The Legacy of Bruce Bryer," p. 12.

Chapter 7: Two Hits

1. Webster K. Cavenee et al., "Prediction of Familial Predisposition to Retinoblastoma," *New England Journal of Medicine* 314:1201–7, 1986.

2. Fred Gilbert, "Retinoblastoma and Cancer Genetics," *New England Journal of Medicine* 314:1248, 1986.

3. Alfred G. Knudson, "Mutation and Cancer: Statistical Study of Retinoblastoma," *Proceedings of the National Academy of Sciences* 68:820–23, 1971.

4. Alfred G. Knudson and Anna T. Meadows, "Chromosomal Deletion and Retinoblastoma," *New England Journal of Medicine* 295:1120–23, 1976.

5. Ibid.

6. Ibid.

7. William F. Benedict et al. (including Robert S. Sparkes), "Patient with 13 Chromosome Deletion: Evidence That the Retinoblastoma Gene Is a Recessive Cancer Gene," *Science* 219:973–75, 1983.

8. Robert S. Sparkes et al., "Gene for Hereditary Retinoblastoma Assigned to Human Chromosome 13 by Linkage to Esterase D," *Science* 219:971–73, 1983.

9. Webster K. Cavenee et al., "Expression of Recessive Alleles by Chromosomal Mechanisms in Retinoblastoma," *Nature* 305:779–84, 1983.

10. Robert A. Weinberg, "Finding the Anti-Oncogene," *Scientific American* 258:44–51, 1988.

11. Webster K. Cavenee et al., "Expression of Recessive Alleles."

12. Webster K. Cavenee et al., "Genetic Origins of Mutations Predisposing to Retinoblastoma," *Science* 228:501–3, 1985.

13. Thaddeus P. Dryja et al., "Molecular Detection of Deletions Involving Band q14 of Chromosome 13 in Retinoblastoma," *Proceedings of the National Academy of Sciences* 83:7391–94, 1986.

Chapter 8: Cancer Unleashed

1. Bert Vogelstein et al., "Genetic Alterations During Colorectal-Tumor Development," *New England Journal of Medicine* 319:525–32, 1988.

2. Lisa A. Cannon-Albright et al. (including Mark Skolnick), "Common Inheritance of Susceptibility to Colonic Adenomatous Polyps and Associated

Colorectal Cancers," *New England Journal of Medicine* 319:533–37, 1988.

3. Peter Nowell, "Molecular Events in Tumor Development," *New England Journal of Medicine* 319:575–77, 1988.

4. Mark Skolnick, "The Utah Genealogical Data Base: A Resource for Genetic Epidemiology," *Banbury Report* 4: *Cancer Incidence in Defined Populations* (Cold Spring Harbor, New York: Cold Spring Harbor Laboratory Press, 1980), 285–97.

5. Mark Skolnick et al., "Mormon Demographic History I. Nuptiality and Fertility of Once-Married Couples," *Population Studies* 32:5–19, 1978.

6. Eldon Gardner, "Breast Cancer in One Family Group," *American Journal of Human Genetics* 2:30–40, 1950.

7. Eldon Gardner, "Inherited Susceptibility to Breast Cancer in Utah Families," *Encyclia* 57:27–46, 1980.

8. D. Timothy Bishop et al., "Segregation and Linkage Analysis of Nine Utah Breast Cancer Pedigrees," *Genetic Epidemiology* 5:151–69, 1988.

9. Randall Burt et al., "Dominant Inheritance of Adenomatous Colonic Polyps and Colorectal Cancer," *New England Journal of Medicine* 312:1540–44, 1985.

10. Cannon-Albright et al., "Common Inheritance."

11. Eric Fearon, Bert Vogelstein et al., "Somatic Deletion and Duplication of Genes on Chromosome 11 in Wilms' Tumour," *Nature* 309:176–78, 1984.

12. Stanley Orkin et al., "Development of Homozygosity for Chromosome 11p Markers in Wilms' Tumour," *Nature* 309:172–74, 1984.

13. A. Koufos, Mark F. Hansen et al. (including Webster Cavenee), "Loss of Alleles at Loci on Human Chromosome 11 During Genesis of Wilms' Tumour," *Nature* 309:170–72, 1984.

14. Anthony E. Reeve et al., "Loss of Harvey Ras Allele in Sporadic Wilms' Tumour," *Nature* 309:174–76, 1984.

15. M. Muleris, "Consistent Deficiencies of Chromosome 18 and of the Short Arm of Chromosome 17 in 11 Cases of Human Large Bowel Cancer: A Possible Recessive Determinism," *Annales du Génétiques* 28:206–13, 1985.

16. Eric Fearon et al. (including Bert Vogelstein), "Clonal Analysis of Human Colorectal Tumors," *Science* 238:193–97, 1987.

17. Ibid.

18. Leslie Roberts, "The Race for the Cystic Fibrosis Gene," *Science* 240:142, 1988.

19. Fearon et al., "Clonal Analysis."

20. Lemuel Herrera et al. (including Avery Sanberg), "Brief Clinical Report: Gardner Syndrome in a Man with an Interstitial Deletion of 5q," *American Journal of Human Genetics* 25:473–76, 1986.

21. Walter S. Bodmer et al., "Localisation of the Gene for Adenomatous Polyposis on Chromosome 5," *Nature* 328:614–16, 1987.

22. Ellen Solomon et al., "Chromosome 5 Allele Loss in Human Colorectal Carcinomas," *Nature* 328:616–19, 1987.

23. Harold Schmeck, "Cancer of Colon Is Believed Linked to Defect in Gene," *New York Times*, p. 1, 13 Aug. 1987.

24. Ellen Solomon et al., "Chromosome 5 Allele Loss."

25. Marc Leppert et al. (including Ray White), "The Gene for Familial Polyposis Coli Maps to the Long Arm of Chromosome 5," *Science* 238:1411–13, 1987.

26. Bert Vogelstein et al., "Genetic Alterations."

27. D.P. Lane et al. (including Arnold Levine), "T-Antigen Is Bound to a Host Protein in SV40 Transferred Cells," *Nature* 278:261, 1979.

28. D. Linzer et al., "Characterization of 54K Dalton Cellular SV40 Tumor Antigen Present in SV40 Transformed Cells and Uninfected Embrional Carcinoma Cells," *Cell* 17:43–52, 1979.

29. Suzanne F. Baker et al. (including Bert Vogelstein), "Chromosome 17 Deletions and p53 Gene Mutations in Colorectal Carcinomas," *Science* 244:217–21, 1989.

30. Michael Waldholz, "Gene Linked to Colon Cancer Identified," *Wall Street Journal*, p. B2, 14 April 1989.

31. J. Mackay et al., "Allele Loss on Short Arm of Chromosome 17 in Breast Cancer," *Lancet* 8625:1384–85, 1988.

32. Janice M. Nigro et al. (including Bert Vogelstein), "Mutations in the p53 Gene Occur in Diverse Tumour Types," *Nature* 342:705–8, 1989.

33. John D. Minna et al., "Recessive Oncogenes and Chromosomal Deletions in Human Lung Cancer," *Current Communications in Molecular Biology: Recessive Oncogenes and Tumor Suppressors*, ed. Webster Cavenee, Nicholas Hastie, and Eric Stanbridge (Cold Spring Harbor, New York: Cold Spring Harbor Laboratory Press, 1989), 57–65.

34. Eric R. Fearon et al. (including Bert Vogelstein), "Identification of a Chromosome 18q Gene That Is Altered in Colorectal Carcinoma," *Science* 247:49–56, 1990.

Chapter 9: Genes and the Heart

1. Robert A. Norum et al., "Familial Deficiency of Apolipoproteins A-I and C-III and Precocious Coronary Artery Disease," *New England Journal of Medicine* 306:1513–19, 1982.

2. Pamela Lyon, "The Great Adventure," *Biologue* 6:1 (University of Texas Health Science Center, Dallas, 1985). The story of the Brown-Goldstein discovery in the succeeding paragraphs is based on this source.

3. Ibid.

4. Norum et al., "Apolipoprotein A-I."

5. Ibid., 1318.

6. Jan L. Breslow, "Apolipoprotein Genetic Variation and Human Disease," *Physiological Review* 68:85–132, 1988.

7. Ibid., 124.

8. Sotirios K. Karathanasis, "Apolipoprotein Multigene Family: Tandem Organization of Human A-I, C-III, and A-IV Genes," *Proceedings of the National Academy of Sciences* 82:6374–78, 1985.

Chapter 10: The Map

1. Daniel J. Kevles, *In the Name of Eugenics* (New York: Alfred A. Knopf, 1985), 245–47.
2. Victor A. McKusick and Thomas H. Roderick, "Medical and Experimental Mammalian Genetics: A Perspective," *Birth Defects Original Articles Series*, vol. 23, no. 3 (White Plains, New York: March of Dimes, 1987).
3. Victor A McKusick and Frank H. Ruddle, "The Status of the Gene Map of the Human Chromosomes," *Science* 196:390–405, 1977.
4. Roger P. Donahue et al., "Probable Assignment of the Duffy Blood Group Locus to Chromosome 1 in Man," *Proceedings of the National Academy of Sciences* 61:949–55, 1968.
5. Ibid., 949.
6. Victor A. McKusick, "The Anatomy of the Human Genome," *Hospital Practice* (April 1981), 82–100.
7. Torbjorn O. Caspersson, "The Background for the Development of the Chromosome Banding Techniques," *American Journal of Human Genetics* 44:441–51, 1989.
8. McKusick and Ruddle, "The Status of the Gene Map," *Science* 196:390–405.
9. Victor A. McKusick, *Morbid Anatomy of the Human Genome* (Bethesda, Maryland: Howard Hughes Medical Institute, 1988), 25.
10. Richard Severo, "A Geographer of the Gene World," *New York Times*, 3 Sept. 1980.
11. Maya Pines, *Mapping the Human Genome* (Bethesda, Maryland: Howard Hughes Medical Institute, 1987), 3.
12. Ibid., 4.
13. Ibid., 6.
14. Jean L. Marx, "Putting the Human Genome on the Map," *Science* 229:150–51, 1985.
15. Pines, 6.
16. David Stipp, "Genetic Map That Could Speed Diagnosis of Inherited Disease Touches off Dispute," *Wall Street Journal*, 8 Nov. 1987.
17. Collaborative Research Inc., Bedford, Massachusetts, *Annual Report for 1987*, 3.
18. David Stipp, "Genetic Map."
19. McKusick and Ruddle, "The Status of the Gene Map," 390–405.
20. Stephen S. Hall, "The Sequel," *California Magazine*, July 1988, 63.
21. Ibid., 65.
22. Robert Kanigel, "The Genome Project," *New York Times Magazine*, p. 44, 13 Dec. 1988.
23. Ibid., 98.
24. Pines, 10.

25. Victor A. McKusick, "Mapping and Sequencing the Human Genome," *New England Journal of Medicine* 320:910–15, 1989.

26. Renato Dulbecco, "A Turning Point in Cancer Research: Sequencing the Human Genome," *Science* 231:1055–56, 1986.

27. Christopher Joyce, "The Race to Map the Human Genome," *New Scientist*, p. 5, 5 Mar. 1987.

28. Ibid.

29. Hall, "The Sequel," 69.

30. Christopher Joyce, "The Race," 5.

31. Robert A. Weinberg, "The Case Against Gene Sequencing," *The Scientist*, 16 Nov. 1987.

32. McKusick, "Mapping and Sequencing," 910–915.

33. Ibid., 913.

34. Larry Thompson, "A Big Ticket, Low-Profile Science Venture," *Washington Post*, 4 Jan. 1989.

35. McKusick, "Mapping and Sequencing," 913–14.

36. Robert A. Weinberg, "The Human Genome Sequence: What Will It Do For Us?," *BioEssays* 9:91–92 (Aug.-Sept. 1988).

Chapter 11: "Siss im Blut"

1. "An Interview with Janice A. Egeland, Ph.D.," *Currents in Affective Illness* 6–6:5–12, 1987.

2. Janice A. Egeland and Abram M. Hostetter, "Amish Study, I: Disorders Among the Amish, 1976–1980," *American Journal of Psychiatry* 140–1:56–61, 1983.

3. Janice A. Egeland and James N. Sussex, "Suicide and Family Loading for Affective Disorders," *Journal of the American Medical Association* 254:915–18, 1985.

4. "An Interview with Janice Egeland."

5. Ibid.

6. Ibid.

7. Janice A. Egeland, Daniela S. Gerhard, and David L. Pauls, "Bipolar Affective Disorders Linked to DNA Markers on Chromosome 11," *Nature* 325:783–87, 1987.

8. S. Hodgkinson et al., "Molecular Genetic Evidence of Heterogeneity in Manic Depression," *Nature* 325:805–6, 1987.

9. S.D. Detera-Wadleigh et al., "Close Linkage of C-Harvey Ras-1 and the Insulin Gene to Affective Disorder Is Ruled Out in Three North American Pedigrees," *Nature* 325:806–8, 1987.

10. Claudia C. Wallis, "Is Mental Illness Inherited?," *Time* magazine, p. 67, 9 March 1987.

11. Miron Baron et al., "Genetic Linkage between X-Chromosome Markers and Bipolar Affective Illness," *Nature* 326:289–92, 1987.

12. Julien Mendlewicz et al., "Polymorphic DNA Marker on X Chromosome and Manic Depression," *Lancet* 8544:1230–31, 1987.

13. David Van Biema, "In Search of a Killer," *Washington Post Magazine*, 25 Sept. 1988.

14. Linda E. Nee et al., "A Family with Histologically Confirmed Alzheimer's Disease," *Journal of the American Medical Association* 40:203–8, 1983.

15. Peter H. St George-Hyslop, Rudolph E. Tanzi et al., "The Genetic Defect Causing Familial Alzheimer's Disease Maps to Chromosome 21," *Science* 238:664–66, 1987.

16. David E. Comings, "Presidential Address: The Genetics of Human Behavior—Lessons for Two Societies," *American Journal of Human Genetics* 44:452–60, 1989.

17. Anne S. Bassett, "Partial Trisomy Chromosome 5 Cosegregating with Schizophrenia," *Lancet* 8589:799–801, 1988.

18. R. Sherrington et al., "Localization of a Susceptibility Locus for Schizophrenia on Chromosome 5," *Nature* 366:164–67, 1988.

19. J.L. Kennedy, "Evidence Against Linkage of Schizophrenia to Markers on Chromosome 5 in a Northern Swedish Pedigree," *Nature* 366:167–70, 1988.

20. Eric S. Lander, "Splitting Schizophrenia," *Nature* 366:105–6, 1988.

21. Herbert Pardes et al., "Genetics and Psychiatry: Past Discoveries, Current Dilemmas and Future Directions," *American Journal of Psychiatry* 146–4:435–43, 1989.

Chapter 12: Genes in a Bottle

1. Sheila B. Gilligan, Theodore Reich, and C. Robert Cloninger, "Etiologic Heterogeneity in Alcoholism," *Genetic Epidemiology* 4:395–414, 1987.

2. C. Robert Cloninger, Theodore Reich, and Shozo Yokoyama, "Genetic Diversity, Genome Organization, and Investigation of the Etiology of Psychiatric Diseases," *Psychiatric Developments* 3:225–46, 1983.

3. Eric C. Devor and C. Robert Cloninger, "Genetics of Alcoholism," *Annual Review of Genetics* 23:19–36, 1989.

4. Ibid., 19.

5. Robert C. Cloninger et al., "Psychopathology in Adopted-out Children of Alcoholics, the Stockholm Adoption Study," in *Recent Developments in Alcoholism*, vol. 3, ed. Marc Galenter (New York: Plenum Publishing Corp., 1985), 37–51.

6. Daniel J. Kevles, *In the Name of Eugenics* (New York: Alfred A. Knopf, 1985), 339.

7. Donald W. Goodwin, "Alcoholism and Genetics," *Archives of General Psychiatry* 42:171, 1985.

8. C. Robert Cloninger et al., "Psychopathology," 37–51.

9. C. Robert Cloninger, Theodore Reich, and Samuel B. Guzeman, "The Multifactorial Model of Disease Transmission: III. Familial Relationships

Between Sociopathy and Hysteria (Briquet's Syndrome)," *British Journal of Psychiatry* 127:23–32, 1975.

10. Soren Sigvardsson et al., "An Adoption Study of Somatoform Disorders I, II, and III," *Archives of General Psychiatry* 41:853–78, 1984.

11. Ibid., 869–70.

12. Gilligan, Reich, and Cloninger, "Etiologic Heterogeneity," 395–414.

13. Ibid., 409.

14. C. Robert Cloninger, "Neurogenetic Adaptive Mechanisms in Alcoholism," *Science* 236:410–16, 1987.

15. Ibid., 413.

16. C. Robert Cloninger, Soren Sigvardsson, and Michael Bohman, "Childhood Personality Predicts Alcohol Abuse in Young Adults," *Alcoholism: Clinical and Experimental Research* 12:494–505, 1988.

17. David E. Comings, "Presidential Address: The Genetics of Human Behavior—Lessons for Two Societies," *American Journal of Human Genetics* 44:452–60, 1989.

18. Ibid., 455.

19. Ibid., 456.

20. Devor and Cloninger, "Genetics of Alcoholism," 29.

Chapter 13: Predictive Medicine

1. Marion Steinmann, "In the Shadow of Huntington's Disease," *Columbia, the Magazine of Columbia University* 13–2:15–19, 1987.

2. Michael Waldholz, "Probing the Cell: A Project to Identify All Human Genes Sets New Medical Horizons," *Wall Street Journal*, p. 1, 3 Feb. 1986.

3. Raymond L. White, "DNA in Medicine: Human Genetics," *Lancet* 8414:1257–62, 1984.

4. Michael Waldholz, "Probing the Cell: The Diagnostic Power of Genetics Is Posing Hard Medical Choices," *Wall Street Journal*, p. 1, 18 Feb. 1986.

5. Alan Newman, "The Legacy on Chromosome 4," *Johns Hopkins Magazine* 20:30–39, 1988.

6. Ibid.

7. Ibid.

8. Peter Gorner, "Out of the Shadow: A New Genetic Test Can Foretell Agonizing Death," *Chicago Tribune*, 4 Aug. 1988.

9. O.W.J. Quarrell et al., "Exclusion Testing for Huntington's Disease in Pregnancy with a Closely-Linked DNA Marker," *Lancet* 8545:1281–83, 1987.

10. Marlene Huggins et al. (including Michael Hayden), "Ethical and Legal Dilemmas Arising During Predictive Testing for Adult Onset Disease: The Canadian Experience of Huntington Disease."

11. M. Loch et al., "Predictive Testing for Huntington Disease," *American Journal of Medical Genetics* 32:217–24, 1989.

12. Huggins et al., "Legal Dilemmas."

Chapter 14: A Niche in Society

1. Gina Kolata, "Rush Is on to Capitalize on Testing for Gene Causing Cystic Fibrosis," *New York Times*, p. C3, 6 Feb. 1990.
2. Leslie Roberts, "The Race for the Cystic Fibrosis Gene," *Science* 240:142, 1988.
3. Robert G. Knowlton et al. (including Lap-Chee Tsui, Manuel Buchwald, and Helen Donis-Keller), "A Polymorphic DNA Marker Linked to Cystic Fibrosis Is Located on Chromosome 7," *Nature* 318:380–82, 1985.
4. Ray White et al., "A Closely Linked Genetic Marker for Cystic Fibrosis," *Nature* 318:382–84, 1985.
5. Brandon J. Wainwright et al. (including Robert Williamson), "Localization of Cystic Fibrosis Locus to Human Chromosomes 7 cen-q22," *Nature*, 318:384–85, 1985.
6. Peter Newmark, "Testing for Cystic Fibrosis," *Nature* 318:309, 1985.
7. Leslie Roberts, "Race for Cystic Fibrosis Gene Nears End," *Science* 240:285, 1988.
8. Jean L. Marx, "The Cystic Fibrosis Gene Is Found," *Science* 45:925, 1989.
9. Ibid, 925.
10. Robert Pokorski et al., "The Potential Role of Genetic Testing in Risk Classification," *Report of the Genetic Testing Committee to the Medical Section, American Council of Life Insurance* (June 10, 1989).
11. Charles Cantor, "AAAS Observer Roundtable," *The AAAS Observer*, p. 8, 3 Nov. 1989.
12. Tabitha M. Powledge, "Toward the Year 2005," *The AAAS Observer*, p. 1, 3 Nov. 1989.
13. Neil A. Holtzman, *Proceed with Caution* (Baltimore, Maryland: Johns Hopkins Press, 1989), 152–53.
14. Ibid.
15. UJA–Federation of New York, press release (October 1989).
16. Arno G. Motulsky, "Impact of Genetic Manipulation on Society and Medicine," *Science* 219:135–40, 1983.
17. Ibid.
18. Marc Lappé, "The Predictive Power of the New Genetics," *Hastings Center Report* 20:18–21, 1984.
19. Robert Pokorski, cited above.
20. Ben Webb, "Genetic Screening: UK Physicians Demand Action," *Nature* 341:91, 1989.
21. Ibid.
22. Kathleen Nolan and Sarah Swenson, "New Tools, New Dilemmas: Genetic Frontiers," *Hastings Center Report* (Oct./Nov. 1988), 40–46.
23. Motulsky, "Impact of Genetic Manipulation."
24. Holtzman.

25. Pokorski et al., "Role of Genetic Testing."

26. Holtzman.

27. Marc Lappé, "Ethical Issues in Genetic Screening for Susceptibility to Chronic Lung Disease," *Journal of Occupational Medicine* 30–6:493–501, 1988.

28. Dorothy Nelkin and Laurence Tancredi, *Dangerous Diagnostics: The Social Power of Biological Information* (New York: Basic Books, 1989).

29. Ibid., 102.

30. Alan L. Otten, "Probing the Cell: Genetic Examination of Workers Is an Issue of Growing Urgency," *Wall Street Journal*, p. 1, 24 Feb. 1986.

31. Robin Marantz Henig, "High-Tech Fortunetelling," *New York Times Magazine*, pp. 20–22, 24 Dec. 1989.

32. William Saletan, "Genes 'R' Us," *The New Republic*, pp. 18–20, July 17 and 24, 1989.

33. National Center for Human Genome Research, National Institutes of Health, *Report of the Working Group on Ethical, Legal and Social Issues Related to Mapping and Sequencing the Human Genome*, Bethesda, Maryland, 1989.

34. Daniel E. Koshland, Jr., "Sequences and Consequences of the Human Genome," *Science* 246:189, 1989.

35. S.E. Luria, "Letters," *Science* 246:873, 1989.

36. Ari Berkowitz, "Letters," *Science* 246:874, 1989.

Chapter 15: Choices

1. Cathy Holding and Marilyn Monk, "Diagnosis of Beta-Thalassaemia by DNA Amplification in Single Blastomeres from Mouse Preimplantation Embryos," *Lancet* II:532–35, 1989.

2. John Kelly, "Pre-Embryos" letter, *Lancet* I:335:116, 1990.

3. Yuri Verlinsky et al., "Genetic Analysis of Polar Body DNA: A New Approach to Preimplantation Diagnosis (abstract)," *American Journal of Human Genetics*, supplement, 45:A272 (1072), 1989.

4. Kathleen Nolan and Sarah Swenson, "New Tools, New Dilemmas: Genetic Frontiers," *Hastings Center Report* (Oct./Nov. 1988), 40–46.

5. John D. West et al., "Sexing the Human Pre-Embryo by DNA-DNA Insitu Hybridisation," *Lancet* I:1345–47, 1987.

6. "Fetal Sexing: Is It Ethical? When?," *Medical World News*, p. 18, 28 Sept. 1987.

7. James P. Grigsby, "Learning Disabilities in Fragile X Syndrome," *Fragile X Foundation Newsletter* (Fall 1986).

8. L. Bertilsson et al., "Debrisoquine Hydroxylation Polymorphism and Personality," *Lancet* I:555, 1989.

9. J. Kagan, J.S. Reznick, and N. Snidman, "Biological Bases of Childhood Shyness," *Science* 240:167–71, 1988.

10. R. J. Herrnstein, *I.Q. in the Meritocracy* (Boston: Atlantic–Little, Brown, 1975).

INDEX

abortion, 20, 119, 277, 278, 281–
   284, 294, 308, 321
  amniocentesis and, 309, 311
  sex determination and, 313
acetylcholine, 238
adopted children, 275
  in genetic studies, 253–59, 262–
   263
age:
  hemochromatosis and, 55
  Huntington's disease and, 31,
   82
alcoholism, 249–66
  genetic susceptibility to, 17, 18,
   19, 249–66, 300
  hemochromatosis and, 55, 56
  in men vs. women, 255–60,
   264–65
  Tourette's syndrome and, 263–
   265
  Type I, 257, 260–62, 265
  Type II, 256–62, 264–65
Allen, Cleona, 232–33
alpha-one-antitrypsin (AAT), 301–
  302
Alta, Utah, meeting at, 49–51, 54,
  60–64, 70, 71–72, 181

Alzheimer, Alois, 237
Alzheimer's disease, 17, 19, 236–
  243, 245, 300
American Council of Life
  Insurance, 297, 299–300
American Journal of Human Genetics,
  67–68, 111
American Psychiatric Association,
  243
American Society for Human
  Genetics, 122, 311
amino acids, 26, 194, 195
  genetic code for, 24–25
  in proteins, 24, 125–26
Amish, manic-depression of, 102,
  225–34
amniocentesis, 309, 311, 313, 321
amyloid, 238, 240–41
angina pectoris, 181
animal fat, in diet, 183, 184
animals, DMD gene in, 123–24
Anne (Belgian girl), muscular
  dystrophy of, 108–10, 115,
  119–120, 125
Annual Review of Genetics, 252
antibodies, 126–27, 130
"anti-cancer" genes, 134

apolipoprotein (apo) A-I gene, 179, 195–99
apolipoprotein (apo) A-IV gene, 198–99
apolipoprotein (apo) C-III gene, 180, 195–98
apolipoprotein (apo) E genes, 192–194
Ashe, Arthur, 185
atherosclerosis, 183–84, 185, 191, 192, 193, 196–98
Atomic Energy Commission, 218
autosomes, 105, 109, 119, 212
autosome 21, 119–20
Avila Giron, Ramon, 44, 46, 83, 84

bacteria:
  DNA libraries in, 75
  gene transference in, 62
  restriction enzymes in, 62–63
Baltimore, David, 221
Banbury Center, 67
Baron, Miron, 234–36, 241
Barranquitas, 83–84, 95
bases (base nucleotides), 23–25, 210, 216, 219
  in DMD gene, 125–26
  polymorphisms in, 61–65
Bassett, Ann, 242–43, 244, 246–47
Baylor College of Medicine, 281, 282
Beadle, George W., 23, 24
Beaudet, Arthur, 270–71, 281, 284–89, 292
Becker, P. E., 104
Becker-type muscular dystrophy, 104, 105, 128
Beckman, Lars, 254
behavior, genetic determinants of, 20, 318–19
Benzer, Seymour, 40
Berkowitz, Ari, 305
beta-hemoglobin, 307–8
beta-thalassemia, 307–8
Bias, Wilma B., 208
Bilheimer, David W., 190
Billings, Paul, 303–4
BioEssays, 224
biological underclass, 21, 302
bipolar disorder, see manic-depression

Bird, Ted, 83–84
Bishop, Tim, 158, 162
Blackburn, Henry, 185
blacks, G6PD deficiency in, 204–5
bladder cancer, 166
blood groups, 208–9, 211, 212
Bodmer, Sir Walter, 70, 170–71
Bohman, Michael, 253–59, 263
Bonilla, Ernesto, 83
Boston University School of Medicine, 271
botany, 22
Botstein, David, 48, 60–75, 217, 221, 288, 304
  at Alta meeting, 50–51, 54, 60–64, 70, 71–72, 75, 181
  at Banbury meeting, 67
  Cavenee and, 146–47
  Gusella-Housman discovery and, 101
  Housman's rift with, 89
  at NIH meeting, 87–89
  White and, 65, 73, 87–89
Botstein-White-Skolnick-Davis paper (1980), 66–69, 71, 80, 111, 128–29, 133, 195, 214, 287
Boyer, Herbert, 72–73
brain cancer, 175–76
Brandt, Jason, 274–75
breast cancer:
  chromosome 17 and, 175
  of Mormons, 158–59
  risk of, 52, 54
  Skolnick's studies of, 158–59, 160
  susceptibility genes and, 52, 54, 60, 138–39, 158, 163, 295
Brenner, Sydney, 223, 224
Breslow, Jan, 179–80, 185, 191–199, 284
  background of, 191–92
Briquet's syndrome (hysteria), 258
Broder, Samuel, 134
Brown, Michael, 45, 46, 188–91
Bryer, Bruce:
  background of, 104
  death of, 131
  Francke's work with cells of, 107, 108, 111, 115–16
  Kunkel's work with cells of, 115–19

reverse genetics and, 103–20
X chromosome of, 107–8, 111,
  116, 122
Buchwald, Manuel, 287–89
Bulfield, Graham, 129
Burghes, Arthur H. M., 125
Burt, Randall, 155–56, 159–62

Canadian Muscular Dystrophy
  Association, 114
cancer, 17, 132–78, 313–14
  bladder, 166
  brain, 175–76
  breast, see breast cancer
  colon, see colon cancer
  drugs for, 18
  inheritance of, 132–34, 138–39,
    140
  lung, 154, 163, 176–77
  oncogenes and, 145–46, 149
  p53 and, 173–78
  progression to, 174
  skin (melanoma), 163
  supressor genes and, 150–51,
    168–69, 174–78
  Vogelstein's study of, 156, 163–
    169, 171–78
  Wilms' tumor, 141, 164–65,
    166, 168
  see also retinoblastoma
Cannon-Albright, Lisa, 155–56,
  158, 162
Cantor, Charles, 220, 293
carcinoma, 152–53
Carmen (Huntington's disease
  sufferer), 85–86
Cartwright, George E., 59
Caskey, Thomas, 285–87, 290–91
Caspersson, Torbjorn O., 209–11,
  213
Cavalli-Sforza, Luca, 53
Cavenee, Webster, 132–35, 145–
  150, 169
  p53 gene and, 175–76
  Vogelstein compared with, 164
  White and, 133, 134, 135, 145–
    147, 149, 161
cDNA probes, 74–76, 78, 195
Cell, 127, 216
cell division, DNA replication
  during, 25

cells:
  cytogenetics and, 202–3
  membranes of, 186–87
  replication and growth of, 177–
    178
Centre d'Etude du
  Polymorphisme Humain
  (CEPH), 215, 216
CF, see cystic fibrosis
Charcot-Marie-Tooth (CMT),
  303–4
Chase, Tom, 83–87, 94
Chicago Tribune, 274
Children's Hospital (Boston), 115,
  197–98
Children's Hospital (Los Angeles),
  144
Children's Hospital (San
  Francisco), 131
Childs, Barton, 204
cholesterol, 17, 179–99
  familial hypercholesterolemia
    and, 188–91
  functions of, 186–87
  HDL, 180, 187–88, 195, 197,
    198
  IDL, 187
  LDL, 187–91
  levels of, 183–84
  liver's manufacture of, 186
  transportation of, 187–88, 194
  Type III disorder and, 192–94
  VLDL, 187
chorionic villus sampling (CVS),
  309, 313, 321
chromosome 1, 206–8
chromosome 3, retinoblastoma
  and, 135, 148
chromosome 4, 101–2, 273
chromosome 5:
  deletions of, 172
  FAP gene and, 170–71
  schizophrenia and, 243–44
chromosome 6, 60, 230, 239
chromosome 7, 281, 288–89
chromosome 9, 211, 212
chromosome 11, 231–33, 246, 266
  Vogelstein's work with, 164–65,
    166, 167, 171
chromosome 13:
  mental retardation and, 142
  retinoblastoma and, 135, 142–

chromosome 13 *(cont.)*
  144, 150, 167
  riflips and, 147–48, 150
chromosome 14, 80
chromosome 16, 209, 220
chromosome 17, 171–78
  breast cancer and, 175
  colon cancer and, 167, 168, 171,
    172
  p53 gene on, 173–78
chromosome 18, colon cancer and,
    167, 171, 172–73, 177
chromosome 19, 220
chromosome 21, 220, 246
  Alzheimer's disease and, 240–41
chromosomes:
  autosomes, 105, 109, 119, 212
  banding patterns of, 210–11,
    213, 214
  crossover of, 149, 150
  cytogenetics and, 202–4, 211
  cytology of, 107–8
  DNA in, 24, 210
  Donahue's study of, 206–9
  etymology of word, 23
  extra, Down's syndrome and,
    203–4
  in germ cells, 57–58
  identification of, 210
  number of, 203
  physical mapping of, 220–21
  sex determination and, 105–6
  see also X chromosomes; Y
    chromosomes
chromosome walking, 117–18,
    120
chronic granulomatous disease,
    104, 107
Churchill, Winston, 185
cigarette smoking, heart disease
    and, 181, 183
Cincinnati, University of, 150
City of Hope, 138
cloning:
  of DNA, 75
  of genes, 37, 90, 115
Cloninger, C. Robert, 249–66
  background of, 252
CMT (Charcot-Marie-Tooth),
    303–4
Cohen, Stanley, 72–73
Cold Spring Harbor Biological

Laboratory, 50, 67, 112–15,
    223
Collaborative Research Inc., 215–
    216, 281, 286, 288, 290, 292–
    293
Collins, Francis, 289, 290, 292
colon cancer, 154–75, 295
  in cancer families, 138–39, 155,
    157, 159–60
  chromosome 17 and, 167, 168,
    171, 172
  familial adenomatous polyposis
    and, 169–72
  p53 gene and, 173–75
  polyps and, 154–56, 160–61,
    163, 167, 168
  Skolnick's research on, 155–63
  stages of, 166–67
  testing for, 156, 160–61
  Vogelstein's research on, 166–
    169, 171–75
color blindness, 58, 106
  G6PD and, 204–5
  manic-depression and, 227, 234
Columbia, 268
Columbia University, 220, 247,
    271
Comings, Brenda, 263–64
Comings, David E., 242, 263–65
Committee to Combat
    Huntington's Chorea, 37, 39–
    40, 41, 43
competition, in science, 114, 118
computer analysis, of riflips, 99–100
Congress, U.S., 38, 41, 45, 220
Conneally, Michael, 41, 42, 92,
    95, 101–2, 245, 275
  G8 probe and, 99–100
coordinate system, genetic, 211,
    213
corn plants, chromosomes of, 201–
    202
coronary heart disease, 16, 18, 294
  cholesterol and, 179–99
  defined, 181
  epidemiology of, 182–84
  in historical perspective, 182
  hypercholesterolemia and, 45
  incidence of, 181, 182, 184
  increase in, 182, 183
  in women, 180, 184–85, 196–98
  see also heart attacks

Correns, Karl, 22
Crick, Francis, 23, 50
criminal behavior, 17, 252, 258
CVS (chorionic villus sampling), 309, 313, 321
cystic fibrosis (CF), 16, 19, 38, 285–94
    testing for, 279–83, 285–87, 290–93, 312
Cystic Fibrosis Foundation, 287–288
cytogenetics, 202–4, 211
cytology, of chromosomes, 107–8

Dangerous Diagnostics (Nelkin), 302
Darwin, Charles, 22
Dausset, Jean, 215
Davies, Kay E., 111–14, 121, 124, 125, 288
Davis, Ronald W., 60–62, 65–69
    at Alta meeting, 50–51, 54, 60–61, 71–72, 75
    as 1980 landmark paper coauthor, 66–69, 71, 80, 111, 128–29, 133, 195, 214, 287
Deleted from Colorectal Carcinoma (DCC), 177–78
DeLisi, Charles, 217–18, 220
deoxyribonucleic acid, *see* DNA
depression, 17, 19
    see also manic-depression
Devor, Eric J., 251, 266
De Vries, Hugo, 22
diabetes, 17, 18, 55, 294
    insulin and, 26
diet:
    colon cancer and, 155, 156
    fat in, 183–86
    heart disease and, 17, 18, 181, 183–86
    hemochromatosis and, 55, 56
    phenylketonuria and, 19, 299
discrimination:
    genetic, 21, 299, 301–2
    "hereditary meritocracy" and, 320–22
Disteche, Christine, 107
Dixon, Laurene, 104
Dixon, Robert, 104
DMD (Duchenne muscular dystrophy) gene, 109, 111–31
    in animals, 123–24

chromosome walking and, 117–118, 120
    as damaged normal gene, 121
    Davies's discovery of second riflip and, 113, 121
    discovery of, 124–25
    pERT87 probe and, 117–18, 120, 124
    protein produced by, 123, 125–127
    riflips and, 111–16, 118–23, 128
    rRNA genes and, 119–20
    size of, 123–26
    XJ probe and, 120, 123, 124
DNA (deoxyribonucleic acid), 23–26, 37
    of apo E gene, 194
    CEPH samples of, 215, 216
    in chromosomes, 24, 210
    DOE sequencing of, 216–23, 293
    enzymes and, 25–26, 50, 307–8
    of globin genes, 70–71
    heart disease and, 197–99
    of mice, 120, 124, 126–27
    physical structure of, 23–24, 50, 56
    "reading" of, 61
    replication of, 25
    of yeasts, 70
DNA library, 74–77, 96–97
    Kunkel's compiling of, 116
DNA polymorphism, *see* riflips
DOE (Department of Energy), 216–23, 293
dominant genes, 32, 33
    in Alzheimer's disease, 239
    in hemochromatosis, 55, 59
    in Huntington's disease, 32–33, 51, 83, 90, 99
    neurofibromatosis and, 138
    retinoblastoma and, 139
Donahue, Roger P., 205–9
Donis-Keller, Helen, 288
dopamine, 233, 261, 262, 265–66
Dore, Guy, 200
Dor Yeshorim, 294–95
double helix, 23–24
Down, John Langdon, 203
Down's syndrome (mongolism), 203–4, 240, 309
Dozy, Andrees, 71, 115

drugs, 62, 226, 265, 318
   for heart disease, 181
   susceptibility genes and, 17–18
Dryja, Thaddeus, 150–52, 283
Duchenne, G. A. B., 104–5, 106
Duchenne-type muscular
      dystrophy, 104–5, 109, 128
   gene for, see DMD gene
Duffy-a blood groups, 208–9
Dulbecco, Renato, 218–19, 221
dystrophin, 121–31
   discovery of, 121–27
   naming of, 127
   normal muscle and, 129

eating disorders, 245, 264–65
Edinburgh experiment, 312
Eduardo (Huntington's disease
      sufferer), 85–86
Edwards, Corwin Q., 59
Egeland, Janice, 102, 225–36, 241,
      246
Ehlers-Danlos syndrome, 202
Eldridge, Roswell, 101
Elephant Man, 138
El Mal (the bad), 81–86
   legendary origins of, 81–82
   see also Huntington's disease
emphysema, 301
Energy Department, U.S. (DOE),
      216–23, 293
environment:
   cancer and, 156
   genes vs., 225–26, 304
   heart disease and, 182
   susceptibility genes and, 17, 18,
      52
enzymes, 16, 23, 209
   DNA and, 25–26, 50, 307–8
   esterase D, 144–45, 150
   G6PD, 204–5, 235
   restriction, see restriction
      enzymes
epidemiology:
   of heart disease, 182–84
   of Romans, 182–83
esterase D, 144–45, 150
ethics, of genetic testing, 21, 282–
      284, 298–306, 314–15
eugenics, 305, 314–15
evolution, theory of, 22

eye cancer, see retinoblastoma
eye color, as genetic marker, 57

familial adenomatous polyposis
      (FAP), 169–72
"Familial Deficiency of
      Apolipoproteins A-I and C-III
      and Precocious Coronary
      Artery Disease" (Norum et
      al.), 180
FAP (familial adenomatous
      polyposis), 169–72
fat, in diet, 183, 184
Fearon, Eric, 166–68, 171–72, 177
Fernbach, Susan, 282
fertilization, in vitro (test-tube),
      310–12, 314
FH (familial
      hypercholesterolemia), 188–91
"50/50" (Nancy Wexler), 46–47
Fletcher, John C., 270
Folstein, Marshall, 245–46
Ford, Gerald, 45
Fost, Norman, 286
"founder's effect," 227–28
Fox, Maurice, 65, 72
fragile X syndrome, 316–17
Francke, Uta, 107, 108, 111, 115–
      116
Frederickson, Donald, 192
fruit flies, 136
   chromosomes of, 201–2
   "drop-dead," 40

Gallo, Robert, 165–66
Gardner, Eldon, 158
Gauthier, Marthe, 203
G8 probe, 99–100
gel electrophoresis, 220
GenBank, 218
Genealogical Society of Utah, 53,
      157
gene cloning, 37, 90, 115
gene discoveries and
      identification, 15–22
   conquest of genetic diseases
      and, 16–17
   for cystic fibrosis, 287–90
   of DMD gene, 124–25
   ethical issues raised by, 21,
      282–84, 298–306, 314–15
   for Huntington's disease, 99–

102, 133, 267
lack of media attention to, 16
media attention to, 16, 37, 102, 124
of p53 gene, 173–78
rate of, 15–16
of suppressor gene, 150–51
susceptibility to disease and, 17–20, 52
see also mapping of the genome
gene jumping, 289
genes:
"anti-cancer," 134
apo A-I, 179, 195–99
apo A-IV, 198–99
apo C-III, 180, 195–98
apo E, 192–94
coining of word, 23
DCC, 177–78
deletion of, 117–18, 121–23
DMD, *see* DMD gene
DNA and, 23–26, 70–71, 194
DNA distribution in, 121–22
dominant, *see* dominant genes
early searches for, 23–26, 56–57
FAP, 169–71
FH, 188–91
globin, 70–71
loss of, 117
oncogenes, 145–46, 149, 164, 171, 172, 174, 218–19, 221
p53, 173–78
polymorphic, 208
recessive, *see* recessive genes
rRNA, 119–20
splicing of, 26, 62, 72–73
suppressor, 150–51, 168–69, 174–78
susceptibility, *see* susceptibility genes
unlinked, 57
GeneScreen, 286
gene therapy, 37
genetic code, breaking of, 24–25, 26
"Genetic Defect Causing Familial Alzheimer's Disease Maps to Chromosome 21, The," 241
genetic engineering, 26, 37
restriction enzymes and, 62
genetic labeling, 299
genetic markers, 57–61, 230–31,

281–82, 289
hemochromatosis and, 58–60
for retinoblastoma, 144
riflips as, 64–67, 88–89, 93–94, 101–2, 111–20
genetics:
emergence of, as science, 203
medicine's relationship with, 201–2
see also specific topics
genetic tests, 267–306
ethics of, 21, 282–84, 298–306, 314–15
pre-embryo selection and, 308–322
prenatal, *see* prenatal testing
social uses of, 19
time needed for, 18
see also specific diseases
Gerhard, Daniela, 231–33, 241
germ cells, recombination of, 57–58, 108–9
Gershon, Elliot, 234, 246
Gilbert, Fred, 133
Gilbert, Walter, 217–19
Gilles de la Tourette, Georges, 263
Gilligan, Sheila B., 259
globin genes, 70–71
Goldstein, Joseph, 45, 46, 188–91
Gomez, Fidela, 97, 98
Goodwin, Donald W., 255
Green, Howard, 209
G6PD enzyme, 204–5, 235
Gurling, Hugh, 244
Gusella, James, 92–93, 98–102, 225, 231, 232, 271
Alzheimer's disease and, 239–41
Guthrie, Marjorie, 37, 39–40, 41, 43, 45
Guthrie, Woody, 37
Guze, Samuel B., 252

Hall, Stephen S., 216
Hamburger, Rahel, 235
Hamilton, Stanley, 167, 168
Hammersmith Hospital, 314
Hasidic Jews, Tay-Sachs disease and, 294–95
Hastings Center Report, 296
Hayden, Michael, 276, 277–78

HDL (high-density lipoproteins)
cholesterol, 180, 187–88, 195,
197, 198
heart attacks, 181, 182
genetic susceptibility to, 17, 18,
19
heart disease, see coronary heart
disease
hemochromatosis, 54–56
genetic markers and, 58–60
HLA genes and, 58–60
of Mormons, 59–60
hemoglobin, 70, 307–8
hemophilia, 106, 283–84, 294
color blindness and, 58
Henry Ford Hospital, 196–97
Hereditary Disease Foundation,
41, 44, 47–48, 92, 93
"hereditary meritocracy," 320–22
heredity and inheritance:
Mendelian laws of, 22, 32, 51,
158, 159, 188
summary of early research on,
22–25
of traits, pattern of, 22
X-linked, 105–6
see also specific topics
Heritable Disorders of Connective
Tissue (McKusick), 202
Herrera, Lemuel, 170
Herrnstein, R. J., 320, 322
heterozygotes:
familial hypercholesterolemia
and, 188
hemochromatosis and, 55, 60
Huntington's disease and, 45
high-density lipoproteins (HDL)
cholesterol, 180, 187–88, 195,
197, 198
"Highly Polymorphic Locus in
Human DNA, A" (Wyman
and White), 80, 111
Hill, Jon, 60
HLA (human lymphocyte or
leukocyte) antigens, 58–60,
230, 239
Hoffman, Eric P., 126–27, 130
Hoffmann-La Roche, 175
Hogeness, David, 72
Holding, Cathy, 307–10, 312
Holtzman, Neil A., 294, 295, 299,
300

homozygotes:
familial hypercholesterolemia
and, 188–91
hemochromatosis and, 55, 59–60
Huntington's disease and, 45–
46, 83
Wexler's search for, 83–86, 91
Hood, Leroy, 217, 220
Hospital for Sick Children, 108,
281
Housman, David, 87, 89–95, 98,
225, 231–33, 241, 314
background of, 90
Botstein's rift with, 89
grant sought by, 92, 93
Tobin and, 48, 49
Venezuela project and, 91–95
Howard Hughes Medical Institute,
80, 87, 147, 161, 170, 214–16
HUGO (Human Genome
Organization), 224
Human Gene Mapping Workshop
(HGM), 213–14, 224
Human Genome I, 293
Human Genome Project, 21–22,
304
human lymphocyte (or leukocyte)
antigens (HLA), 58–60, 230,
239
Huntington, George, 31–32
Huntington's disease, 16, 29–51,
81–102, 232, 268–78
age and, 31, 82
discovery of gene for, 99–102,
133, 267
financing of research on, 37, 41,
42, 45, 91
homozygous children and, 45–46
as ideal test of Botstein's
theory, 90–91
incidence of, 31, 38, 51
lack of interest in, 38
loneliness of, 97–98
testing for, 267–78, 312
third-party test requests and,
275–76
in Venezuela, 44–48, 81–89; see
also Venezuela project
Huntington's Disease
Commission, 45, 47
hybridization, 74–80, 116, 124
mouse-human, 209, 213

hydroxylators, poor, 318
hypercholesterolemia (FH), 45
hyperlipoproteinemias, 192–94
hysteria (Briquet's syndrome), 258

Illinois Masonic Medical Center, 311
Imperial Cancer Research Fund, 170
inheritance, *see* heredity and inheritance; *specific topics*
insulin, human, 26
insurance, 297–300, 303, 309
    denial of, 268, 271–72, 304
Integrated Genetics, 286, 292–93
intermediate-density lipoproteins (IDL) cholesterol, 187
IQ tests, 317–18, 320, 322
iron, hemochromatosis and, 54–56, 59–60
Isaacs, Susan, 247–48
Israel, manic-depression in, 235

Jackson Laboratory, 129
Japan, 222
    heart disease in, 182, 183
Jeffries, Alec, 71
Jerusalem Mental Health Center, 235
Johannsen, Wilhelm, 23
Johns Hopkins Hospital, 271, 273–275
Johns Hopkins Magazine, 272
Johns Hopkins School of Medicine, 156, 200–207, 246
    McKusick at, 200–205
    Vogelstein at, 165, 166, 172–73, 177, 295
Journal of the American Medical Association, 266

Kaback, Michael, 285–87, 290–92
Kan, Y. W., 71, 115
Kanigel, Robert, 217
Karathanasis, Sotirios K., 179–80, 191, 195, 197–99
Karolinska Hospital, 132, 150
karotyping, 203
Kazazian, Haig, 292
Kennedy, James, 244
Kidd, Kenneth, 232, 244

kidney, Wilms' tumor of, 141, 164–65, 166
Kindred 107, 158–59
Kindred 1001, 159
Kindred 1002, 160, 162
Kingsbrook Jewish Medical Center, 294–95
Klinefelter's syndrome, 204
Knorring, Anne-Liis von, 258
Knudson, Alfred G., 135–45, 150–151
    background of, 136–37
    childhood cancer studies of, 137–38
    neglect of cancer theories of, 135–36
    see also two-hit theory
Koenig, Michel, 125
Konopka, Ronald, 40
Koshland, Daniel E., 305
Kravitz, Kerry, 59, 60
Kunkel, Henry, 115
Kunkel, Louis, 114–19, 121–30, 290
    background of, 115
    experimental procedures of, 116–17
    Kunkel, Louis O., 115

Laguneta, 84–86, 95
Lalouel, Jean-Marc, 215
Lancet, 175, 270, 307–10
Lander, Eric, 244–45
Lappé, Marc, 296–97, 301–3
Latt, Sam, 116
Lawrence Berkeley Laboratory, 220
Lawrence Livermore Laboratory, 220
LDL (low-density lipoproteins) cholesterol, 187–91
Lederberg, Joshua, 53–54, 64–65
Lejeune, Jerome, 203
leucine, 24, 25
leukemia, 138, 313–14
Levine, Arnold, 173
Lewis, Jerry, 113
liver:
    cholesterol manufactured by, 186
    hemochromatosis and, 54–55, 59
    LDL processing in, 190

Los Alamos National Laboratory, 218, 220
low-density lipoproteins (LDL) cholesterol, 187–91
Ludwig Institute for Cancer Research, 175
lung cancer, 154, 163
  Minna's views on, 176–77
  p53 genes and, 176
Luria, S. E., 305
lymphocytes, 58

McKusick, Victor A., 200–207, 218, 227
  background of, 200–201
  mapping project of, 200, 210–214, 216, 222–24
malaria, 182–83
Mandel, Batsheva, 235
Maniatas, Tom, 70, 75
manic-depression, genetic cause of, 102, 225–34, 245, 247, 300
mapping, 22, 58, 200–224
  Cavenee's work with, 134–35
  CEPH and, 215, 216
  Collaborative's claims and, 215–216
  disagreements over use of, 89
  DMD gene and, 113
  DOE sequencing project vs., 216–23
  extra-large families needed for, 87–89, 94, 95
  goal of, 89
  hemochromatosis and, 60
  McKusick's interest in, 200, 210–14, 216, 222–24
  riflips and, 66, 68, 70, 71, 73, 80, 87–89, 134–35, 146–47, 161, 214–16
  Ruddle's work with, 212–13
  sequencing compared with, 220–23
  start of, 293–94
Maracaibo (city), 81–84, 95
Maracaibo, Lake, Huntington's disease in villages on, 44–48, 81–89, 93–98
Marfan's syndrome, 202
Marina del Rey meeting, 48
Martin, Joseph, 91

Massachusetts, University of, 49, 73
Massachusetts Eye and Ear Infirmary, 150, 152
Massachusetts General Hospital, 91, 98–99, 271
Massachusetts Institute of Technology (MIT), 21, 48, 49, 75, 80, 90, 150, 216, 221
  globin gene research at, 70–71
Maxam, Allan, 217
Mazzocchi, Anthony, 302
M. D. Anderson Hospital and Tumor Institute, 135, 136, 139, 140
mdx mouse, 130
Meadows, Anna, 141–42
Medical World News, 312–13
melanoma, 163
men, males:
  alcoholism in, 255–60, 264–65
  heart disease in, 184
  hemochromatosis in, 56
  hemophilia-color blindness link in, 58
  muscular dystrophy in, 104–5, 113, 115–19, 121, 125, 130–31
Mendel, Gregor, 22, 32
Mendelian Inheritance in Man (McKusick), 211–12, 214
Mendelian segregation, 51, 52
Mendel's laws of inheritance, 22, 32, 51, 188
  breast cancer and, 158, 159
Mendlewicz, Julien, 234
mental retardation, 170, 299, 316–317
  retinoblastoma and, 142
mental traits, selection of, 316–18, 320
Meyer, Laurence, 163
mice, 307–8
  DNA of, 120, 124, 126–27
  human hybrid cells, 209, 213
  p53 gene in, 173
  X chromosome, of, 129–30
Migeon, Barbara, 210
Miller, C. S., 210
Minna, John, 176–77
Monaco, Anthony P., 117–18, 123–24, 126

mongolism (Down's syndrome), 203–4, 240, 309
Monk, Marilyn, 307–10, 312
Morgan, Thomas Hunt, 23, 136
Mormons, 52–54
    cancer of, 156–63, 169
    hemochromatosis of, 59–60
    riflips and, 66, 67, 80, 215
    Southern blot of blood samples of, 77–80
Motulsky, Arno G., 203, 296, 297, 299, 319–20
multiple hit theory, *see* two-hit theory
multiple sclerosis, 16, 17, 18
Murray, J. M., 111–12
muscular dystrophy, 16, 19, 103–120, 294, 312
    Becker-type, 104, 105, 128
    cure for, 130–31
    Duchenne-type, 104–5, 109, 128; *see also* DMD (Duchenne muscular dystrophy) gene
    in females, 108–12, 115, 119–20
    misdiagnosis of, 128
    research on, 38, 103–31
    unexpectedness of, 38
Muscular Dystrophy Association, 112, 113, 115, 124
"Mutation and Cancer" (Knudson), 141
mutations, 22, 38, 53–54, 198–99
    of apo E, 193
    bacterial gene transference and, 62
    good, 54
    Huntington's disease and, 82
    radiation and, 217–18
    retinoblastoma and, 140, 141, 142
Myers, Richard, 269, 271–72

National Academy of Sciences, 222
National Cancer Institute, 176
National Foundation-March of Dimes (now March of Dimes Birth Defects Foundation), 213
National Health Service, British, 298
National Institute of Mental Health, 245, 253

National Institute of Neurological Disorders and Stroke (NINDS), 45, 87, 91, 93
National Institutes of Health (NIH), 38, 44, 45, 65, 146, 188, 192, 214
    DOE vs., 220, 221, 223
    1979 meeting at, 87–89
    1990 meeting at, 291–92
    Office of Human Genome Research of, 223
    Vogelstein's work at, 165–66
    White's proposal to, 66, 67, 75
National Research Council (NRC), 222, 223
National Science Foundation, 220
natural selection, 54
Nature, 101–2, 122, 124, 145, 149, 170, 233–34, 244, 245, 289
Nedda (schizophrenic), 39
Nee, Linda, 225, 236–40
Negrette, Americo, 82–84, 95
Nelkin, Dorothy, 302, 303
nervous system, 260–62, 319
neuroblastoma, 141
neurofibromatosis, 138
Neurological Institute and Hospital, 131
newborns, genetic tests for, 18–19
New England Journal of Medicine, 128, 132–33, 142, 179, 180, 196, 197, 201
    colon cancer research in, 154–155, 161, 163, 172, 173
New Haven Human Gene Mapping Library, 213–14
New Republic, 304
New York Psychiatric Institute, 247–48
New York Times, 102, 170, 286
New York Times Magazine, 217
Noble, Ernest, 266
Nolan, Kathleen, 298–99, 312
nondisclosing exclusion tests, 276–277
norepinephrine, 262
Norum, Robert A., 180, 196–97
Nowell, Peter, 155
nucleotides, base (bases), 23–25, 61–65, 125–26, 210, 216, 219
Nussbaum, Alexander L., 195

obsessive-compulsive disorders, 245, 247–48, 264–65
occupational choice:
  genetic tests and, 19, 300–302
  restrictions on, 19, 21, 276, 300–302
Ochs, Hans, 107, 116
"On Chorea" (Huntington), 31–32
oncogenes, 145–46, 149, 218–19, 221
  ras, 164, 171, 172, 174
ophthalmoscopes, 139
Otten, Alan, 302

Pagon, Roberta, 107
Pardes, Herbert, 245, 246
Partridge, Terry, 130
Pauls, David, 232
pAW101 probe, 79, 80
pAW1016–18 probe, 77
Pearson, Peter, 113
Pedigree 110, 231–32
Pedigree 214, 230
Pedigree 265, 230
personality, alcoholism and, 260–261
pERT87 probe, 117–18, 120, 124
Peutz-Jeghers syndrome, 201
p53 gene, 173–78
phage, 75
phenylketonuria (PKU), 299
  testing for, 18–19
physical traits, selection of, 315–16
Pines, Maya, 214, 215
Pokorski, Robert, 297–98
polymerase chain reaction, 307–8, 311
polymorphism:
  chromosome 1 and, 208
  DNA, *see* riflips
  genetic markers and, 57
polyps, colon cancer and, 154–56, 160–61, 163, 167, 168
population genetics, 51–54
  genetic markers and, 57–60
Porter, Ian, 204–5
"Potential Role of Genetic Testing in Risk Classification, The," 297, 299–300
pre-embryo selection, 308–22
  of mental traits, 316–18, 320
  of physical traits, 315–16

preimplantation diagnosis, 308–11
"Preliminary Communication" (Holding and Monk), 307–10
prenatal testing, 19–20, 93, 269, 309–10
  for cystic fibrosis, 280–83, 289, 294
  for Huntington's disease, 276–278
  for muscular dystrophy, 119, 294
Proceedings of the National Academy of Sciences, 80, 111, 141, 209
Proceed with Caution (Holtzman), 294
proteins, 23, 289–90
  amino acids in, 24, 125–26
  amyloid, 238, 240–41
  apo A-I, 179, 195–99
  apo A-IV, 198–99
  apo C-III, 180, 195–98
  apo E, 192–94
  cholesterol processing and, 186, 187, 192–96
  DCC gene and, 177–78
  DMD gene and, 123, 125–27
  HLA, 58–60
  in muscle development and function, 107
  supressor genes and, 150–51
  see also dystrophin
Psychiatric Developments, 250
psychiatry, psychotherapy, 225–26, 232, 242–45, 247, 265
pueblas de agua (water villages), 81–82, 84–86, 93

Quaid, Kimberly, 274
Quinlan, Bill, 151–53
Quinlan, Bonnie, 151–52
Quinlan, Mrs., 151–53
Quinlan, William Francis, 152–53

ras, 164, 171, 172, 174
Ray, Peter N., 120
Reagan, Neil, 155
Reagan, Ronald, 146, 155
recessive genes, 38
  cystic fibrosis and, 286
  in hemochromatosis, 56, 59
  in Huntington's disease, 32, 51
  inheritance and, 32, 33

retinoblastoma and, 139
Recombinant DNA Advisory
    Committee (RAC), 66
recombination of germ cells, 57–
    58, 108–9
Regier, Darrel, 234
Reich, Theodore, 226, 250, 259
Reilly, Philip, 282–83, 290–92
Renwick, James H., 208–9
replication of DNA, 25
restriction enzymes, 62–64, 70, 71,
    76, 111, 198
    Kunkel's use of, 116
    in Southern blot, 77, 78, 79
restriction-fragment-length
    polymorphisms, see riflips
retinitis pigmentosa, 104, 300
retinoblastoma, 132–53
    Cavenee's work with, 132–35,
        145–50
    chromosome 13 and, 135, 142–
        144, 150, 167
    colon cancer compared with,
        168, 169
    esterase D and, 144–45
    incidence of, 134
    inheritance of, 139, 140
    Knudsen's study of, 135–45
    manifestation of, 139–40
    mental retardation and, 142
    non-inherited, 140
    ophthalmoscope and, 139
    treatment of, 132, 139, 140
    two-hit theory of, 140–45, 147–
        150
reverse genetics, 103–20, 125–28,
    186, 282
    defined, 103–4
Ridenour, Dorothea, 283
riflips (restriction-fragment-length
    polymorphisms; RFLPs;
    DNA polymorphisms), 61–80,
    245
    alcoholism and, 250–51, 265–66
    Banbury meeting and, 67
    Botstein's Alta meeting
        explanation of, 61–64
    in Botstein-White-Skolnick-
        Davis paper, 66–69
    cancer and, 159, 161, 163–73
    chromosome 13 and, 147–48
    DMD gene and, 111–16, 118–

123, 128
    G8 probe and, 99–100
    as genetic markers, 64–67, 88–
        89, 93–94, 101–2, 111–20
    in globin genes, 70–71
    Gusella's work with, 98–102
    hybridization and, 74–80
    maps, 66, 68, 70, 71, 73, 80,
        87–89, 134–35, 146–47, 161,
        214–16
    naming of, 66
    retinoblastoma and, 135, 146–
        147, 150
    Wyman and, 69–70, 73–80
Roberts, Leslie, 287
Romans, ancient, epidemiology of,
    182–83
Roswell Park Memorial Institute,
    170
Roth, John, 50
Rowland, Louis P., 128–29, 131
Royal College of Physicians, 298
Royal Victoria Hospital, 175
rRNA genes, 119–20
Ruddle, Frank H., 211–13, 216,
    219, 223–24

Sabin, Jessie, 34–36
Sabin, Leonore, see Wexler,
    Leonore Sabin
Sabin, Mr. (father), 34, 36
Sabin, Paul, 35
Sabin, Seymour, 35
St George-Hyslop, Peter, 240
Saint Vitus' dance, 32
Sanberg, Avery, 170
Sanger, Frederick, 217
San Luis, 82–83, 95
sarcolemma, 129
Schaffer, Arthur, 249–51, 265
schizophrenia, 17, 19, 225, 242–
    245, 300
    drugs for, 18
    Wexler's work with, 39
Schmickel, Roy D., 119–20
Schulze, J., 204–5
Science, 145, 150, 168, 169, 174,
    177–78, 216, 218, 241, 293,
    305, 319
Scientific American, 146
Scientist, 221
Scottsdale meeting, 112, 115

Scully, Brian, 283
senility, 225, 236–43
serotonin, 261, 262
sex determination, 105–6, 312–13
Sherrington, Robin, 244
Shows, Thomas, 213–14
shyness, biological bases of, 318–319
sickle-cell anemia, 70–71, 115, 294, 299
Sigvardsson, Soren, 258, 262–63
Sinsheimer, Robert, 216–17
skin cancer, 163
Skolnick, Mark, 51–54, 59–61, 64, 70, 101, 135, 270, 284
  background of, 53–54
  breast cancer studies of, 158–59, 160
  colon cancer research of, 155–163, 169
  Lederberg and, 53–54, 64–65
  Mormon genealogies studied by, 53–54, 59–60, 87
  as 1980 landmark paper coauthor, 66–69, 71, 80, 111, 128–29, 133, 195, 214, 287
  White and, 66, 161
smoking, heart disease and, 181, 183
Solomon, Ellen, 70, 170
Southern, Edward M., 77
Southern blot, 77–80
  making of, 77, 78
Sparkes, Robert, 144–45, 147, 150
splicing of genes, 26, 62, 72–73
Squibb Corp., 175
Stahl, Frank, 72
Stanbridge, Eric, 176
sterilization, Huntington's disease and, 83
Stockholm Adoption Study, 255–259
Streisinger, George, 72
Strong, Louise, 135, 141, 142, 147
suicides of Amish, 228, 230
suppressor genes, 150–51, 168–69
  p53 as, 174–78
susceptibility genes, 17–20
  alcoholism and, 17, 18, 19, 249–266, 300
  breast cancer and, 52, 54, 60, 138–39, 158, 163, 295

colon cancer and, 138–39, 154–157, 159–60, 295
drugs and, 17–18
environment and, 17, 18, 52
heart disease and, 185–86, 188–191
lung disease and, 301–2
prenatal testing for, 19–20
Sutton, Walter, 23
Sweden, 253–59, 301–2
  alcoholism in, 249, 253, 255–59, 262–63
Sweeney, Karen, 272–75
Sweeney, Paul, 272–74
Swenson, Sarah, 298–99, 312

Tancredi, Laurence, 302–3
Tanzi, Rudolph, 240
Tatum, Edward L., 23, 24
Tay-Sachs disease, 290, 294–95
Tennessee, University of, 130–31
tests, see genetic tests; prenatal testing; *specific diseases*
thrombosis, coronary, 182
thymidine kinase, 209, 210
Time, 234
Tobin, Allan, 42, 48–49, 67, 87, 90
Tourette's syndrome, 263–65
traits:
  inheritance of, pattern of, 22
  pre-embryo selection and, 315–318, 320
Tschermak, Erich von, 22
Tsui, Lap-Chee, 287–90, 292
Tufts University-New England Medical Center, 131
Turner's syndrome, 204
"Turning Point in Cancer Research, A" (Dulbecco), 218–19
twins:
  alcoholism in, 255
  Huntington's disease in, 274–75
  manic-depression in, 226
two-hit theory, 140–45, 147–50, 164
  Cavenee's testing of, 148–49, 150
  colon cancer research and, 154, 156
  p53 gene and, 174, 176

Sparkes's testing of, 144–45
Type III disorder, 192–94

United Jewish Appeal-Federation of New York, 294–95
U.S. Air Force Academy, 299
Utah, University of, 49–54, 80, 147
 colon cancer research at, 155–63

Van Biema, David, 237
variable number tandem repeat (VNTR), 80
Venezuela, Huntington's disease in, 44–48, 81–89
Venezuela project:
 G8 probe in, 100
 Housman and, 91–95
 pedigree from, 99
 problems in, 95–97
 Wexler and, 83–89, 91–98, 231
Verellen-Dumoulin, Christine, 108–11, 119
Verlinsky, Yuri, 311
very-low-density lipoproteins (VLDL) cholesterol, 187
Vogelstein, Bert, 156, 163–69, 171–78, 284, 295
 awards of, 176
 background of, 165–66
 p53 gene and, 173–78
Vovis, Gerald, 293

Wallace, Peggy, 100
Wall Street Journal, 174, 216, 270, 302
Ward, Patricia, 269–70, 283–84
Washington Post, 223, 237
Watson, James, 23, 50, 223, 293–294
Weinberg, Robert, 20–21, 146, 174, 221–22, 224
Weiss, Mary, 209
Wexler, Alice, 29–31, 33–37, 42, 47, 86, 268
Wexler, Leonore Sabin, 29–31
 background of, 34–36
 death of, 46–47
 suicide attempt of, 42–43
Wexler, Milton, 29–31, 34–43, 45, 46, 82, 87, 100, 268

research efforts of, 34, 35, 37–42, 86, 94
Wexler, Nancy, 29–31, 33–37, 42–48, 100–102, 191, 267–73, 304, 322
 frustration of, 86–87
 Housman's collaboration with, 91–94
 loss of confidence of, 98
 at NIH meeting, 87–89
 on riflip technique, 88
 Venezuela research of, 83–89, 91–98, 231
White, Ray, 65–80, 99, 214–16, 219, 236, 270, 284, 288, 289
 background of, 65, 72–73
 at Banbury meeting, 67
 Botstein and, 65, 73, 87–89
 Canevee and, 133, 134, 135, 145–47, 149, 161
 colon cancer studies of, 169–72
 grant proposal of, 66, 67, 75
 at NIH meeting, 87–89
 as 1980 landmark paper coauthor, 66–69, 71, 80, 111, 128–29, 133, 195, 214, 287
 Skolnick's relationship with, 66, 161
 Vogelstein compared with, 164
 Wyman, 69–70, 73–80, 214
Whitehead Institute, 21, 150, 216, 221
Williamson, Robert, 111, 288, 289
Wilms' tumor, 141, 164–65, 166, 168
Wilson, Barbara, 279–83
Wilson, John, 279–83
Wilson, Richard, 279–83
Wilson, Sarah, 279–83
Winston, Robert, 314
women, females:
 alcoholism in, 255–60, 264–65
 breast cancer in, 52, 54, 60, 138–39, 158–59, 160, 163, 175, 295
 heart disease in, 180, 184–85, 196–98
 hemochromatosis in, 56
 muscular dystrophy in, 108–12, 115, 119–20
 resistance to X-linked inherited diseases in, 106, 109

women, females *(cont.)*:
  as unknowing carriers of X-
    linked inherited diseases, 106
Wood, Donald S., 113, 114, 118,
  122, 125, 126
World Federation in the
  Neurology of Huntington's
  Chorea, 44
Worton, Ronald G., 108, 110, 111,
  114, 121, 124, 125, 129
  Schmickel's collaboration with,
    119–20
  XJ probe of, 120, 123
Wyman, Arlene, 69–70, 73–80, 89,
  111
Wyngaarden, James B., 223

xanthomas, 196–97
X chromosomes, 204, 205, 212,
  314
  in Anne's case, 109–11, 115,
    119–20
  of Bryer, 107–8, 111, 116, 122
  fragile, 316–17
  hemophilia and, 283–84

inheritance and, 105–6
manic-depression and, 227, 230,
  234–35
of mice, 129–30
muscular dystrophy and, 103,
  105–13, 115, 116, 119–20,
  122, 125
sex determination and, 105–6
translocations on, 108–12
XJ probe, 120, 123, 124
Xp21 band, 110–14, 121
Xp21 breakpoint, 110–11, 114

Yale University, 235, 244, 318–19
Y chromosomes, 312, 314
  defects in, 106
  sex determination and, 105–6
yeasts, riflips of, 70
Yokoyama, Shozo, 250
Young, Anne, 96–97
Young, Brigham, 53

Zannis, Vassilis I., 179, 191, 193–
  195, 197–98
Zech, Lore, 210

The Age of the Gene

ten years after this book was first published, molecular biologists the world over were fully consumed by the international race to map the human genome. The effort, whose twin goal is to identify all the 100,000 or so human genes and then understand the critical roles each of these genes play in the exquisitely complex dance of human biology, had truly matured into one of the most—if not perhaps the most—momentous scientific exploration of our time.

A cascade of gene discoveries throughout the 1990s was making it a certainty beyond doubt that no other pioneering scientific enterprise–not the all-pervasive advances in computer technology or even the tumultuous Internet communications revolution–will ultimately have as great a long-term impact on human progress. That's because, by the decade's end, researchers had begun producing evidence, as many had suggested in the 1980s, that unlocking the mysteries of the human genetic machinery was providing the single most powerful tool for solving the world's vexing public health puzzles, such as, but not limited to, cancer, Alzheimer's disease, asthma, diabetes, depression and arthritis. Researchers, doctors and even politicians who in 1990 had never even heard the word "genome" were lauding, with much justification, that the new gene based science ushered in by many of the discoveries described earlier in this book, will quite likely lead to healthier, more productive, and, perhaps, even longer lives.

And no wonder. Throughout the 1990s the public was bombarded, sometimes almost daily, with awe-inspiring and often hard-to-believe or fathom news emerging from human biology labs. The Human Genome Project had gone from being a relatively-small, somewhat controversial, government-funded operation born at the National Institutes of Health in 1991, to a colossal, hotly-competitive commercial chase among entrepreneurial scientists in biotech and drug-making corporations. For them, discovering, and then patenting a gene or a new gene-solving technique carried the promise of palpably huge financial awards. Entirely new companies, whose publicly-owned shares traded explosively in the midst of the 1990s bull market, were created to help finance ambitious, increasingly wealthy and well-regarded molecular biologists and their high-profile gene-searching laboratories.

Indeed, by 1999, one of these companies, a publicly-traded venture named Human Genome Sciences, (or HGS) based in Washington, D.C., was claiming, quite brazenly but with ample evidence, that it already had uncovered large chemical portions of almost every human gene. Another company, Millennium Pharmaceuticals Inc. in Cambridge, Mass. was filling its financial coffers with gene-hunting contracts worth hundreds of millions of dollars each being paid by giant pharmaceutical companies convinced that understanding the genetic basis of diseases would lead them to a mother-lode of powerful new medicines. And a dozen major drug makers, in part worried they were losing out in the commercial gene race to HGS, Millennium and other new so-called "genomics" companies, formed their own consortium to quickly uncover hundreds of thousands of riflip-like gene markers, which, the companies now believed, would expedite the discovery of all the major new medicines likely to be developed in the first few decades of the new century.

Examples abounded to back these audacious-seeming claims. For instance, scientists reported they had found genetic switches that controlled the life of cells. They suggested that drugs acting upon these genes might someday actually inhibit cell-senescence, keep tissue and organs composed of these cells functioning longer and better, and thereby postponing disease and even death. Other scientists unmasked all the genes inside HIV, the AIDS-causing virus. This, in turn, led drug makers to create a growing menu of innovative anti-AIDS medicines. And these therapies quickly helped dramatically reduce the death rate of AIDS in the U.S.

and other western nations, and even gave some hope that AIDS might someday be vanquished.

One of the most fertile areas of gene research involved cancer. By 1999, for example, researchers had uncovered dozens of genes that, when somehow damaged, allow cells to grow aberrantly into tumors. Bit by bit, these genes, and the proteins they produce, helped the scientist begin to chart, with surprising detail, the previously unknown biochemical pathways that transform a healthy cell into a tumor. As a result, numerous large and small drug makers, most of whom never before ventured into the daunting challenges of cancer research, were spending growing portions of their research budgets attempting to design safer and more effective anti-cancer therapies to interrupt these newly-visible cancer-causing chemical networks. And several of these new-age cancer drugs were already in human testing.

It was no surprise then that researchers at academic, pharmaceutical and biotechnology company laboratories were now fully in the thrall of thousands and thousands of gene-hunting races like those first few early ones in the 1980s that had won acclaim for Nancy Wexler, Francis Collins, Bert Vogelstein and Mark Skolnick. Indeed, spurred in large part by the unprecedented economic and public health potential, scientists now were convinced that by 2001, years ahead of schedule, they will have plotted out the precise sequence of the entire three billion chemical letters, and the genes they constitute, that together make up the architecture of the most marvelous of all Earthly molecules, human DNA. It is a good bet that this accomplishment alone will give scientific historians in the future reason to hail the last decade of the 20th century as the beginning of the age of the gene.

In the years since GENOME was first published, the gene hunting stories chronicled in chapters here led to numerous scientific surprises, successes, and, of course, disappointments and failures, too.

None reflected this new era more than the race to find the so-called breast cancer gene. Not surprisingly, however, this endeavor not only captivated the public but it consumed the efforts of Utah's Mark Skolnick and Ray White, of Francis Collins, and of a dozen or so other labs around the world.

It began in October of 1991, when Mary-Claire King, a previously unheralded researcher, startled the worlds of medicine and biology when she reported at a sparsely-attended scientific meeting in Cincinnati that her tiny, under-financed lab at the University of California at Berkeley had pinpointed the approximate location of a gene that was likely the cause of about 5% or so of all breast cancers. Up until that time, it was a hotly-debated issue in oncology circles as to whether breast cancer, or for that matter, any cancer, could be triggered by the passing, from parent to child, of an inherited genetic defect. Sure, doctors had seen scores of families, much like the large colon cancer kindreds among the Mormons unearthed in Utah by Skolnick, in which breast cancer seemed rampant. But these clusterings of cancer, scientists argued, could as likely be due to some shared environmental factors, such as years of exposure to carcinogens in the air or water, to high-risk lifestyles, such as fat-laden diets, or, more likely, to simple coincidence. Over a lifetime, a woman's risk in the U.S. of developing breast cancer was about one in eight, meaning that millions of Americans will develop the disease, and that, from a purely statistical point of view, it was likely that the cancer would strike more than one woman in a family. What King and others saw as genetic, the skeptics argued, was just the inevitable misfortune of the cancer accumulating densely in some families, while not arising at all in others.

But by 1991, King and her crew had amassed the largest collection anywhere of families in which at least two, and sometimes as many as five or more, relatives across several generations were haunted by the specter of breast, and sometimes, ovarian cancer. King, at that time 46 years old, had dedicated the previous 15 years to solving breast cancer, and it had consumed her professional and personal life. When RFLP markers were developed in the early 1980s, King set her lab to using them to pick through genetic material in the DNA specimens she had culled from hundreds of women in the U.S. and overseas in which cancer appeared to be inherited. Finally, after more than five years of experiments, using an RFLP identified in Ray White's Utah lab, King and her young research colleagues found that one shared spot in chromosome 17—most likely the site of a single gene–that seemed to be altered in many of the relatives with breast and ovarian cancer, but was not common among those relatives who were disease-free.

"It was a stunning, life-altering, discovery," Francis Collins remembers. Then director of genetics research at the University of Michigan, he was fresh from helping identify the gene for cystic fibrosis. No one among the young band of gene-hunters was more famous or respected. Collins had also helped develop a powerful technique for isolating a gene once its approximate location in a chromosome was found. Sitting in the Cincinnati audience, the 41-year old Collins decided to team up with King, and together he proposed the two would prove, once and for all, that it was possible to inherit a risk for breast cancer. But the King and Collins collaboration had competition. Soon afterward, White, Skolnick, and others in the U.S., Canada, Great Britain and France, were enmeshed in a high-stakes, much publicized competition to get to the gene first. The scientists were convinced that by finding the gene and the protein it expressed, they would finally have a handle on the previously hidden biochemical mechanism underlying most, if not all, breast cancers. It would be a finding, they proposed, that would lead them to new ways to diagnose and treat the disease.

Skolnick, however, soon found his efforts hampered by lack of funds, despite his access to an unusually large number of Mormon families his lab had identified in and around Salt Lake City who were afflicted by what seemed to be an inherited form of the disease. But Skolnick did not have technical wizardry of a Francis Collins or Ray White, but instead his skill was more akin to an epidemiologist, whose talent was identifying what appeared to be powerful genetic trends in large populations. His efforts to gain NIH funding to finance a hunt for King's gene were rebuffed. Frustrated, he turned to some local entrepreneurs who helped him create a company called Myriad Genetics Inc., with the expressed goal of building the lab capacity and hiring the scientists with the lab know-how he lacked. And quietly, without much notice, he built a giant-sized operation. It was the first of many gene-hunting companies to soon emerge in the early 1990s. Meanwhile, King and Collins were attracting a flood of news articles documenting their assault on the gene and highlighting the gene's potential for helping women learn if they were born with a higher than normal susceptibility to cancer

One news story, first published in December 11, 1992 of the *Wall Street Journal*, more than any other, proved that King's gene existed and that its identification in a family had powerful consequences no one had expected. Soon after linking up with King, Collins's associates in Michigan began seeking their own families in which to search for the gene. One family that came to their attention involved two sisters for whom breast cancer had become especially tragic. Two other sisters in the family already had recently succumbed to the disease, and their mother and their aunt also previously had cancer, although they survived the disease. The family seemed to be haunted by a genetic defect, likely King's long-sought after gene, and the Michigan researchers recruited them as part of their effort to track down the gene.

In 1992, when one of the remaining sisters, known in subsequent medical research reports as Janet, also developed breast cancer, her younger sibling, known as Susan, in her early 30s, married and with two young children, became certain she was next. In fact, the King discovery was leading many oncologists to believe that, for some families, a gene disorder was at work. And Susan's doctor took the extraordinary step of advising her that to prevent her from developing disease she should consider having her own breasts surgically removed in a pre-emptive strike to thwart what seemed in her family was inevitable. Indeed, as the gene hunters uncovered hundreds of families where the cancer seemed inherited, more and more women were having their breasts removed. A fifth sister in the family of Susan and Janet already had become so fearful of the destiny she seemed to face she already undergone the breast-removal surgery.

At the same time Susan was preparing to have the surgery at University of Michigan's medical center, Collins's lab was in the final stages of using an RFLP marker to identify whether the family indeed had the gene defect on chromosome 17. By analyzing DNA from the women the scientists were able to pick out which women in the family carried a gene defect, and which women didn't. While neither Susan nor Janet were aware of the lab's breakthrough, Janet's doctor, Barbara Weber, who was part of Collins's team at the medical center, was aware the lab was testing the family's DNA. When Janet happened to tell Weber of Susan's imminent surgery, the doctor suddenly realized that the breast cancer gene hunt was providing her an extraordinary opportunity to make use of the unfolding predictive powers just emerging from the new gene-based science.

"Right then, it hit me," Dr, Weber recalls. "We had set out to find the gene, and to use it to help uncover new ways to treat the disease. But all of a sudden I realized, even before we had isolated the exact identity of the gene, that we already had the ability to see who in the family likely carried the defect. Even though we hadn't expected to use if for this purpose, I realized we had a tool that could tell (Susan) if she really did have the same genetic risk that had caused disease among her sisters."

Weber called down to the lab, got the results of their analysis and then called Janet and Susan into her office. "It was an extraordinary moment," says Weber, recalling that it took her some time to explain the lab's new-found ability to identify which family member was at risk and which was not. Finally, she asked Susan if she wanted to know the answer. "I could-n't believe what I was hearing," says Susan. "It sounded more like science fiction than medicine." Indeed, at that time it was. Weber told Susan that the lab had found that her stricken sisters, mother and aunt all harbored a similar gene defect on chromosome 17. But, Weber told her, Susan did not. "I was just days away from having my breasts removed," says Susan, "and what Dr. Weber told me was that I didn't have to do it. I was stunned."

It was, indeed, the first widely-reported instance in which the new gene technology was able to allow cancer doctors to peer into a science-based crystal ball and help them make a predictive diagnosis that previously was beyond human knowledge or know-how. But others in the Michigan women's family weren't as lucky. The scientists found that Susan's uncle–her mother's brother–had inherited the gene defect and had passed it along to some of his daughters. None of these daughters had previously been ill, nor had any of them had any idea the inherited risk of cancer might also be a threat to them. "Out of nowhere, it seemed to these women, a research study involving their cousins that they knew nothing about was providing them information about their future that they didn't want, and, of course, were totally unprepared to deal with," remembers Collins. "It was, for all of us doing gene research, quite a sobering time."

In the months that followed, the press was filled with stories of other women in other families in the research studies who were now able to find out about their future risk. And, perhaps more than any other set of news stories, it gave "the public a sense of what was going on in genetics," recalls Mary-Claire King. "It also gave our efforts to find the gene an even greater urgency. We were now able to alert some women of their higher risk, but there was no medicine, and no treatment, other than monitoring them closely, or offering the surgery, that we could do about their heightened genetic risk."

But by the summer of 1994, the scientists still hadn't found the gene, and King and Collins were growing frustrated. Collins, by this time, had been named to head the new U.S. Human Genome Project, and moved his lab to the NIH research campus in Bethesda, Md. Quietly, unbeknownst to the King-Collins team, or other labs, Skolnick's colleagues at Myriad were closing in on the gene. "We had built a very strong group," Skolnick says.

In early September, 1994, Myriad Genetics reported it had found the gene, called BRCA1. The Myriad scientists found that just one change in the thousands of chemical letters that constitute the gene was enough to produce a disease-causing defect. Although the precise nature of the gene, how it worked in the body, or even how the defect arises, was still not known, the research provided powerful evidence to scientists that a seemingly innocuous alteration was enough to upend the normal activity of a protein. This discovery, that a tiny change was sufficient to put a person at high risk of disease was in itself one of the most influential insights to arise from the new gene-hunting technology. "It told us that the difference between being healthy and being at risk of disease was often such a slight variation in someone's genetic makeup," says Collins. "It was a remark-able discovery and has really changed the way we think about disease and its biological causes."

The BRCA1 discovery produced several other immediate effects. For one, it catapulted Skolnick into the ranks of the nation's premier gene hunters. Myriad also patented the discovery and used its exclusive com-mercial rights to the gene to develop a marketable gene test that millions of women could use, not just women in research studies, to determine if they, too, carried a gene that put them at risk of cancer. Myriad began offering the test to doctors and the company's fortunes soared.

But the company soon came under criticism because it was selling a test that many doctors and their patients found difficult to interpret. That's because scientists examining the breast cancer families found that inher-iting the defect didn't cause a woman to develop cancer for certain. Instead, only 50% to 85% women who harbored an altered version of BRCA1 developed the disease. Moreover, it was unclear when during a woman's lifetime the cancer might arise, how serious the illness would be, and, whether the gene defect might cause cancer of the breast, ovaries or both, since women with the altered version of BRCA1 were also at high risk of ovarian cancer, too. As a result, doctors were uncertain as what action was best to prevent disease.

"We were worried that women took the test thinking it would give them a clear solution," says Fran Visco, an advocate for patient rights. "We wanted to make certain that Myriad and the doctors provided women with adequate counseling to go along with the test, to make certain that someone who found out she had the gene (alteration) could decide what options she had, or that a positive result might not give them any useful options at this time. We know it left a lot of women very confused." Visco and others also were concerned that if a health insurer paid for the test, the insurer might use the information to deny future health coverage to women or their daughters found to carry the altered version of BRCA1.

Despite the ethical concerns of some, Myriad soon signed lucrative agreements with several large drug companies who wanted Myriad to use its new gene hunting skills to find genes involved in other diseases, too. Myriad was the first of a slew of new so-called genomics company; its success soon encouraged other gene hunters and venture capitalists to cre-ate other genomics firms to conduct similar gene hunts of their own. Soon, Millenium outside Boston, HGS outside Washington, D.C., Incyte

Pharmaceuticals in California, and Genset in Paris were collecting families from all around the globe, in whom, like the breast cancer families recruited by King, a common disease seemed to have a strong genetic component. Researchers at these and other new companies launched campaigns to find genes involved in cancers of the skin, prostate and brain, Alzheimer's disease, and a host of other maladies including diabetes, asthma, and even depression, psoriasis, schizophrenia, baldness and obesity. And the race was on to unearth families in places like Iceland, Israel or Sardinia, where scientists often can find large families with homogenous genetic backgrounds, a situation which allows scientists to better target their gene-hunting techniques

"It was now an accepted certainty, using the techniques uncovered by King and Skolnick, and newer ones emerging from our labs and others, that we could indeed find the genetic or biological aspect of disease and that, in doing so, we now had a very powerful new way to explore the causes of disease," says Eric Lander, who directs the genome research labs at Massachusetts Institute of Technology's Whitehead Biomedical Research Institute in Cambridge, Mass. and who helped found Millenium. "This newfound ability to dissect the basic underlying causes of disease wasn't lost on the big pharmaceutical makers, who were searching for ways to provide newer, more potent, more effective and safer drugs."

One of the first contracts signed by Millenium was with Hoffmann-La Roche, the Nutley, N.J. unit of the giant Swiss drug maker, Roche Holding Ltd. It's goal was to find a gene that predisposed some people to being severely overweight. "It was a pretty good chance that being overweight had a strong biological component, and that this was driven by the inheritance of certain genetic traits," Dr. Lander says. "Millenium and others felt that finding one or more genes involved in obesity would finally give us a new understanding as to what was going on. That, in turn, might lead us to new drugs."

That lead soon came from a researcher at Rockefeller University in New York City who that obesity in a special breed of mice was related to a defect in a newly-identified gene that makes a protein called leptin. The scientist also reported that humans contained the same gene and that some obese people did, indeed, have alterations or changes in their leptin gene that differed from people who weren't overweight. Rockefeller patented the gene discovery, by now a common tactic, and licensed it to Amgen, a California biotech outfit, that believed giving healthy versions of the leptin protein might be a useful anti-obesity medicine. But at Millenium, scientists used the same discovery to reveal the biochemical pathways that may make some people more prone to being overweight. Soon afterward, they found that leptin, a hormone produced by cells from the leptin gene, played a critical role in controlling the intake of nutrients into cells. The Millenium scientists found that leptin did this by activating another protein on the cell's surface. Millenium and Roche then decided if they could find a chemical to interact with this activity, they might someday find a drug that blocked excessive weight gain in a manner different than any previous medical therapies.

In fact, several of the biggest drug makers undertook efforts of their own to find new genes or the pathways in which newly-found genes worked. Until the mid-1990s, most pharmaceutical companies were content to leave most basic biology research to academics. When a new biological understanding was reported by a university scientists, the drug maker often jumped on it, using their renowned ability to synthesize and test new chemicals to interact with the newly-uncovered disease pathway But the new gene technology suddenly gave the drug makers reason to believe they also could conduct their own basic research, too. At the same time, the researchers at the drug makers were under intensifying pressure from their top management to develop drugs that provided significant advances over existing medicines. The cost of research and development at the companies was exploding, and the only way to justify the expanding costs was through higher revenues. But the increasingly powerful managed care health insurance business was putting pressure on the drug companies to keep prices of existing drugs stable. The only medicines that could be carry premium price tags were those that differed significantly from existing products, and provided a measurably better way of treating illnesses.

Not surprisingly, the big pharmaceutical makers turned to the gene-hunting technology for answers. "If we are going to produce the kind of medicines that will have a significant impact on public health, on diseases that still defy adequate treatment, it became clear we were going to have go about our business differently," says George Poste, a leading science executive at SmithKline Beecham Plc., the London-based drug maker. Poste set off a land-rush stampede for new gene research among competing drug companies when, in 1993, he decided to spend $125 milllion for the rights to develop new medicines based on the gene discoveries being made by Human Genome Sciences. The move was especially bold because HGS was only a year old at the time, and had yet to show that any of the genes it was uncovering would lead to new medicines. But Poste was certain he was right, and other companies listened.

In fact, HGS's technology worried some big drug-makers. HGS had been formed to exploit a technology breakthrough by Craig Venter, a scientist at the NIH. By the early 1990s, Venter had found a way to quickly pick out a gene within a large segment of DNA by identifying just a partial length of the gene's sequence. Venter claimed that these partial sequences, or gene tags, were so likely to lead to the discovery of the genes themselves that he prodded his bosses at NIH to patent the tags. But others in the gene research community were concerned that Venter's proposal would mean NIH might own the commercial rights to tens of thousands of genes, thereby giving the federal government a financial stake in the drugs and diagnostics eventually created as a result of the new gene finds. Frustrated by the controversy, Venter left the NIH and with the support of a venture capitalist, licensed his speedy gene discovery technology to HGS. Soon afterward, HGS cut its deal with SmithKline, and in quick succession, Pfizer Inc., Merck & Co. and other companies made similar deals with other new gene hunting "genomics" companies.

One of the earliest indications that the new gene technology was changing the way drug makers carried out their jobs occurred at the venerable drug making giant, Merck. Despite its size and reputation, throughout its history, the company had steered clear of cancer research. Says Edward Scolnick, director of the company's research and development: "Merck prides itself in going after drugs to interact with biology that we understand. Cancer, until very recently, was just not well understood." But in the 1990s no disease was yielding its secrets quicker as a result of the gene-based science assault than cancer. And no scientist epitomized this effort to use gene technology to unlock cancer's secrets than Bert Vogelstein at Johns Hopkins. By the early 1990s, Vogelstein's early work in understanding the role of the gene p53 was proving to be a model for other cancer research at Merck and elsewhere.

Indeed, new research by Vogelstein and others helped show why cancer may best help explain the power of understanding the genetic basis of disease and why genes are so powerfully involved in disease development. But to understand that, one must understand why genes are damaged and why those genetic wounds play such a vital role in the evolution of every disease that afflicts humans.

Each time a cell divides, in order to replenish itself or grow new tissue, it must make an exact duplicate of its full complement of genes. The entire human genome is composed of three billion nucleotides or chemical letters. That means that during cell division the cell must reproduce an exact replica of its genetic sequence of letters. But cells sometimes make mistakes. And often these nucelotide letters in newly created cells are reproduced in jumbled manner. This shuffling of the order of the genetic letters is actually essential, a trick nature uses to differentiate one person from another. And genes for particular proteins often vary in large and small ways from one species to another, creating traits and characteristics that distinguish one animal from another. Indeed, it is this altering of genetic structure through this process of copying mistakes that is at the heart of evolution.

But sometimes these mistakes during cell replication can create alterations so serious they lead to mutated or mutant genes that can cause troubling and even lethal health problems. If these serious mistakes occur in the cells that are female eggs or male sperm, they can create a so-called germ-line mutation in which an altered version of a gene is transmitted in defective form to a newly-born individual. Sometimes these mutations can occur in cells within existing individuals, causing problems in the functioning of specific cells during that person's life, but not to succeeding generations. Genes can also be damaged by exposure to some environmental factors in the air, water, or food.

Cancer, put simply, is an unwanted copying or cloning of a single cell. It is a cell dividing in an unchecked manner. Specifically, gene hunters like Vogelstein are uncovering a set of genes—perhaps numbering several hundred—whose job it is to produce protein molecules that manage the life cycle of a human cell. These molecules are command and control agents that transmit signals to cells, telling them when to divide and when to stop dividing. Cell division signals emanate from outside or inside a cell as a result of a wide number of developments. And when enough copies have been produced, new signals are transmitted by other protein molecules also produced by specific genes.

The breakthrough scientists made in the 1990s is learning that cancer is mostly caused by defects in genes that normally produce proteins that order a cell to stop dividing. In other words, cancer is simply the result of a missing or damaged protein that is unable to carry out its task of turning off cell growth. Scientists have come to call these important genes "tumor suppressors" because one of their primary functions is keeping cell division under control. But when the suppressor genes are altered, either because they were inherited in an aberrant form or damaged during life, cells usually under their watchful eye become renegades, massing bit by bit into armies of cells that can eventually mass into deadly tumors.

"The understanding that cancer results when genes are damaged or mutated is the single most important step we have made in advancing the battle against cancer, in developing better ways to prevent, diagnose and eventually treat the disease," says Richard Klausner, the director of the federal government's National Cancer Institute. "If there are cures in the coming decades, they will all arise from this new understanding of the genetics of cancer."

The power of this new genetic model of cancer was first made most explicit in an unusual 1993 lab experiment involving a tiny slice of tissue long-ago removed from a tumor taken from the 1968 U. S. presidential candidate and former U. S. Senator, the now-deceased Hubert H. Humphrey. When the experiment's results were reported in the New England Journal of Medicine it signaled to many cancer doctors that a powerful change was occurring in the field of cancer care. "We did the experiment and published it in a highly-visible respected journal precisely because we wanted to alert the world to the power of the new genetics,." says David Sidransky, a former student of Vogelstein who now runs his own lab at Johns Hopkins.

One day in 1967. Sen. Humphrey was worried to find some spots of blood in his urine. Soon afterward, his doctor sent off a urine specimens on a glass microscope slide to several pathologists. doctors who analyze the shape of cells to see if they are cancerous. All the pathologists except one decided the urine sample contained no evidence of trouble. But one of them, John Frost at Johns Hopkins Hospital, believed the cells he saw under a microscope were I the very earliest stages of cancer. Having no way to prove his case. Dr. Frost was out-voted by doctors who believed the politician was in no danger. Sen. Humphrey went on to lose the presidency against Richard Nixon and then returned to the Senate. A few years later, more blood in the urine led to further exams showing that the politician indeed had bladder cancer. And in 1977, he died from the disease. Meanwhile, Frost had retained the original urine sample, a dried bit of cells sitting on a glass microscope slide and locked it away in drawer hoping he might someday prove that he had seen the first signs of cancer 10 years before the Senator died.

One day in 1992 Frost turned to a young doctor training in his lab who had worked with David Sidransky, one of the young lab researchers who helped find p53. Frost showed the slide to his student and asked him if it looked like the cells were cancerous. When he told the student that he was looking for a way to prove he'd been right the student turned to Sidransky. Several weeks later, the Hopkins researchers were able to get hold of a slice of Humphrey's tumor that had been saved at the hospital where his cancer was treated with surgery. Following days of extensive testing, the researchers found that the slice of tissue contained cells with a p53 alteration.. They then returned to the slide and found that p53 genes in the urine sample slide contained a alteration in the exact spot in the gene where it was altered in the cancer slice removed from Humphrey in 1976. It turned out that the urine contained some few cells shed from the beginnings of a tumor along the inner walls of Humphrey's bladder.

"The detective work showed that indeed Frost was right," Sidransky says. "We had the evidence. Humphrey had the very early beginnings of a cancer when he ran for presidency."

As important, however, the experiment showed for the first time that scientists could detect from a urine sample a cancer tumor so small it cannot be identified and caught. Like many cancers, a bladder cancer treated early can be cured. "If we knew in 1967 what we know now, Humphrey would never have succumbed to a cancer," Sidransky says.

These days, Sidransky is developing a kit doctors can use to scan urine samples for p53 defects, a test that certainly will save thousands of lives when it goes into full use. Indeed, cancer doctors are just now beginning to analyze the genetic make-up of cancer tumors biopsied and removed from patients, seeking to determine what kind of genes have gone awry. P53 defects turn out to be signals that a cancer is particularly fast-growing, suggesting to doctors that the cancer they are treating should be attacked especially aggressively with surgery, chemotherapy and radiation.

This kind of "predictive medicine," only a dream a decade earlier, was now expected to become more widespread as a early-warning system to treat numerous cancers, as well as other diseases from Alzheimer's disease to asthma and diabetes. In the future, researchers say, doctors will not only take a family history of patients, but they will use the gene discoveries to identify a person's in-born risk for illness. With a simple blood sample, researchers will be able to produce a genetic profile suggesting what maladies a person may get down the road.

By decade's end numerous drug makers, including researchers at Merck, were developing drugs to interact with the cell-dividing signals uncovered by the cancer gene researchers. Merck, Schering-Plough Corp., Johnson & Johnson and others were testing new drugs that inhibit the excessive cell dividing signals sent out by a defective version of a gene called Ras. In 1999, Warner-Lambert Co. acquired a small biotech company called Sugen because of the smaller company's pioneering work in understanding how overactive Ras genes and proteins fuel cancer growth. Still other companies were developing therapies that involved injecting healthy versions of p53 into tumors, hoping that the proteins produced by the p53 genes will act as a brake deep inside a cell.

Indeed, by the tail end of the 1990s decade drug makers had become convinced that the kind of gene alterations underlying cancer was likely at work in every disease. These gene variations don't necessarily cause disease outright, the scientists believed, but instead make people especially prone or susceptible to disease. In other words, the very slight, often single nucleotide letter, differences within genes that make one person different from another, also appeared to be the kind of variability that makes one person more likely to develop an illness, such as diabetes or asthma, than another.

One of the first of these type of disease-susceptibility genes was found, by luck it now seems, by Allen Roses, the researcher at Duke University who in the late 1980s was chasing after a gene for Alzheimer's disease that he believed was located on chromosome 19. Roses's lab was analyzing DNA from a large population of families where two or more members had Alzheimer's disease when they honed in on a region of chromosome 19 that was consistently altered in those relatives with the illness. T their surprise, the Duke scientists found that the alteration lay in the middle of a gene called Apoe, that produced a protein well-known to scientists. But the protein didn't seem to have anything to do with brain or neurological damage. Instead, Apoe's role was to ferry fats throughout the body. Researchers knew that the gene and its protein came in several versions, and that variability often affected how well cells made use of fats.

But Roses found that people in the families he studied who harbored one of these versions, Apoe4, had a much higher than usual risk of developing Alzheimer's. "Apoe4 doesn't cause Alzheimer's outright," Roses explains, "but something about having that particular version of the protein does increase the chances of getting the disease." Other Alzheimer's disease researchers were skeptical because they believed the disease resulted from the accumulation of thick plaques inside the brain that damage and kill neurons. But when these scientists examined other sets of Alzheimer's disease patients they found that Roses was right. Yet, to them, the Apoe link simply confused matters. T Roses, however, it made things clearer. "If you believe most diseases have a genetic aspect, that it results from some genetic variation, then Apoe makes perfect sense," he says. "The protein is a clue, its simply part of a biological pathway that, when affected in certain way, can promote disease."

Under this theory, every major common illness has at its root genetic variation. T Roses, the key to solving Alzheimer's and a host of other common illnesses is finding the slight alteration in a gene that increases some people's susceptibility to disease. In 1997, Roses was hired by the drug maker, Glaxo Wellcome Plc., to help the company transform its drug discovery operations into one that used gene-based research as the key to solving disease. The central technology Roses decided to use was a new type of gene RFLP-like gene marker called "SNPs," which stands of single nucleotide polymorphism. These SNPs, like RFLPs, are distributed throughout the genome. Their power, however, was their ability to hone in on the single letter changes in a gene that produced the biological difference between a person being susceptible to disease or not.

Under his guidance and prodding, Glaxo decided to get aggressively involved in producing a map of hundreds of thousands of SNPs sprinkled throughout the genome, and to use them as signposts to guide scientists to where, among the jumble of DNA, family members with disease shared a gene alteration that differed from those members who were healthy. Within two years of coming to the company, and by scouring dozens of families' DNA with the SNP markers, Glaxo had found genes that seemed to confer susceptibility to migraines, diabetes and even psoriasis. Other companies using the same technology claimed they were close to getting genes involved in baldness, in prostate cancer and even genes that seemed to play previously unknown roles in causing high blood pressure and heart attacks.

Perhaps even as important, the SNP map is expected to herald an entirely new field of "personalized medicine." Drug researchers soon became convinced that genetic variability detected by SNPs would help them understand why some people respond better to existing drugs than do other people. "We've always suspected that different reactions to drugs depended on a person's biology," says Michael Silver, a research executive at Pfizer Inc., who joined with Roses and other drug company leaders in forming a consortium in 1999 to create a genome-wide SNP map. Pfizer and the other companies plan to use the SNPs to create tailor-made drugs, medicines created specifically to target specific segments of the population whose genetics suggest they would respond best to particular drugs. By personalizing medicine, the drug makers hope, they will avoid the kind of dangerous side effects that often inhibit the success of medicines because the drugs cause serious problems for some people, but not others. Using genes to guide the use and development had become known by this time as a new area of drug discovery: pharmacogenetics. And every major drug maker now embraced it fully as the way to carry out research into the new century. Truly, the gene had come of age.

"If we can figure out, by analyzing a person's genetic makeup, which drug will serve them best and which drug might cause them side effects, we will produce an extraordinarily powerful new way to treat people with common illnesses," says Pfizer's Silver. "It's the smart way to make and use medicines, and it's the way genetic research will become part of the way every doctors practices medicine in the future.

—v—